从两种
文化中学习

LEARNING FROM TWO CULTURES

欧亚城市发展、更新、保护及管理
理论与实践

Urban Development,
Renewal,
Preservation and
Management
in Europe and Asia

德国ISA意厦国际设计集团 编著

中国建筑工业出版社

目录

关于 ISA　　　　　　　　　　　　　　　6

人即是城市　　　　　　　　　　　　　10
继承过去，规划未来

理论文章　　　　　　　　　　　　　　**22**

设计塑造生态：　　　　　　　　　　　24
不使用高科技的生态新城
Dita Leyh

从经济起飞到生活质量：　　　　　　　34
实践与研究中城市规划的经验和任务
Michael Trieb

我们应以何种风格建造？　　　　　　　50
——21 世纪初建筑学定位的探讨
Phillipp Dechow

从数量到质量：　　　　　　　　　　　60
韩国城市规划控制的经验
Seog-Jeong Lee

新城：　　　　　　　　　　　　　　　70
有关城市的梦想
张亚津

项目实践　　　　　　　　　　　　　　**82**

城市发展

01　世界文化遗产城市波茨坦城市总体规划　　85
02　丽江玉龙新城规划　　　　　　　　　　　99
03　泉州市城市新区建筑天际轮廓控制规划方法研究　111
04　泉州市区和环湾地区空间发展战略研究　　121
05　广州市南沙光谷地区城市发展规划　　　　129
06　阳宗海旅游度假区西北部概念规划设计　　137
07　东山蝶岛发展战略规划　　　　　　　　　143
08　十堰市东部新城概念规划及城市设计　　　151
09　福州市东部新城中心城市设计　　　　　　161
10　杭州天堂鱼生活：杭州市运河新城概念规划　171
11　唐山市南湖生态城概念性总体规划设计　　183
　　及起步区城市设计
12　大同市御东新区概念性总体规划　　　　　191
　　及核心区概念性城市设计
13　文昌市抱虎角概念规划　　　　　　　　　199
14　昆明市绿地系统概念规划　　　　　　　　205

城市更新

01	伦茨堡控制性规划	**217**
02	"霍费尔天空"： 提升城市品质的范例项目	**229**
03	埃尔旺根步行区	**239**
04	埃斯林根总体规划和公共空间城市设计导则	**247**
05	"粉红岛"：韩国首尔方背洞街区广场	**261**
06	**Gangseogu** 城市更新规划	**269**
07	北京市昌平区奥运自行车训练馆 及周边地区城市设计	**285**

城市保护及管理

01	世界文化遗产城市施特拉尔松：城市景观规划	**293**
02	新勃兰登堡城市景观规划	**305**
03	世界文化遗产城市波茨坦瑙恩郊区 及耶格尔郊区的设计导则	**315**
04	泉州市法石街区保护 与整治规划及环境设计	**325**
05	泉州市城市新区公共空间控制规划方法研究	**333**

图片来源及参考文献   **340**

项目概览   **342**

CONTENTS

ABOUT ISA   6

The Human Being Is the City   10
Learning from the Past, for the Future

ARTICLES   22

Ecology by Design: 24
Ecological New Towns without High-Technology
Dita Leyh

From Economic Miracle to Quality of Life: 34
Urban Development Experiences &
Tasks of Practice and Research
Michael Trieb

In Which Style should We Build?   50
Architecture at the Beginning of the 21st Century
-A Positioning Approach
Phillipp Dechow

From Quantity to Quality: 60
Experiences from City Supervision in South Korea
Seog-Jeong Lee

New Town: 70
Dream City
Yajin Zhang

PROJECTS   82

URBAN DEVELOPMENT

01	The Urban Development Planning of World Cultural Heritage City of Potsdam	85
02	Masterplan of Yulong New Town, Yunnan	99
03	Planning of the Skyline of the City of Quanzhou, Fujian	111
04	Strategic Study on Development of the Urban Districts and General Area Around Quanzhou Bay, Fujian	121
05	Urban Development of Nansha Guanggu in Guangzhou	129
06	Conceptual Planning of the Northwestern Part of Yangzonghai Resort, Yunnan	137
07	Dongshan Butterfly Island in Fujian: the Strategic Planning of Urban Development	143
08	Conceptual Planning of the East New Town in Shiyan, Hubei	151
09	City Planning for the Center of the Eastern New Town of Fuzhou, Fujian	161
10	A Fishing Life Paradise in Hangzhou: Concept Planning for the Canal New Town in Hangzhou	171
11	Conceptual Planning and Urban Design for an Ecological Zone in Tangshan, Hebei	183
12	Conceptual Planning of Yudong District and Urban Design of the Core Area, Datong	191
13	Conceptual Planning for Baohujiao Area in Wenchang, Hainan	199
14	Concept Planning for Kunming Green Space System	205

URBAN RENEWAL

01	Framework Development Planning of Rendsburg	**217**
02	"Heaven of Hof": an Initiative-project for Upgrading the City	**229**
03	Pedestrian Zone of Ellwangen	**239**
04	Master Plan and Design Guidelines for the Public Spaces of Esslingen am Neckar	**247**
05	"Pink Islands": a Street Square in Bangbae-dong, Seoul, South Korea	**261**
06	Urban Renewal Gangseogu	**269**
07	Urban Design of the Area Surrounding Changping Olympic Cycling Training Center, Beijing	**285**

URBAN PRESERVATION AND MANAGEMENT

01	World Cultural Heritage City of Stralsund: Townscape Planning	**293**
02	Townscape Planning Neubrandenburg	**305**
03	Design Code for the Nauener Vorstadt and the Jaegervorstadt in World Cultural Heritage City of Potsdam	**315**
04	Urban and Landscape Design for the Fashi District in Quanzhou, Fujian	**325**
05	Research of the Planning Methods for Open Space in Quanzhou, Fujian	**333**

IMAGE SOURCE & REFERENCE   340

PROJECT LIST   342

关于ISA
ABOUT ISA

ISA 意厦国际设计集团是一个由城市规划师、建筑师、景观建筑师、地理学家和室内设计师组成的国际性、多领域设计团体,通过多年在城市规划领域的工作积累,它已发展成为一支技术完善的、经验丰富的团队,贯穿常规工作的国际协作为其发展提供了强大动力。

The ISA Stadtbauatelier is an international, interdisciplinary planning group, consisting of urban planners and architects, and it can, based on the many experiences within the field of urban planning, count on a well-established and experienced team that receives new stimuli on a daily basis from international composition and cooperation with experts in landscape architecture, geography and interior design.

人是集团工作的核心，不论性别、年龄、种族与文化，这包括几千年以来所有文化背景下的人。人被视为个体，视为独特的自我，他不仅有生理和心理特性，也有精神的特性。
The human being is the objective of the work in our office. The human being, independent of sex, age, race and culture. The human being, as it has been recognized for thousands of years by all cultures. The human being is understood as individual, unique in itself, not only consisting of body and soul, but also containing a spiritual mind.

集团理念要求我们将城市视为艺术品，就像过去几千年中地球上的所有文化在漫长的岁月中对其的塑造成果。
The philosophy of the office demands to recognize cities as works of art, like the cities of great personalities than have been linked inseparably for thousands of years, in all cultures of the world.

人即是城市
继承过去，规划未来
The Human Being Is the City
Learning from the Past, for the Future

1. Introduction

On the way to better cities

This publication is a report of a group of people, on the way to better cities. Of its point of origin, of the wanderer's current point of view, of its final goal – and the long way to go. Of the goal to rightfully re-establish the human being as true objective and purpose of the city. Of the trials to bring quantity and quality back into balance in urban planning. Of the efforts to satisfy the needs, not of one, but of all citizens in the city.

It is the report of one stage of the long journey in the real world, to plan cities based on an integral picture of man, not only on single-sided economic, technical or ideological points of view. And it is also the report of a group of people of different ages, different sexes and different cultures, that have found a way together to commonly live their individual dreams, to work, all on their own schedule, on a better urban world for the people.

1. 引言

迈向更美好的城市

本书是这样一个团队的记录——他们不断追求更美好的城市，对他们行动的初衷、当前状况、目标愿景以及将来要走的漫漫长路进行的记录。他们的目的是让人们再次享有自己的权力——人是城市的本质所在。他们努力尝试，追求城市规划中数量和质量的平衡。他们尽力满足城市中人类全面的而不是单一的需求。

这是一篇关于追求真实世界的漫漫长路中一个阶段的报道，他们努力从整个人类的层面对城市进行规划，而不只是单方面的从经济、科技或者思想的层面。同时，这是一篇关于不同年龄、不同性别、不同文化的人们共同追求自己梦想的回溯，他们以自己的方式去创建一个更适合人类生存的城市环境。

图 01　一起来对话：迈向更美好的城市
Fig. 01　Together in dialogue – on the way to better cities

关于 ISA ABOUT ISA

从思想到行动

本书开篇将向您介绍ISA集团专业"智囊团"是怎样建立、如何发展的，以及在今天的研究和实践中处于怎样的位置，还有其遵从的理论。

接下来的章节介绍了ISA的合伙人们从学术和城市建设角度对城市规划经验和任务的看法，内容涵盖城市生态、城市识别性、城市发展及人文城市哲学。思想和行动之间还有一个章节，将详述ISA工作的整体多样性、设计草图作为城市规划思考工具的重要性。

规划项目案例阐述部分也就是对行动的详细介绍。根据全球城市规划实践所涉及的三大领域，这些项目被分为了城市发展、城市控制和城市更新三大类。对每个工作领域而言，这是一种经验积累：有哪些工作任务，人们应该怎样着手，可以怎样工作，可以找到怎样的解决办法。

我们并不认为这里所介绍的思想和项目可以作为模范样板，而是希望借此推动在城市规划的实践中，针对每一个规划任务去寻找自主、创新、灵活的解决方式，帮助读者实现自己的追求。

变难以实现为实际任务

因此，这也是一篇反映我们一直以来探索而未曾达到的目标的故事。全世界对城市的文化和独特识别性的追求，以及为将城市建设为一个整体的生态有机体作出的努力都体现了这一点——在当前和接下来几个世纪各种各样的城市规划任务中，克服片面的城市规划，意识到建立务实的人类形象；城市界面是建筑和城市规划之间的联系点，考虑其角色时不能忽略这一点；作为中心的、还没实现的任务，新城建设的实例尤其展现了将来的梦想："城市将会成为和谐的人类社会"，而不是相反，像现在普遍存在的那样，在单独的规划中，所有的这一切仍然在探索中，社会必须去适应城市规划图纸。

然而，这种目前尚未实现的平衡并不是悲观的。相反，它对所有人来说都是一种敦促、鼓励和引导，正如这个群体对其思想和行动所秉承的那样，继承过去，展望未来——为未来城市中人们的愿景服务。

From thinking to acting

The first part of the book portrayed how the "think and act tank" – the ISA group – was founded, has developed, and where it stands today, in research and practice, and on which philosophies it is based.

This following section reflects on the thinking of significant, individual personalities of the ISA, about their experiences and objectives on fundamental aspects of urban planning, ranging from urban ecology, to city identity, to urban development, to the human urban philosophy. The stage between thinking and acting is what creates this collage-like section demonstrating the entire range of the ISA works, and presents the role of the concept draft as urban planning instrument.

The acting becomes clear through the examination of selected projects, organized in the three major areas of practice of the global urban planning practice: urban development, urban supervision, and urban regeneration. In every one of these working areas, they reflect experience: what are the existing tasks, how can they be approached, how can they be worked on, and what are the possible solutions.

The ensuing published thoughts and projects should stimulate - not to copy blindly – but to develop individual methods from their basis, to encounter every urban planning task freshly, creatively and originally. To support the reader's particular efforts, is the true concern of this publication.

Transforming the unmatched to become the task

This book therefore is also a story of all that has not been achieved. It is obvious everywhere, in the worldwide search for the cultural and individual identity of cities, and the efforts of cities to become an integral ecological organism. It becomes clear in the various undertakings that do now and will soon occur in urban planning, from the decrease of single-sided planning to the realization of a pragmatic conception of man. As a central, but not yet achieved, objective the dream of the future appears in the example of the new towns. "A harmonious human society is to become a city," and not the opposite, as it still is in the reality of so many of our plans: the city, into which the society must fit.

But the outlook of the so far unachieved is not pessimistic. Quite the contrary, it is the invitation, encouragement, and suggestion that everyone – as this group does - reflect and report their thoughts and actions, learning from the past and working for the future – for the future of the people in the future cities.

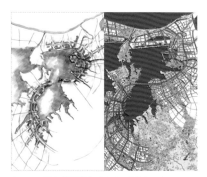 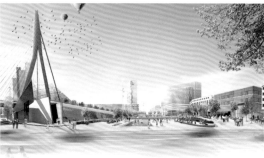  

图 02　从头脑风暴的初步方案到最终的总体规划
Fig. 02　From brainstorming with first ideas to the final masterplan

图 03　临汾市总体规划：以人为本的整体环境规划
Fig. 03　Masterplan for the city of Linfen: overall environmental planning with man as the focus

2. ISA : an international think- and act tank

ISA International Stadtbauatelier is an intercultural think- and act-tank in the areas of regional and urban planning, landscape planning, and architecture. The team is cast internationally and consists of practitioners, researchers, teachers and consultants from over seven different nationalities. It is comparably small, concerning the number of personnel, but highly qualified, competent and responsible. The office is closely linked to a network of highly specialized experts that can be consulted if necessary. It is being operated inter-generationally and inter-culturally, in Europe, Asia and Southern America.

Fundamental way of thinking

The think-tank is based on the philosophy that planning does not only concern the material needs of the society, but also the immaterial needs. It is bound to the principles of true creativity that always asks, "what should happen," and does not simply copy other solutions. A particular emphasis is directed to the traceability of the works, to make these comprehensible for everyone and to pave the way for substantive civil participation.

Diverse range of activities

This thinking and planning institution is working in the areas of regional and urban development planning, land use planning, master- or framework-planning, detailed development planning and design policies. It develops plans for small cities as well as for metropolitan areas, for new towns as well as world heritage sites. Urban development, supervision and rehabilitation are part of its everyday objectives; be it for historical cities, green cities, eco cities or high tech cities.

International planning group

The ISA International Stadtbauatelier planning group is a company over 30 years of age. It was founded in Germany, but has developed more and more to be an international and intercultural planning group. Because of its background – as an outsourced institution of the University of Stuttgart – it was responsible from its very beginnings for work on special tasks in practice, research and teaching, in all areas of architecture, landscape planning, urban- and regional planning within its home country as well as abroad.

2. ISA：一个国际化智囊团

ISA 意厦国际设计集团是一个致力于区域规划、城市总体规划、景观规划和建筑领域的跨文化智囊团。这是一支由来自多个国家的实践者、研究者、教育者和评论者组成的国际队伍。尽管从成员数量上来说，ISA 的规模相对较小，但是每位成员都有很高的专业技能和丰富的经验。欧洲、亚洲和南美洲遍布他们跨时代、跨文化的工作身影。

基本理念

我们的"城市智囊"以这个理念为基础：规划不仅要满足社会的物质需求，而且要满足其非物质需求。集团遵循着真正创新的原则，不断提出"应该实现什么"，而不是单纯地照搬其他解决办法。集团特别重视工作的可实施性，让每个人都能理解，使真正的市民参与成为可能。

多样的工作领域

作为一家咨询与规划机构，我们主要致力于区域性规划和城市发展规划、城市用地规划、城市总体规划或控制性规划、详细规划和法定规划层面的工作。既服务于小城市，又服务于大都市，既服务于新城，又服务于世界文化遗产城市。城市发展、城市控制和城市更新就像历史城市、城市绿地系统、生态城市或者科技新城一样，是 ISA 集团日常工作组成部分。

国际性的设计团队

它是一个有着三十多年历史的集团，成立于德国，时至今日已发展成为一个跨文化的国际设计集团。追寻其起源——斯图加特大学的资源衍生从一开始它就致力于实践、研究与教育的综合任务，工作领域涵盖德国与国际建筑、景观规划、城市和区域规划所有领域。

图 04　与大学间的紧密联系
Fig. 04　The close contact to the university

研究、教学与实践
今天，ISA 成员不仅进行规划实践，也在德国、亚洲和南美洲的重要学术机构进行研究和教学科研工作。在城市规划与建筑领域，三种不同文化的融合成为集团设计团队的重要特征。面对日新月异的实践任务，积累了专业经验的三代设计者同时兼顾国际学术资历、学术研究和教学任务，形成了 ISA 集团的"智库"。

国际协同工作
ISA 意厦国际设计集团的高品质工作是以不同文化的特性和功能在同一项目中的协同配合为基础的。很多项目都是来自不同大洲的团队成员，在日常工作中借助现代科学手段共同完成，从而使一种文化形态下的项目经验和知识，与其他文化形态下的理念和才智紧密结合、互相激发。

3. ISA 意厦国际设计集团的发展史

ISA 意厦国际设计集团于 20 世纪 70 年代末成立于德国斯图加特，其时称为 "Stadtbauatelier 设计事务所"，至今已有超过 30 年的历史。起初，它被定位为一个城市规划、建筑设计和城市改造的规划和实践事务所。它受城市或者政府的委托，进行城市规划领域的工作，解决相关问题。成立之初即与斯图加特大学建筑和城市规划系紧密联系，如今又与欧洲规划协会（EIP）、斯图加特大学文化和工艺研究国际中心机构（IZKT）密切合作。由于大多数的城市规划任务兼具理论与实践的自然特性，ISA 意厦国际设计集团在设计和实施中始终贯彻着研究和实践两条线索。事务所的发展可分为三个阶段：20 世纪 80 年代、20 世纪 90 年代和 21 世纪。

Research, Teaching and Practice
Nowadays the members of the ISA are not only practicing, but are also working in research and teaching in leading universities in Germany, Asia and Southern America. The integration of international qualifications and academic research and teaching in three different cultural spheres is one of the special characteristics of the ISA group. With the professional experiences of three generations of urban planers and architects, the ISA often faces new types of practical assignments.

International synergies
Through the synergy of the characteristics and abilities of different cultures collaborating on common projects, a special quality is derived from the work of the ISA International Stadtbauatelier. On a daily basis, international employees work together on many undertakings.Everyday, projects are worked on by international employees on different continents, with the aid of state-of-the-art technology, bringing new ideas and skills based on their different cultures into the project.

3. The development of the ISA International Stadtbauatelier

The ISA International Stadtbauatelier was founded at the end of the 1970s in Stuttgart, Germany as "Stadtbauatelier," and can look back over more than 30 years of practice. It was designed from its very beginnings as a design and realization office for urban planning, cityscape and urban design. The office is working on tasks and problems in those fields, commissioned by municipal or federal authorities. Being closely linked to the faculty of the the University of Stuttgart architecture and urban planning institute, in its very beginnings, the office nowadays is closely linked to the European Institute for Planning (EIP), an institute of the International Center for Culture and Technology Research (IZKT) of the University of Stuttgart. Based on that fact, most of the ISA Stadtbauatelier's urban planning tasks, in a theoretical or practical manner, combine research and practice in planning and realization. From this, the office developed in three temporal chapters: the 1980s, the 1990s and currently in the 21st century.

图 05 一个国际化的工作团队在项目中采用多种技术——从草图到 3D 动画
Fig. 05 An international team working in a variety of echniques on a project - from sketch to 3D animations

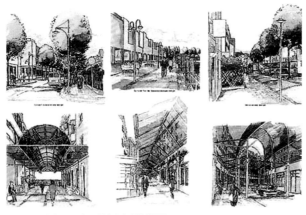

图 06　手绘草图和透视图转变为专业的效果图
Fig. 06　The development of hand sketches and perspectives to professional renderings

Consulting, Planning and Realization in German Cities
From its very beginnings, the office was working on a broad range of planning tasks, starting with the design of urban spaces, e.g. for Leonberg, Ulm, Luebeck, to detailed development plans, framework and urban development strategies for cities like Esslingen am Neckar, Flensburg and Rendsburg. The emphasis of all planning projects was directed towards the three-dimensional development of the village- and cityscape. Design concepts for villages and entire cities, like Moelln or Ludwigsburg, served as a foundation for design recommendations and guidelines.

Having served the German reunification on all levels
In the 1990s, with the reunification of Germany and the experiences of the 1980s, ideas for solutions to new urban development problems were widely sought after. The ISA Stadtbauatelier was commissioned by various state-governments, municipalities, and semi-public urban development institutions to work on the giant task of developing a method of urban development-, structuring- and design planning for former East German communities. During this period, many new projects were added to the office portfolio, especially in the field of integral urban development planning, e.g. in Potsdam, Germany; planning for world heritage cities like e.g. Wuerzburg or Stralsund; and urban design and architectural concepts for prefabricated slab settlements in cities like Jena or Gera. These tasks were carried out simultaneously with ongoing projects in the fields of urban space design, development of industrial areas, urban planning, and urban design in Germany.

Learning from the old cities, to develop new ones
During the 21st century, activities in the fields of public space and industrial planning intensified, complemented by new working disciplines of new town planning for up to 500,000 inhabitants and plans for industrial areas, mostly in Korea and China. Furthermore, innovative street and square designs were planned and realized for several German cities. Guidelines for public space as well as for urban street furniture and public lighting plans broadened the working range of the ISA Stadtbauatelier. A number of these projects were developed in a new form of public-private partnership, and extensive participation processes together with the local authorities. The working fields of commercial area planning, detailed development planning and design guidelines are still an essential component of the works of the Europe offices, in research as well as in practice.

在德国城市中进行的咨询、规划和实施
自奠定初期，事务所就进行了大范围内的规划，从莱昂贝格、乌尔姆和吕贝克(世界文化遗产城市)进行的公共空间规划，到在埃斯林根、弗伦斯堡和伦茨堡进行的建筑设计规划、控制性规划和城市发展规划，所有规划项目的重点是城市形态和村镇的三维发展。我们同时开始进行城乡总体规划，如在 Moelln 或路德维希堡进行的规划，是设计准则和设计条约的基础，也是当地建筑法的基础。

德国统一影响到了各个层面
20世纪90年代，随着两德统一，80年代的经验以及新的城市发展理念遭到了质疑。ISA 意厦国际设计集团应不同州政府、团体和半国家化的城市开发机构要求，为过去的东德也制定了统一的城市发展、城市结构和城市规划条约。在这个时期，事务所承接了新领域的任务。特别是城市总体发展规划，如德国波茨坦、智利的塔尔卡瓦诺，世界文化遗产城市与历史城市整体景观规划，如维尔茨堡或施特拉尔松，以及重要城市更新项目，例如板材建筑房屋的城市规划和建筑规划方案，如耶拿和格拉。这些任务同时代表了德国城市建设同时期的重点，大量公共空间规划、工业区规划、城市规划和城市设计平行进行。

总结旧城经验，建设新城
21世纪，公共空间和产业规划的项目不断增多，还有很多主要在中国和韩国进行的、居民超过50万的新城规划以及工业区规划。德国城市规划工作也在进一步拓展中，例如在德国的埃斯林根和斯图加特进行的系列公共空间导则，包括街道和广场规划，这些项目均得以实施。公共空间的概念规划以及街道小品和城市照明概念规划拓展了 ISA 意厦国际设计集团的工作领域。这些项目部分是通过公私合作伙伴这样一种全新的形式以及社区积极参与得以实施的。工业园区的规划、建筑设计以及规划导则的研究与实践也是集团欧洲总部工作的重要组成部分。

关于 ISA　ABOUT ISA

国际项目的发展

在此期间，ISA 意厦国际设计集团在欧洲和亚洲的项目逐渐展开，特别是在法国巴黎的 Cergy-Pontoise 新城规划以及在亚洲的大量新城规划项目。在韩国首尔开始进行"Sang Am 媒体城"及大型商务中心和居住区规划项目，并在首尔和釜山陆续完成了从室内设计、建筑单体设计到建材区设计等各种工作，随后承担了首尔中心区恩平新城整体规划项目。受市政府的委托，ISA 还承担了 Seoul-Gangseogu 城区（人口约 100 万）的控制规划示范性项目，作为其他城区规划的典范。此后，ISA 在中国和韩国展开了大量建筑设计、公共空间规划、城市更新、居住区规划、新城规划、城市发展专项规划和区域性规划项目。如今在亚洲的工作延展到历史文化保护、旅游度假区规划等更多领域。项目遍布北京、上海、广州、福州、泉州、厦门、重庆、昆明和丽江等城市。旅游规划、新城规划、城市改造和生态规划成为了 ISA 意厦国际设计集团工作的新重点。

Development of international activities

At the same time, the ISA Stadtbauatelier extended its international activities in Europe and Asia – more precisely projects in France, for the new town of Cergy-Pontoise in Paris, as well as larger scale projects in Asia. Starting with the urban expansion project "Sang-Am Media City," including working and living in Seoul, South Korea, followed by various planning projects from hardware store areas to new towns in Seoul and Busan. Additionally, at that time the international work of the ISA Stadtbauatelier began consultations and projects in South America. Urban development plans and urban design plans for Santiago, Chile were developed, and in Talcahuano, Chile urban design plans were developed in cooperation with the Bio-Bio University there. On the commission of the city of Seoul, an exemplary development framework plan was developed for the district of Seoul-Gangseogu (approximately 1 mil. inhabitants) as well as test patterns for other parts of the city. Even now, there are planning projects to follow in China and Korea, in the fields of architecture, public spaces, urban renewal, large-scale settlements, planning of new towns, specific urban development planning and regional planning. At the moment, the range of activities is being broadened once again by planning projects for world heritage sites and holiday regions. Work has been done for the cities of Beijing, Shanghai, Guangzhou, Chongqing, Xiamen and Kunming, along others. Regional planning for the economic sector of tourism, the planning of new towns, urban rehabilitation, ecological landscape planning and city ecology have consequently become new focal points in the working areas of the ISA Stadtbauatelier.

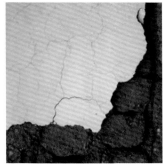
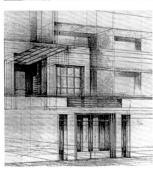
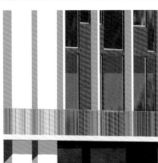

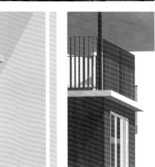

图 07　耶拿高密度住宅区的建筑设计及实施

Fig. 07　Architectural design and implementation of high density housing area in Jena

图 08　对世界遗产城市丽江历史建筑形态的现代诠释。设计师同样可以从立面设计中了解过去，开创未来

Fig. 08　The structure of the existing World Heritage town of Lijiang is a modern interpretation. Even in facade design you can learn from the past to create the future

4. The ISA International Stadtbauatelier today

From Stadtbauatelier to ISA International Stadtbauatelier

Since its founding in the year 1978 until the year 2000, the name of the office was Stadtbauatelier. From 2000 to 2007, the name was changed to SBA Stadtbauatelier, to ensure/assure the memorability of the name in foreign countries. Since the year 2007, the projects of the office are being worked on under the name of ISA International Stadtbauatelier, both to avoid confusion of names, but also to take into account the increasing number of international activities and new partners. The ISA Stadtbauatelier is an international, interdisciplinary planning group, consisting of urban planners and architects, and it can, based on the many experiences within the field of urban planning, count on a well-established and experienced team that receives new stimuli on a daily basis from international composition and cooperation with experts in landscape architecture, geography and interior design.

ISA is working in research and practice in the working arenas of urban planning, architecture, spatial planning and interior design in Europe, Asia and Southern America. It combines long-term experience with new concepts in the areas of urban and regional planning, urban renewal, and city image planning, new towns and new quarters, public space, urban lighting, architecture, industrial architecture, interior design, landscaping and city ecology.

The team of the Stadtbauatelier

The Stadtbauatelier is a team of individuals with particular qualities and a thoroughly personal interest for urban design and architecture. This attitude results in efficient group dynamics, and a shared motivation based on the teamwork of the individuals. The working teams are cast internationally and interdisciplinary. For special tasks external consultants are called in.

4. 今天的 ISA 意厦国际设计集团

从 Stadtbauatelier 设计事务所到 ISA 意厦国际设计集团

考虑到事务所不断增加的国际业务和新增的国际合作伙伴，组建于 1978 年的 Stadtbauatelier 设计事务所逐步转化为 ISA 意厦国际设计集团。自 2007 年起的工作均以此命名进行。ISA 意厦国际设计集团是一个由城市规划师、建筑师、景观建筑师、地理学家和室内设计师组成的国际性、多领域设计团体，通过多年在城市规划领域的工作积累，它已发展成为一个设施完善、经验丰富的团队，贯穿常规工作的国际协作为其发展提供了强大动力。

ISA 致力于欧洲、亚洲和南美洲的城市规划、建筑设计、公共空间和室内设计的实践和研究工作。30 年经验同时仍然在不断汲取与城市和区域规划、城市更新和城市景观规划、新城和城市区域规划、公共空间规划、城市照明设计、建筑设计、工业建筑设计、室内设计、景观规划和城市生态规划领域中的新理念。

Stadtbauatelier 设计事务所团队

Stadtbauatelier 是一个由专业技能突出和对城市规划和建筑设计有浓厚兴趣的人组成的团队。成员之间的协作保证了团队的高效率工作，为团队注入了能量。这是一个跨学科的国际团队，且聘请专业咨询人士担任特殊项目的咨询顾问。

图 09 ISA 意厦国际设计集团北京和斯图加特的工作团队
Fig. 09 Team of ISA InternationalenStadtbauateliers in Beijing and Stuttgart

这样可以通过与其他专业规划人员的协作和交流来汲取更多经验。总体项目负责人组织团队，协调团队的技术与设计协作，所有团队成员之间密切配合。这不仅有利于针对问题找到各自合适的解决办法，也利于形成一种有效的解决方式。这是工作中多样性创意的来源之一。作为一家以规划和研究为任务方向的设计机构，Stadtbauatelier 的工作仍然努力保持多领域发展。

实践工作领域

集团的实践工作主要包括两个领域。一个是区域性规划、城市发展规划、城市总体规划、城市用地规划、概念性规划、建筑设计规划、城市设计导则与法规等；另一个是公共空间、街道与广场的设计与实施，从设计方案到特殊的街道照明和街道小品的布置。工作范围还涵盖独栋建筑、酒店、办公楼、旅馆、高品质住宅区和特殊建筑的设计，甚至室内设计。

研究工作领域

集团成员一直致力于不同课题的研究工作，主要研究课题涵盖"中国历史文化名城保护与建筑更新设计"、"城市夜景规划"、"城市空间序列规划"、"高科技与工业企业园区规划"、"城市建筑设计导则的制定"、"不同文化差异下的城市发展比较"及"中国新城规划"等各个领域。此外，ISA 意厦国际设计集团还进行各项研究工作，例如与 Schlaich Bergermann 及其合伙人事务所（SBP）协同工作，在联邦德国政府交通、建筑与住房部的支持下，完成了"德国工业区气候防护罩"研究项目，该项目荣膺美国"2010 绿色优秀设计奖"。

By this manner of working, the long-time experience in coordinating and exchanging thoughts can be accessed, and complemented by the experts involved in the planning. The responsible project managers organize the team and are accountable for the conceptual and technical coordination, but as much as possible, the team is working together in an equal partnership. This does not only help to find individual solutions for every particular task, but also helps find a specific approach to both a method of resolution and an adequate definition of the problem. The creative variety of projects is partly rooted here. The working field of the ISA Stadtbauatelier as a planning office and research institution embraces a wide range of activities.

Practical working areas

The Stadtbauatelier is working on different scales of project, reaching from regional plans, urban development plans, master plans for entire cities and land use plans, to framework plans, detailed development plans, urban design plans including design guidelines and regulations, to the planning and realization of public spaces, streets and squares. The conception and development of specific street lamps and the street furniture belongs in this field of work as well. In some cases, even singular houses, restaurants, office buildings, hotels, high quality settlements and special buildings are being designed by the ISA.

Research tasks

Members of the Stadtbauatelier are working on different research projects like "Gestaltungsgrundsaetze der Lichtplanung" (design principles in lighting design), "Planung von Sequenzen im staedtischen Raum" (planning of sequences in the public space), "Lebensqualitaet in Gewerbegebieten" (quality of life in trade areas), "Staedtische Gestaltungsrichtlinien fuer die Architektur" (urban design guidelines for architecture), "Kulturelle Unterschiede in der Erfahrung der Stadt" (cultural differences in the experience of the city) or "Methoden der New Town Planungen" (methods of new town planning). Furthermore, the Stadtbauatelier worked on the research project "Klimahuellen fuer Gewerbegebiete" (climate shells for trading areas), in cooperation and under the leadership of the office Schlaich, Bergermann und Partner (SBP), funded by the Federal Ministry of Transport, Construction and Housing (BMVBW).

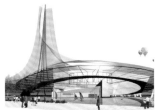

图 10　ISA 意厦国际设计集团各种规模的项目：从城市街道单体设计到区域规划
Fig. 10　The range of the different scales of ISA work: from the individual object in the urban street design to regional planning

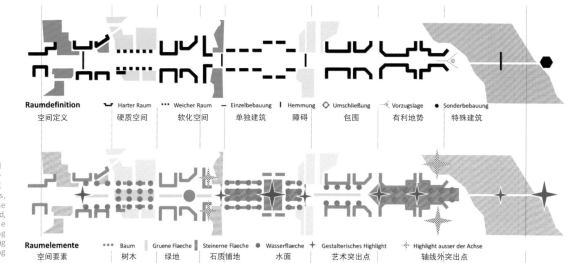

图 11 当前项目的研究内容通常都得以应用，例如大同新城规划——城市空间序列性研究

Fig. 11 In current projects, the contents of the research will be applied, for example the Newtown Planning for the city of Datong – sequence planning in urban spaces

Key personnel of the ISA Stadtbauatelier

The key personnel of the ISA Stadtbauatelier are internationally and interdisciplinary working urban planners, architects and interior designers: Prof. Dr.-Ing. habil. M. Trieb (University of Stuttgart, Germany), Prof. Dr.-Ing. S. J. Lee (Seoul National University, Seoul, Southern Korea), Dipl.-Ing. Dita Leyh (ISA Stuttgart, Germany), and Dr.-Ing. M. Arch. Yajin Zhang (ISA Beijing, China), sharing their responsibility with the associated partners Dipl.-Ing. Phillipp Dechow (Stuttgart) and Dipl.-Ing. Ximi Lu (Beijing).

Awards and competition successes

The ISA Stadtbauatelier has won public prizes such as the award "Beispielhaftes Bauen" (exemplary building for the planning and realization of public streets and squares in Esslingen am Neckar and Ellwangen), or the prestigious "Green GOOD DESIGN Award 2010" for the research work "Klimahuellen fuer Gewerbegebiete" (Climate Shells for trading areas) in cooperation with Schlaich, Bergermann und Partner, consulting engineers (SBP) and the BSVI (Bundesvereinigung der Strassenbau- und Verkehrsingenieure e.V. (federal association of road construction and traffic engineers) for the draft and realization of the pedestrian street in the city center of Ellwangen. The ISA Stadtbauatelier has furthermore achieved several first placements in many international competitions.

ISA 意厦国际规划集团核心成员

集团是一个由城市规划师、建筑师、室内设计师组成的多领域国际性设计团体。发展至今，集团合伙人包括 M. Trieb 教授（德国斯图加特大学规划学教授）、S.J. Lee 教授（韩国首尔大学）、硕士工程师 Dita Leyh（ISA 德国斯图加特总部）、博士工程师张亚津（ISA 驻中国地区总代表）以及联合合伙人硕士工程师鲁西米（中国北京）、硕士工程师 Phillipp Dechow（斯图加特）。

所获公共嘉奖及竞赛获奖

ISA 意厦国际设计集团获得了很多公共嘉奖，例如埃斯林根和埃尔旺根的公共街道和广场的规划与实施被评为"典范建筑"，与 Schlaich Bergermann 及其合伙人事务所（SBP）协同工作完成的"德国工业区气候防护罩"研究项目荣膺美国"2010 绿色优秀设计奖"，埃尔旺根内城的步行街区规划和实施获得联邦街道和交通工程协会 BSVI 嘉奖。

图 12 研究项目"德国工业区气候防护罩"荣膺绿色优秀设计奖
Fig. 12 Green Good Design Award for the research project "Climatic Covers for Commercial Areas"

5. ISA 意厦国际设计集团理念

人是规划的出发点和目标

自 ISA 意厦国际设计集团建立，其工作就一直以人性化的理念和城市设计理论为基础。人是设计工作的核心，不论性别、年龄、种族与文化，这包括几千年以来所有文化背景下的人。人被视为个体，视为独特的自我，他不仅有生理和心理，也有精神。躯体、性别和种族只是人类临时的服饰。人不仅仅有物质需求，也有非物质，即精神和灵魂的需求，其周边的环境必须满足这些需求。

城市是使命

城市就是环境，被视为人类的第三件服饰。衣服是第一件，房子是第二件，城市，特别是公共空间是人的第三件服饰。城市被视为一种缩影，一个小的合成世界，它的任务是满足人类的所有需求，包括人类从出生到去世的所有身体的、精神的和情感的需求。城市是人类生存社会的物质基础和保护外壳，是其存在的缩影。

城市绝非各种建筑物的随意组合，尽管很多城市都存在这样的问题。城市是每个社会的物质存在、社会群体和文化发展的可视性表达。因此，城市可以理解为空间和时间上有创造力的整体设计任务。每个城市都需要自己独特的城市意象，这可以通过城市设计和城市结构来实现，就像设计一座建筑一样。

5. The philosophy of the ISA International Stadtbauatelier

The human being as origin and objective

From the beginnings of the ISA Stadtbauatelier, the work was based on a human philosophy and the urban planning design theory that developed from that philosophy. The human being is the objective of the work in our office. The human being, independent of sex, age, race and culture. The human being, as it has been recognized for thousands of years by all cultures. The human being is understood as individual, unique in itself, not only consisting of body and soul, but also containing a spiritual mind. Body, sex and race are understood as temporarily clothing of the human being. The human being therefore does not only have material but also immaterial, psychological and spiritual, needs that must be satisfied by its environment.

Cities are the task

The cities are this environment. They are considered the third dress of the human being. Clothing is the first, housing the second, and cities – especially the public space – are the third dress of the human body. Cities are considered as microcosms, small synthetic worlds that have the task of satisfying all human needs, its physical, mental and spiritual needs, from birth to death. Cities therefore are nothing less than the material foundation and cover of the human urban society, the microcosm of its existence.

Consequently cities are not to be seen as a coincidental collection of buildings, as becomes apparent in an overwhelming number of cities. They are the visible expression of material existence, the social society and the cultural development of a particular urban community. This is the reason why cities should be seen as creative, integral design tasks, embedded in space and time. Every city needs its own image, a design derived from within, a cityscape that is designed like a building.

图 13　从规划到实施的中国丽江新城
Fig. 13　A city emerges from planning to realisation, Lijiang, China

LEARNING FROM TWO CULTURES

Serving the human needs

The fundamental task of the ISA Stadtbauatelier therefore is not simply the fulfillment of physical human needs – this is only the prerequisite – but additionally and especially, it is the psychological and spiritual-intellectual needs of the human being. That is to say: emotional sensations, qualities of diversification, stimulations or astonishments in the public space, as well as intellectual and spiritual experiences of the human evoked by artistic beauty in the public space are just as important for the quality of life as the physical living conditions, work, communication, etc.

To design spiritual-intellectual experiences, to create an attractive atmosphere with the aid of psychological creative urban guidelines and visualizations is the objective of the projects of the ISA Stadtbauatelier. To achieve this, by the means of the most innovative developments and concrete planning, is always the basis of every project developed by the office. Consequently, urban planning design strategies – like the development of urban design images or the development of design guidelines like unity and diversity, continuity and change, or the sequence of particularity and generality – are nothing more than the urban planning perception of the basic principle of the synthesis of yin and yang, which are the basic design instruments of the ISA Stadtbauatelier.

Cities are to be recognized as art

The philosophy of the office demands to recognize cities as works of art, like the cities of great personalities that have been linked inseparably for thousands of years, in all cultures of the world. Cities as works of art, not to fulfill the aesthetic demands of an over-cultivated minority, not simply to establish a functional and interesting environment, but to give the human being the possibility to promote its own development, through the help of its surroundings.

All of this, because the timeless truth – first the human influences its environment, then the environment influences the human – is valid in the past as well as today. Modern, lively and beautiful cities are the best precondition to form, cultivate and develop humans and their culture, in an emotional and intellectual way as well as in a moral and spiritual way.

满足人类需求

ISA 意厦国际设计集团的根本任务不仅是满足人类的身体需求——这只是前提，特别还要满足人类的心理、精神和智力需求。也就是说，对于生活质量来说，精神实现、生活品位、在城市空间中的刺激与惊喜，以及城市的艺术美给人的智力和精神体验与物质生活条件、工作和交流一样重要。

满足人的精神和智力体验，借助城市规划与城市设计的心理行为学原则和视觉效果来创造吸引人的环境是 ISA 意厦国际设计集团的目标。这种以详细规划的不断更新发展为基础的理念是集团进行每个项目的基础。因此，城市景观规划，如"统一性"与"多样性"、"持续"与"改变"乃至"特殊性"和"共同性"，这些城市规划理念的平衡与动态，就像阴阳原则在城市规划中的转变——是 ISA 意厦国际设计集团重要的设计工具。

城市应该被理解为艺术品

集团理念要求将城市视为艺术品，就像过去几千年中地球上的所有文化在漫长的岁月中对其形成的意识一样。城市是艺术，不只是为了满足少数人的审美愿望，不只是为了创造功能性的和有趣的环境，而是让人类借助环境，在此基础上进行进一步发展。

这一永恒的真理——往往是人先创造城市，然后城市也相应会影响人——不仅适用于过去，也适用于现在。现代的、生动的和美好的城市是塑造、养育和延续人类及其文明的最佳前提，无论是从思想和智力层面，还是从道德和精神层面。

图 14　德国 Murrhardt 市 "Am Oberen Tor" 广场喷泉的设计和实施，与 Atelier Dreiseitl 合作
Fig. 14　Design and realization of the square "Am Oberen Tor" with fountain for the city Murrhardt in Germany, project in collaboration with Atelier Dreiseitl

图 15　作为艺术品的城市
Fig. 15　The city as art

规划一个和谐的社会

对我们工作所有要求的目标是，有意识地将我们的城市规划任务和城市视为、理解并规划为社会的"家"。这一工作意义非凡。城市规划——或者是城市发展，城市控制或者城市更新——只是一种工具，它可以为物质基础、社会可能及和谐社会的文化内涵创造前提。

以集团理论的最高标准来衡量，我们的研究和实践工作从一开始就一直在努力追求"和谐社会"。新城的规划开发是建设和谐社会的外壳；城市设计中的空间序列时序是公共空间规划的核心工具，是在这条道路上的一大进步。在所有规划实践中——从城市发展规划到街道空间设计，我们不断追求质量和数量，并努力实现其平衡，这有时是我们工作的重头戏，但大多数时候只是很小的一部分；同时，我们将研究成果融入实践工作中，在这条道路上不断进步。根据我们的判断，这些成果往往是不够的；但是从"过程就是目标"这个道理来看，我们一直在努力。

Planning for a harmonious society

It is the objective of our work and all of these requirements to recognize the central task of urban planning – to see, understand and plan cities as the home of the urban society. Consequently the task becomes more significant. Urban planning – may it be urban development, city management or urban renewal – is to be seen only as an instrument, to create the prerequisites for the development of the material substructures of social possibilities and cultural perspectives of a harmonious urban society.

Measured on the high expectations of our philosophy, we are just at the beginning with our research and practical work, or at most we are in the early stages. Research concerning the development of new towns as cover for a harmonious urban society, sequential planning as essential instrument in the design process of public space, or the realization of a pragmatic conception of man into the everyday urban planning practice in different cultures, are all steps on a long path. The constant efforts on all practical tasks – from urban development planning to street space design – to see quantitative and qualitative goals at the same level, and to bring those into a balance, are sometimes large, but mostly small steps of a working process. During which, we attempt to walk that long path from philosophy to research into practice. In our own perception, the results are always inadequate, but it is from the knowledge that, "the way is the goal," where we derive our motivation.

理论文章
ARTICLES

01 设计塑造生态：
不使用高科技的生态新城
Ecology by Design:
Ecological New Towns without High-Technology
Dita Leyh

02 从经济起飞到生活质量：
实践与研究中城市规划的经验和任务
From Economic Miracle to Quality of Life:
Urban Development Experiences &
Tasks of Practice and Research
Michael Trieb

03 我们应以何种风格建造？
——21世纪初建筑学定位的探讨
In Which Style should We Build?
Architecture at the Beginning of the 21st Century
—A Postitioning Approach
Phillipp Dechow

04 从数量到质量：
韩国城市规划控制的经验
From Quantity to Quality:
Experiences from City Supervision in South Korea
Seog-Jeong Lee

05 新城：有关城市的梦想
New Town: Dream City
张亚津

设计塑造生态：
不使用高科技的生态新城
Ecology by Design:
Ecological New Towns without High-Technology

Dita Leyh

For forty years now, ecology has played a major role in German architecture and urban planning. This has manifested itself both in the form of legislation, and in industrial trends toward a specialization in the manufacturing of ecological products. This tendency is also reflected in the ideology of its people. Politics attempts to influence sustainable urban development in Germany through legislation. To this end, the following three laws are essential:

- The Intervention Regulation: a law that dictates an ecological compensation for urban interventions; also called intervention-compensation regulation
- The Energy Savings Ordinance (EnEV): The Energy Savings Ordinance is a law regulating the energy consumption of buildings
- The Waste Water Ordinance

The "Intervention Regulation" is a law, demanding an ecological compensation for every urban intervention. The ecological consequences of every urban project are observed by environmental studies that also determine the nature of compensations used. Examples of possible compensations could be the creation of biotopes, the planting of trees, or the creation of green rooftops. These measures can be executed within the planning area itself or outside of it. The "Energy Savings Ordinance" dictates the maximum allowable annual energy consumption for a new or rehabilitated building. In the current version of this ordinance, this is 40-80 kWh/(m²a). And, the "Waste Water Ordinance" demands a tax for any kind of wastewater draining on a plot of land, including rainwater. This is intended to promote on-site retention and infiltration as well as the evaporation of rainwater. This can be achieved for example by greening rooftops, water permeable ground coverings and unsealing of the ground.

These laws are supported by various subsidy programs, such as "The Social City" or the program of internal development and redensification of inner cities. Another major support, implemented by sustainability focused politics, are subsidies provided to companies specializing in the production or marketing of ecological products (e.g. solar cells, systems for the use of alternative energy sources, insulation, etc.). With this background in mind, the increase of eco-friendly urban areas in Germany is not surprising. But, the success of such urban areas is determined in no small part by the attitudes and lifestyles of its inhabitants. The foundation of all of these developments is an open minded citizenry who are well educated in ecological matters. Schools, ecological organizations and clubs, and the plethora of ecologically oriented studies at the universities support this from an early age.

近四十年来，生态在德国的建筑设计和城市规划中扮演着重要角色。相关政策法规的制定、生态产业的发展以及市民们世界观的改变体现出了这一点。政策上人们首先试图通过立法，来影响德国城市的可持续发展。以下这些法律起着重要作用：

- 《干预条例》：对城市建设造成的生态损失加以平衡补偿的条例，也可称为干预平衡条例；
- 《节能条例》（《EnEV》）：对建筑物的能源消耗加以规定的法律；
- 《排水处理条例》。

《干预条例》要求所有的城市建设始终实现生态平衡。建设行为的生态影响通过环境报告确定，由此进一步确定生态平衡的措施。相关措施可以是创造群落生境、植树或者屋顶绿化等。这些措施可在规划区内或规划区外加以实施。《节能条例》对一个新建或改建建筑每年的能源消耗量加以规定。根据目前的条例规定，应该是40～80千瓦时/(平方米·年)。《排水处理条例》要求对地块产生的污水(包括雨水)缴纳税收。这些税收用于促进地块中雨水的储存、渗透和蒸发，屋顶绿化改造，以及可透水地面的建设和不可透水地面的改造。

还有一些其他的相关辅助计划，例如"社会化城市计划"和城市内的其他发展计划，例如用于防止城市无序蔓延并且要求加大建成区的城市密度。可持续发展政策的另一个重点是对生态产品相关的产业分支进行援助（例如太阳能电池、替代能源系统、建筑保温隔热材料等）。广大市民对生态问题的积极态度是所有这些相关政策和发展的基础。这些认识从幼年起就开始培养，后期又通过学校教育、生态机构和协会或者是大学里以生态为主题的课程得以深化。在这种情况下，出现越来越多的生态社区也就顺理成章了。社区居民的态度和生活方式在很大程度上决定着这种社区的成功出现。这里不只是考虑一种普遍适用的解决方案，还需要有实施所在地的社会基础和对可持续生活方式的认同。

图宾根的"法兰西社区"是这种"生态住宅区"的典型代表;这是一个位于过去法国营房区的新的城市拓展区,这里有6000居民,2500个工作岗位。营房的历史建筑和新建建筑被整合融入新的城市规划结构中。这里通过一定的措施保障了可持续发展、混合用地和短途城市的实现,例如要求容积率达到3,在同一个建筑中融入各种不同的功能,如办公室、作坊或者是有居住功能的艺术家工作室。

交通方面引入了汽车分享系统,并且建设了集体车库,使得汽车被停放在社区的周边。"法兰西社区"的一个典型特征是业主的直接参与和共同开发的模式——"集体开发模式"。

该地区按照传统高密度的欧洲城市风格被分成小块,每个地块允许建设一座最高5层的城区住宅,这些住宅共同组成一个组团(Block)。每个地块都有由多个业主组成的开发团队,他们共同委托同一个建筑师,同建筑师一起实现各自的住宅设想。

这种业主集体开发实施建设的方式一方面缩减了开支,实现了各个业主的愿望,另外还形成了社区感,使居民在搬进来以前就有了很好的邻里关系。该社区由此获益于丰富多样的建筑和良好的社区互动。这个社区的成功建立在前述因素的基础上,此外居民的协作、市政府的协调组织、公共参与以及开发团队的团结一心也是积极性的促进。

相当一部分城市化进程很快的国家,必须快速规划和建设新城,在这种地方,城市可持续发展的法律、政治、科技和社会基础往往很不健全。没有高科技,没有适用的法律,人们应该怎样规划和建设生态城市呢?为了回答这个问题,我们应该回顾一下过去各个地区建筑和城市建设的区域性传统。

What this also means, however, is that these same concepts cannot simply be applied everywhere and expected to produce similar results. The social basis for acceptance of a certain lifestyle must first exist to ensure the success of eco-friendly urban areas.

One example of an "ecological city" is the "French Quarter" in Tuebingen; a city expansion for 6000 inhabitants with 2500 employment opportunities, situated on the site of a former French Army Base. The historical barrack facilities have been integrated into a new urban structure, complemented by new buildings. Measures have been taken to ensure sustainability, mixed use and short distances. A floor space ratio of 3 is mandatory, buildings have been zoned for a variety of different uses such as offices, workshops and studios with residential use, allocated within the same building. Car sharing systems have also been established with private cars being parked in car parks at the margins of the quarter.

Especially characteristic of the French Quarter is a new kind of direct, owner based and collective realization of construction projects, called the "joint building venture". The area is divided into small construction plots, in reference to the traditional European city with its high density, and offers every plot with a townhouse of up to five stories, creating an urban unit with the neighboring buildings. Every urban unit constitutes a group of eventual building owners, who collectively commission an architect, to develop their vision of a block of individual houses "under one roof". This direct realization without the detour of a real estate developer results in lower costs, realization of individual preferences and creates a sense of community that encourages good neighborhoods even before anyone has moved in. The quarter benefits from lively, diversified architectural presence as well as a positive social environment.

The success of this example is largely thanks to the previously mentioned fundamentals, but perhaps owes even more to the participation of the prospective inhabitants of the quarter, as well as a strong support of the planning process by the urban administration, helping with mediation, participation and issues like the conjunction of building ventures.

In many countries currently experiencing rapid urbanization, where entire cities must be planned and constructed at an extremely fast pace, these legislative, political, technological and social fundamentals of sustainable urbanism are still sorely underdeveloped.

图 01　图宾根洛雷托广场的生态住宅区"法兰西社区"
Fig. 01　Ecological settlement "French Quarter" in Tuebingen, Loretto square

从两种文化中学习 LEARNING FROM TWO CULTURES

But how is it possible to plan and build an ecological city, in the absence of appropriate regulation, and without high technology? To answer this question, we must first examine the past regional traditions of architecture and urbanism in each particular area. In traditional architecture and urbanism, principles have developed over centuries that have adapted each city perfectly to its setting and climate. The white villages in southern Spain, for example, are constructed with predominantly local materials, and in order to protect its citizens from an over abundance of solar radiation, the buildings are built in narrow lanes, producing a lot of shadows. The white color of the buildings also helps by reflecting the sun's rays. Many cities in hotter regions use these principles, culminating in canopied or built-over streets, as can be observed in traditional Moroccan cities. Another popular architectural element used, are water basins in courtyards, whose cooling effect is a result of evaporation. Small openings to the outside, insulating clay used in construction, as well as a favorable surface to volume ratio are also useful in keeping heat outside buildings.

In Islamic cities like Dubai, Yazd/Iran or Hyderabad/Pakistan, natural air conditioners, "wind towers", called "Malqaf" or "Badgir", have been built over the centuries, that lead the cooling night breeze into the houses and push hot air out. Extreme examples of protection from heat and cold are the cave houses in Goreme, Cappadocia/Turkey; Artenara, Gran Canaria/Spain, and the ground houses in Matmata/southern Tunisia, completely dug into the ground, protecting inhabitants from the elements and limiting the consumption of construction resources.

在超过几百年的传统建筑和城市建设发展进程中，城市会自然产生各种规则，让城市可以很好地适应周边环境和自然气候。例如西班牙南部的白色村庄，它一方面利用了本地的材料，一方面又通过狭窄的小巷形成阴凉，避免受到太多的阳光直射，白色的立面还可以把阳光折射回去。许多在炎热地区的城市都会利用类似规则，以形成有着局部或者完全遮阴的街道空间，例如摩洛哥的一些传统城市。它们还在内院设计水景，通过水汽蒸发降温。减小开窗面积，利用隔热的建筑材料以及较小的建筑体形系数（就空间体积而言相对小的建筑表面）可以将热量更好地阻挡在建筑之外。伊斯兰的城市，例如迪拜，伊朗的亚兹德以及巴基斯坦的海德拉巴，在过去的几百年安装的都是天然的"空调"——"风塔"（所谓的Windcatcher温凯驰自然通风系统）：它将晚上的凉风引入房间，并将内部的热空气引出。

位于土耳其卡帕多西亚的格雷梅和西班牙阿特纳拉的达加那利的穴居以及位于突尼斯南部马特马他的地居是对抗酷热和严寒的实例，它们融入地形以抵抗极端的气候并降低材料消耗。城市的建筑密度和建筑体形系数也可为寒冷地区提供更好的建筑保温性能。柏林18到19世纪建设的高密度街区就是一个很好的实例，这里每平方米的供暖能源消耗要远远低于独户住宅区，此外还可使供暖管道更短和基础设施更为紧凑等优点。

图 02 马拉喀什（左图）与柏林（右图）的航拍图比较
Fig. 02 Bird's view of Marrakech (left) in comparison to Berlin (right)

图 03 希巴姆(也门)的黏土高层建筑，格雷梅(土耳其)的穴居和迪拜的风塔
Fig. 03 High-rise buildings built of clay, Schibam/Jemen, cave-houses in Gorem/Turkey, and wind tower in Dubai

在酷热和严寒之外，城市设计还应该考虑其他的天气影响，如强降雨和降雪。位于德国汉堡、瑞士伯尔尼和捷克泰尔奇的拱廊街就是很好的实例。在这种情况下，通过非常简单的方式，就能不受天气影响地购物，而不必建设完全封闭的购物中心。黑森林民居上带有突出的大茅草屋顶，可以在夏天阻挡日晒，冬天（由于日照角度的改变）获得充分的日照。这种屋顶形式（四面倾斜的屋顶）可以减小风阻力的作用面积。在寒冷的冬天，安置在房子卧室下面的圈棚也可以产生热量。在屋顶下储存的干草也可以加强保温效果，将热量保留在屋里。

传统建筑的建造往往已经下意识地应用了可持续发展的原则。涉及的建筑材料大多可以再生，并且一般都来自周边地区，这样避免了长途运输。所有这些措施不仅仅会给城市带来技术生态和气候上的好处，也会赋予城市独特的识别性。

当然，在全球化的时代背景下，我们的生活方式和要求也发生了转变，但是因此就一定要以牺牲城市的识别性和可持续性为代价吗？我们为何不能把过去几百年积累的经验以一种与当今需求相适应的方式加以利用？观察现代国际城市的发展趋势就会发现，我们已经丢弃或忽视了很多经验。当今的城市扩展在世界范围内趋于相同。对完全不同气候区的新城进行比较，就会发现这些地方往往有着相似的城市规划结构、建筑类型和建筑材料。例如，位于莫斯科新型商务区的莫斯科国际商务中心——这里的气温有时低达-20°C（大陆性气候，冷温带气候区），我们会在马尼拉（热带气候，半湿润区，最高可达35°C）、北京（温带大陆性气候，潮湿，夏季炎热，冬季最低-20°C）或者迪拜（亚热带干旱区，40°C）看到类似的建筑和城市结构。

The high density of a city, and a resulting favorable surface-volume ratio, also offer protection in colder regions. This became apparent in the very dense "Gruenderzeit" quarters of late 19th century Berlin which had much lower energy consumption levels per square meter than a detached single family house. Not to mention the shorter service lines and more compact infrastructure, etc. Not only heat and cold, but also different weather phenomena like heavy rain and snowfalls have been considered in the design of cities. Arcades like in Hamburg, Germany, Bern, Switzerland or Telc in the Czech Republic are all fine examples. The largely overhanging roofs of Black Forest Houses are designed to protect the building from a variety of different weather phenomena, as well as provide insulation and act as constructive sunshields, allowing only the light of a low winter sun to enter the house. The form of the roof "Krueppelwalmdach", inclined to all sides, reduces the wind loading of the house. The placement of the stables below bedrooms brings additional warmth in cold winters. The hay supplies traditionally stored under the roof, support insulation as well, helping retain heat inside the house.

In the construction of many traditional buildings, sustainable principles have already been in use unconsciously for centuries. Long transport routes were eliminated by the use of local, renewable or mostly renewable materials. All of these measures not only have technical, ecological and climatic advantages, but have also served to develop the individual identity of each city.

Our lifestyles and needs have obviously changed over time as a result of globalization, but must that necessarily lead to a loss of identity and lack of sustainability? Can we not use the knowledge collected over the centuries and adapt it to the needs of the present day? If one looks at the development tendencies of the modern international city, one cannot avoid the feeling that a great deal of knowledge has somehow been lost, or simply ignored. Contemporary city expansions seem very similar throughout the whole world. If one compares new cities and towns in disparate climatic regions, the same urban structures, building typology and materiality, can be found everywhere. One example is Moskva City; a new business district in Moscow, where temperatures can drop as low as -20° Celsius (continental climate, moderately cold climate). An almost identical architecture and urban structure can be observed in Manila, (tropical climate, alternating wet/dry) where temperatures can climb as high as 35° Celsius), as well as in Beijing (moderate continental climate, humid, hot summers) with temperatures as low as -20° Celsius during winter, or Dubai (subtropical, arid) where temperatures can reach upwards of 40° Celsius.

图 04　捷克共和国的泰尔奇地拱廊、德国的黑森林民居
Fig. 04　Arcades in Telc/Czech Republic, Black Forest House/Germany

从两种文化中学习 LEARNING FROM TWO CULTURES

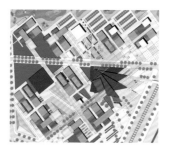

图 05　中国的广州南沙在新行政中心通过一种现代的形式应用了加顶街道的原则
Fig. 05　Guangzhou, Nansha, China: the traditional principle of covered streets to protect from the glaring sun, has been updated into a more contemporary form

In all of these locales, civic identity and reco-gnition value is forged solely by a few landmark buildings, designed by a select handful of star architects, while ordinary urban spaces are often difficult to differentiate.

Which principles of the traditional, sustainable city could be introduced in the contemporary planning of a new town or quarter? And how can sustainable cities continue to emerge using present available resources without using high technology; especially in countries lacking the political and social fundamentals for sustainable urbanism?

Through analysis of the traditional city, it becomes clear that sustainability is not solely determined by urban planning, but is rather the result of a conglomeration of people and institutions involved in the planning, construction, administration and daily use of a city. The diagram below outlines some of the related disciplines (administration, citizens, urban-, landscape-, traffic-planning, architecture, sociology, infrastructure, etc.) that must cooperate to realize sustainability efforts. Only with the coordinated cooperation of the these groups, can a truly sustainable city can be realized.

The focus therefore lies on the following two areas:
- City and nature: How can a city be built, that uses local resources and adapts itself to its own environment?
- City and human being: How can urban design influence the behavior of the citizenry toward sustainability?

The following resources are vital to the first topic area:
- Sun / Wind / Water / Landscape / Local materials

Natural resources such as sun, wind, water, landscape and local materials relate to the former. Concepts such as compact cities, mixed use, transport and public urban space relate to the latter.

The following examples illustrate these principles:

1. The City and Nature

The City and Wind
In planning the new government district of Nansha, China as a new district in Guangshou/Canton and the future center of the Pearl River delta, the abundant summer winds are being utilized to cool the city. All buildings in the city are oriented southwest/northeast, so that the cooling summer breeze from a southeastern direction blows naturally through the streets. A forested area on the northern boundary blocks the interfering winter chill from the north. Regarding the buildings, traditional regional principals of architecture are being newly interpreted – the narrow yard of residential buildings for example, promotes cooling with its stack effect. The new buildings in Nansha are being designed as "twin buildings" surrounding a corridor-like yard. In addition to the double skin facade, the traditional cooling system is enforced and adapted to the needs of the present day.

似乎只有通过明星建筑师设计的地标建筑才能创造一个城市的独特性和识别性。其他的日常街道空间则日益趋同，难以互相区分。

在一个新城或者新城区现代规划工作中应该采用哪些传统的、可持续发展城市的原则？应该怎样在缺乏城市可持续发展的法律、社会和政治基础的国家，利用现有的资源而不是高科技来建设可持续发展的城市？

通过对传统城市的分析，我们应当明确一个原则——"学科交叉原则"。可持续发展不是由城市规划决定的，而是所有参与规划、建设、管理和城市使用的人员和机构相互协调的结果。这些原则来自以下方面：行政管理、市民、城市规划、景观规划、交通规划、建筑学、社会学、基础设施等。只有这些原则相互作用，不同措施相互协调才能建设出一个真正可持续发展的城市。

重点在于以下两个领域：
- 城市和自然：怎样建设一个城市，让它最大限度地利用周边的资源，并与之相适应？
- 城市和人：怎样通过城市规划来影响人们在可持续发展层面上的行为？

以下这些资源对于第一个主题来说起着关键作用：
- 日照、风、水、景观、当地材料

第二个主题以"短途城市"这个原则为基础：
- 紧凑型城市、混合用地、交通、公共空间

下面我们就通过规划设计实例来解释一下这些原则。

1. 城市和自然

城市和风
在中国南沙新行政中心的规划中，广东的新城区和珠江三角洲未来的中心将会成为夏天城区降温的主要风向。所有的建筑都是西南/东北走向，这样夏天凉爽的东南风就可以穿过街道。与之相对，冬天的北风也会被林地阻挡在北部地区的边界。对于这个地区的建筑物的平面布局，结合传统的当地建筑原则进行了重新阐释——传统住宅的窄高型庭院可以通过烟囱效应促进降温。办公区的新建筑被设计成"双子楼"，两座相对的建筑环绕着一个狭窄的走廊式庭院布局。结合双层立面既可以强化传统的降温系统，又可与当今的需求相适应。

城市和日照

在不同的气候条件下，有些地区需要利用太阳的光和热，但是在很多的地区却要阻挡它。在上述南沙项目中，就采用了传统的加顶街道，形成了栅栏屋顶形式的人行通廊，同时把所有重要的建筑连接在一起。

城市和水

南沙位于亚热带湿润季风气候区，这里的强降雨在设计和科技上有很大的应用潜力。屋顶和人行道上的降雨，在流入一个大型蓄水池前将经过水池、梯级和芦苇丛的层层天然净化。通过这种方式，人们可以感受到水的净化，并在设计中将其与景观规划联系在一起。湖里的水达到了沐浴水质，可应用于一些休闲活动。这个系统可以通过雨水的保留、蒸发和渗透减轻现有排水装置的负担，同时冷却水体上方的空气。

城市和绿地

城市可以理解为一个有机体，它也像居住在其中的人类一样有着相同的需求——光、空气、水、太阳、能源。这些需求可以通过城市的周边环境得到满足，这样城市也会与其周边环境和谐共处。但是，这种共处对城市来说常常是单边的索取。功能有机体的目标是在城市和自然间形成一种互补的循环。

中国生态城市陈家镇位于上海附近的崇明岛，在这一规划中就尝试实现城市与周边地区的共处。核心城市嵌入不同的景观环中，这些景观环像洋葱表皮的纹理一样围绕城市向外扩展。每一个环都有一个不同的重点。生态发展第一个环由系列人工湖形成，它可以净化雨水和污水，使其可以再利用。从人工湖挖出的土也可以用来创作小型山地景观，把风引向理想的方向。第二个环由生物农业形成，它可以直接提供食物，避免长途运输，也为居民提供了邻近的工作岗位。长江沿岸最外层的环作为自然鸟类繁殖区受到严格保护。

理论文章 ARTICLES

The City and Sun

Depending on the climatic conditions, in some regions, the light and warmth of the sun can be utilized as a resource, but in many cases steps must also be taken to guard against the damaging effects of the sun in hotter climate areas. In the Nansha project mentioned above, the traditional principle of roofed streets has been applied, to create a network of covered walkways, connecting all major buildings. The street roofs are constructed with movable wooden blades, in order to maintain the highest level of flexibility.

The City and Water

The extreme rainfalls that come with the subtropical monsoon climate in Nansha have also been exploited for design and technological potential. The water collected off building and street roofs is naturally purified through a series of basins, steps and reed areas before it pours out into a larger retention basin. The purification process of the water can be made visible, and is being integrated into the overall landscape design. The resulting lake offers a high water quality, and can be used for bathing and other leisure activities. At the same time, the system relieves the existing canalization through the retention, evaporation and filtration of rainwater, and does so as it cools the surrounding air in the process.

The City and Landscape

The city can be seen as organism that has the same needs as a human being: light, air, water, sun and energy. These needs are being fulfilled by the city's environment; the city exists in a close symbiosis with its surroundings. The aim of a functioning organism should be a mutually complimentary cycle between itself and nature, but unfortunately in the case of the city, this symbiosis appears to be decidedly unilateral.

In the Chinese Ecocity Zhenjiazhen, on the Yangtze-island Chongming, close to Shanghai, an attempt has been made to develop this kind of symbiosis between city and nature. The features of the metropolis are embedded into multiple landscape rings that expand around the city in layers, like an onion. Each ring has its own focus. The first ring is a landscape of lakes, cleaning and collecting the rain and wastewater of the city, preparing it for reuse. The earth moved in the process of creating the lakes has then been used to create a landscape of hills that lead the winds in a favorable direction. The second ring is utilized for biological agriculture that provides the city with its food, and creates jobs for its citizens. The outer ring, at the waterside of the Yangtze-river, is used as nature preserve and a breeding area for birds.

图 06 广州南沙应用了有大型蓄水池的天然雨水管理系统
Fig. 06 Guangzhou, Nansha: the natural rainwater management system with large retention basin in front

图 07 中国陈家镇新城与各具特色的生态景观环形成的外部环境和谐共处
Fig. 07 Chen jia chen, China: a new town in symbiosis with its surrounding of characteristic landscape rings

图 08 德国埃斯林根购物街改造前后
Fig. 08 The changing face of a shopping street in "Esslingen am Neckar", Germany, before (above) and after conversion (bottom)

A city such as this that doesn't just take its landscape into consideration but actually internalizes its functionality, embeds itself in its own landscape in a much more profound way than most. This can have a huge influence on its overall ecological impact and balance. In Zhenjiazhen, parks have been arranged in a linear network, so that they are accessible - roughly 5 to 10 minutes on foot - from every part of town. By creating recreational areas in the city, excess leisure and weekend traffic to more remote natural areas can be minimized.

2. The City and its people

The design of a city has a major influence on the behavior of its inhabitants. One main goal is a "city of short distances"; a concept in which many factors come into play. A well designed public space that offers interesting sequences of squares, greens, contractions and expansions, based on the pedestrian pace, can make distances seem shorter than they actually are, and thereby encourage people to get where they are going on foot instead of using other means of transportation. The network of streets can also be laid out so as to actually shorten transport routes. Diagonal streets for example, shorten the distances within a gridded city by large means.

The redesign of a shopping street in Esslingen am Neckar/Germany into a mixed used street, illustrates this concept. Because of initial protests against its conversion to a pedestrian street, the shopping street has been changed into a mixed-use area, including some parking spaces. A multifunctional construction system develops an independent space, and guides visitors, line of sight away from the heterogeneous postwar facades, toward the newly created urban space, while also managing to connect both sides of the street. This "space within a space" creates a more coherent appearance which gives the street a new identity.

Two years after its completion, the same people that protested initially, expressed wishes that the street be converted into a true pedestrian thoroughfare. Since this second conversion, the "Bahnhofstrasse" has become one of the busiest shopping streets in the city. It completes the complex system of pedestrian walkways in the city, becoming one of the final points in the car free city center.

不仅仅是镶嵌在自然中的城市"景中城","城中景"的布置也影响着城市的生态平衡。陈家镇的城市公园按照线性网络分布，这样，从城市的任何一个地方都可以步行5～10分钟到达休闲区。通过创建邻近休闲区，可以减少通向偏远自然区的额外休闲和周末交通。

2. 城市和人

城市规划对居民的行为有很大的影响。首要的目标是实现"短途城市"。在这其中，很多因素都起着作用。一个设计良好的公共空间应该包含空间次序多变的广场、绿地、紧凑或宽敞的街区，其时序安排的趣味性，使人感觉到的心理距离比实际物理距离要短，促使人选择步行。街道网的安排也应该尽量缩短交通路径。例如在一个网格路网中，对角线路径就可以大大缩短交通距离。

德国埃斯林根的购物街改造实例清楚地表明了这一点。最初地方民众反对将其改造为步行区，因此街道被建造成了有停车位的混合功能区。我们为此设计了多功能的支柱系统，形成一个独立的空间，也将人们的注意力从战后建设的异类立面转移到了这一构筑物创建的新空间上。这种"空间中的空间"联系街道两侧，同时形成了一种完整的意象，赋予街道新的识别性。

建成两年后，使用者就表达了将其完全转变为步行区的愿望。调整后，火车站大街成为城市中最受欢迎的购物街。它完善了网络状的城市步行系统，也构成了无机动车城区的重要主干结构。

短途城市的前提是城市密度。但是仅仅有城市密度是不够的。设计和城市风貌也起着本质的作用。如果达到了高层住宅区的密度，建筑由于大尺度的日照间距而疏离，就会出现适得其反的效果——过大的间距和尺度使得人们即使在高密度下也要求助于汽车。在一些需要高密度的情况下，如在亚洲大都市中，人们需要建设高层建筑满足密度上的需要，但是除了垂直向上的密度，也应该在水平向上建设裙房建筑或者多层的组团Block，形成连续的街道空间界面，创造有变化的街道空间，在广场和组团的转角处仍可建设高层。

韩国首尔的城市更新实例解释了这个原则：

首尔的许多居住区都是小的村庄地块通过后期水平向上的密集化形成的。这种高密度和由此产生的绿地和公共空间的短缺问题目前已经得到了一定程度的解决：将高层住宅楼拆除并建成更高密度的住宅区为城市更新提供了财政支持。这种雷同的公寓楼在一定程度上破坏了已经形成的社会结构，还破坏了街道公共空间。

2010年以来，我们开始进行"首尔新愿景"这一项目，尝试将住宅建筑的水平密集化与竖直密集化相结合。现有的社区应该在市民的参与下重新分块，以创造休闲空间和有趣的公共空间。较低的4～5层建筑应该作为空间界面元素保留，并进行重新组织，通过个别的高楼对其加以补充，以实现期望的密度。单个的住宅区通过开放空间相互联系，这样将逐渐形成遍布整个首尔城市网络结构的步行系统。

中国的历史古城丽江——列入联合国教科文组织世界遗产名录，也应用了传统的公共空间和建筑原则，建成了丽江的新卫星城——玉龙新城。其中主体地块的划分符合机动车的现代交通需求，地块内部街巷和广场则互相联系形成连续的步行网络，传统城市的路径关系在新的城市形态下进行了全新演绎。

One major factor for a "city of short distances" is its density, but it is not only the density that is crucial. Design plays a major role as well. If, for example, density is achieved by high-rises where the buildings are spread out because of lightning, the opposite effect can occur. The undefined empty spaces between buildings can actually encourage people to drive more than walk. As in many Asian metropolitan centers, certain densities can be reached by utilizing high-rise buildings, but such structures should be combined with horizontal densification. This means creating small but interesting urban spaces using neighboring ground floors, or creating city blocks with continuous building alignments, and accentuating certain squares or corners through the positioning of high-rise buildings.

Examples in South Korea illustrate this principle.

Many residential quarters in Seoul have developed as a result of the informal densification of small parcels, originating from traditional village structures. The extreme density and the resulting lack of greens and public space are usually solved through the demolition and construction of residential high-rise buildings with a higher density that finances the renewal. These homogeneous apartment buildings destroy the grown social structures and negate the public space with delimitation towards the street.

Since 2010, a new project called "A new vision for Seoul" has attempted to integrate new forms of horizontal densification in Seoul's residential areas alongside the vertical densification. The existing quarters are to be newly parceled, with participation of the citizens, to secure open spaces and other interesting urban spaces. The existing low rise, 4-5 story constructions, will be kept as space creating volume, newly organized and completed by high rise buildings that produce the necessary density. The residential areas will be linked together by a network of urban spaces, to allow a pedestrian network to develop over time through the city of Seoul.

In the historic Chinese city of Lijiang, a UNESCO world heritage site, the principal public spaces and architecture of historical importance have been transferred to the new satellite city Yulong New Town. The construction sites, adapted in size to the demands of motorized traffic, are also accessible via an internal system of car free alleys and squares that reinterpret the street-layout of a traditional city.

图 09　玉龙新城的社区、新城的远景图和施工现场
Fig. 09 Quarter of Yulong New Town and perspective construction site of the new city

从两种文化中学习 LEARNING FROM TWO CULTURES

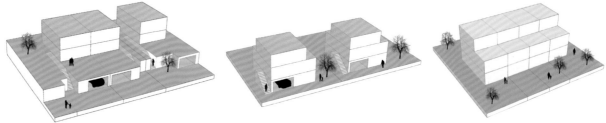

图 10　上图：奥芬堡新建区小巷整体规划图
　　　　下图：根据不同居住类型划分的典型住宅院落
Fig. 10　Above: Masterplan of the new construction area Offenburg Seitenfaden, Germany
　　　　Below: Typical residential yard with group of different residential typologies

The density of a city is the basis for another principle of sustainable mixed-use cities. Without a certain minimum density and number of inhabitants, supply facilities like supermarkets, kindergartens and schools cannot be planned, because they require a catchment area of a certain size. If density is high enough, facilities for daily needs should be planned poly-centrically and allocated to specific areas. The following chart shows the distribution of service facilities within a 10 minute radius on foot in the new satellite city of Shanghai: Cheng Qiao New Town.

Apart from the mixing of services and residential areas, "social mixture" is another important factor for sustainable cities. Different residential typologies should therefore be mixed within each city district to attract various levels of population. Not just the district, but the building as well, should contain varying sizes and floor layouts, to better meet the budgetary needs of potential buyers or tenants. Social housing can be integrated in mixed quarters without attracting attention. Through the social control of the inhabitants, as well as the promotion of mutual tolerance in the neighborhood, many issues and problems of social housing blocks can be avoided.

城市密度也是功能混合的前提。功能混合是可持续发展的另一个重要原则。没有达到特定的人口密度就无法规划临近居民区及步行可达的生活服务设施，如超市、幼儿园或学校，因为这些设施都需要其涉及范围内的居民数达到一定的水平。一旦确定了密度，就应该尽可能以多中心的用地结构去规划这些分属各个社区的日常生活设施（图11展示了上海卫星新城——城桥新城步行十分钟以内可达的服务设施分布）。

除了服务和居住的混合，"社会混合"也是可持续发展城市的重要组成部分。同一个社区内的居住类型应该这样混合：满足不同层次的人群对居住形式的不同喜好。不仅仅是在同一个社区内，同一栋建筑里的住房大小也应随买主或租户的预算有所不同。社会性住宅也可以通过这种方式"悄悄"融入混合型社区中。一方面通过政府的阶层混合控制，另一方面通过邻里之间的互相宽容，避免单纯的社会性住宅社区的许多问题。

城市规划中的建筑物布局可以在很大程度上加强邻里之间的交流。以奥芬堡的新建区规划为例，多层住宅、联排住宅与独栋住宅共同围绕形成一个个共同的"住宅院落"，院落同时是一个尽端路，满足居民的交通需要。住宅院落为邻里交流提供了平台，在停车取车、互相照看儿童或庆祝活动等行为中实现交流。

所有这些例子都体现出了规划与可持续发展之间的紧密联系。几百年来的规划原则和地方建设传统是可持续发展城市的基础，而不必使用高科技产品——合理的设计本身就可以是生态的！统一的规划整合和规划者、建设者与使用者的融合互动是重要的前提之一。未来的目标是使城市成为一个自给自足、与周边环境和居民协调共处的有机体，而不是单方面消耗现有资源的系统。要想再次实现这个目标，在将来还有很长的一段路要走，回顾过去可以让我们更好地在这条路上前进！

The urban layout of buildings can also support communication between neighbors. In an example of a new construction area in Offenburg in Germany, various different residential typologies, from multistory residential buildings, row houses and detached houses, have been grouped around a common "residential yard" accessible to all of the buildings. This yard becomes a communication platform for the neighborhood, whether it be while unloading the car, while children are at play, or during festivities taking place in the yard.

All of these examples illustrate an intimate link between design and sustainability. Design principles and regional building traditions already in use for centuries, can build a new basis for the sustainability of a city, without the necessity of resorting to high tech products. Design is ecology! One of the most important premises for achieving sustainability is the cooperation and integration of all people and disciplines involved in the planning, construction and use of a city. Future goals should be to view the city as autonomous organism, living in harmony and mutual symbiosis with its environment and inhabitants, instead of as an entity that simply exploits existing resources. Achieving that goal may still be a long way off, but perhaps taking a few steps into our past could help us on our path toward the future.

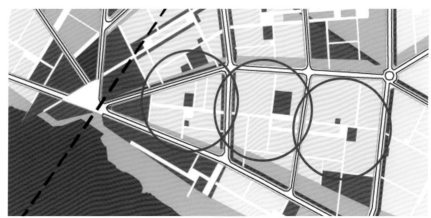

图 11 崇明城桥新城，上海
上图：(红色部分)日常需求设施的分布
下图：十分钟可达的社区中心
Fig. 11 Above : Distribution of service facilities for daily use illustrated in red
Below: 10 minute walk to the centre of the quarter

从经济起飞到生活质量：
实践与研究中城市规划的经验和任务
From Economic Miracle to Quality of Life:
Urban Development Experiences & Tasks of Practice and Research

Michael Trieb

Introduction

Cities are small worlds on their own, carriers of physical life, home of souls and schools of cultural development. They are not only artificially built artifacts, but also lively organisms and living images: small, complete, manmade living environments and urban microcosms.

It is a wonderful task, but a troublesome process, to cultivate urban microcosms through city governance, to revive them via urban renewal or to develop such microcosms in new towns. Even though every place, every project and every era in urban planning demands that the central task be adapted and rethought, similar experiences from another place and time, can still be informative.

For this reason, a selection of urban development experiences gained over the course of thirty years of urban planning practice are illustrated throughout the following. This selection is by no means complete; other urban planning matters, such as the alteration from quantity to quality or the vision and reality of New Towns, have been discussed elsewhere. Naturally, the selection is also subjective and heavily influenced by the tasks the planning team were working on at the time and the author's own perspective.

引言

城市，就自己来说，是小世界，是物质生活的载体，灵魂的家园，文化发展的学堂。城市不仅仅是建造出来的一种艺术品，同时也是活跃的有机体和鲜活的形象——人们所创造的小而完整的生活世界——一个"小宇宙"。

通过对城市的更新来保护这个小宇宙，通过对城市的改造来修整或者在新城中塑造人类新的"小宇宙"是一项奇妙的工程——同时也是一个艰难的过程。但在任何一个地点任何一个项目中，任何时间的城市规划都必须结合其背景通盘考虑和独立计划这项工作。幸运的是，在不同时间、不同地点进行的类似的城市建设经验的研究对规划师不无裨益，可以帮助我们在这个浩瀚如同星辰的信息体系中寻找支点。

因此，接下来将叙述一下在30多年的城市规划工作中总结出来的一些可供选择的经验。对这些经验详尽的阐述被有意识回避——关键点是那些从量到质的转变或者新城当中从"梦想走入现实"的转变——当然从另一个角度来说，这些经验也是主观的。这一方面取决于在这个时期委托进行的任务，另一方面取决于这篇文章作者的观察哲学背景。

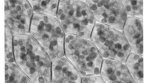

图 01　城市即是一个有机体、小宇宙
Fig. 01 The city as an organism, microcosm

1. 城市意象——城市生命的镜像？

现在，城市意象的意义变得越来越重大。意象是人脑对一种事物的真实、表象和含义的独特反映。相对于现实而言，意象更多地决定了人们的行动和反应。一个城市的形象——城市的感受过程，是城市形象的本质部分，也就是说，一个城市的识别性。

现在，每个城市都在寻找自己的城市识别性，自己独有的"个体性"，力求将其阐释、发展并可感受。这只能通过改善生活质量和强调一些特定的功能、特别的形象和明确的品牌塑造来实现。经验表明，作为城市识别性的镜子，城市意象是城市规划中一个越来越重要的因素。

究竟什么是城市形象呢？一个城市的声誉，是否是当地人与拜访者对其印象的镜子？一种隐藏的真实，以此为基础，我们经历着并触摸着城市——市民生活的一种原本的真实？是个人的生平决定着头脑中城市独特的、不可测的形象吗？或者来自同一个地方更多人的非主观的极具相似性的共识——不管那个地方现在是叫上海、巴黎还是旧金山？

城市意象是生活的基础
城市意象不仅仅是一个城市的视觉表象，还是一个充满经历、经验，有意识或无意识的感觉的复杂整体，由此产生人们的态度和观点[1]。这种部分或者完全被公众认可的城市意象，大大超出人们的想象，是很多经济、社会、文化或者政治决定的基础。大量个人的、职业的和很多时候重要的经济决定不仅仅是在事实，还是在城市意象的基础上作出的。随着全球化的日益剧烈，城市意象也变得越来越重要，因为它可以决定一个城市增长、停滞或者后退。

1. City Image – Reflection of the Lived City?

The image of a city nowadays has become more and more significant. "Image" is each individual's mirror on reality; appearance and meaning interpreted through the subjectivity of the human mind. It is this "image", not the reality itself, which more often than not determines every human action and reaction. The physical appearance of a city, or "cityscape", is a major part of the overall image of a city, and guides its identity.

In recent years every city is seeking its own individuality and struggling to comprehend its "character," in order to refine and transform itself into a tangible experience. This is achieved by promoting the quality of life in a city and accentuating certain features, lending itself a particular appearance and a specific brand. Experience shows that a city's perceived image, as a reflection of the identity of a city, becomes a more and more important factor in urban planning.

But what is the city image? Is it the reputation of a city? The likeness of a city remembered by a visitor or by its citizens? Is it the invisible reality based on how we interact with and experience the city? The actual reality for the city residents? The immeasurable individual picture of the city in one's mind, determined by one's personal biography? Or is it rather an intersubjective memory of multiple individuals converging on the same place, e.g. Shanghai, Paris or San Francisco?

City Image as Basis of Existence

The City Image is not only the visual appearance of a city, but also a complex conglomeration of events and experiences, as well as the resulting conscious and unconscious emotions, attitudes and opinions [1]. More often than one would think, these city images – partially or completely shared by the public – are the foundation of economic, social, cultural or political decisions. Lots of personal, professional, and even profound economic decisions are made not based not solely on objective fact, but on a city's image. With the spread of globalization, the importance of a city's image is growing as well, as it can determine the growth, stagnation or even the downfall of a city.

图 02　弗里德里希港口城市意象：旅游与高科技城市
Fig. 02　Friedrichshafen: tourist and high tech city

New scientific research shows[2] that a city's image is built over several layers, spanning everything from the existing city context, to urban planning, to deliberate city branding. The city context is determined by its geographical setting and historical background: Rome is located on seven hills and reflects European history; London on both sides of the river embodies economic thinking, and Hangzhou, reflecting the advanced Chinese civilization, is situated on the banks of a large lake. The urban planning image is determined by function, design and meaning. A city's image is also often characterized by its various civic functions, e.g. government, economy, industry or higher education. For example, Beijing is often thought of as government city, whereas Shanghai is seen as a city of economy.

Image Carriers – from Opera House to City Skyline
The cityscape is also often an essential element of a city's image. Image carriers can vary from singular buildings like the Sydney Opera House, to public spaces like the Champs Elysée in Paris, to entire city quarters such as the Forbidden City in Beijing. Unmistakable architecture as in Venice and Stralsund, or the unique city skylines of medieval Luebeck or modern Shanghai can also be just as vital.

The importance of a city in terms of its historic, economic, scientific, cultural, social or political impact at the local, regional and national levels, is the second element of urban planning. In this area, size matters far less than the significance of the city itself. According solely to these parameters, the value of the University City of Tuebingen would likely be ranked higher than the industrially City of Stuttgart, the Silicon City Palo Alto higher than San Francisco, and the University City of Cambridge higher than the London Metropolis.

Branding is the third level of these image layers, and represents a conscious effort to enhance a city's image. This can be accomplished through preservation or construction of landmark buildings, like the Sydney Opera House, the Maritime Museum Stralsund, or the Guggenheim Museum in Bilbao; through specific, long-range publicity campaigns; as well as through special events, such as World Expositions, Olympics, Biennials and Festivals. Examples of this would be the well-known Salzburg Festival, the Venice Biennial of Arts, the Shanghai World Expo and the Film Festival in Hollywood.

新的科学研究表明[2]，城市意象有不同的层面，从现有的城市文脉到城市规划直至明确的城市品牌。城市文脉意象是由一个地方的地理位置和历史发展决定的，就像处于7座小山之上的罗马反映了欧洲的历史，伦敦两面临河有很好的经济基础，杭州借助西湖形成中国士大夫文化的高度反映。城市建设的意象是由功能、形态和含义决定的。城市功能是一个较为明确的要素——城市意象受其功能群体影响，比如行政城、经济城、工业城或者大学城，而且经常是受到多重功能的同时影响。北京就不仅仅是行政城，上海也不只是经济城。

意象载体——从歌剧院到城市侧影
城市形态通常是城市意象的核心性因素，意象载体包括单个建筑，如悉尼歌剧院、巴黎香榭丽舍这样的公共空间，北京紫禁城这样的整体街区、威尼斯或施特拉尔松这样不可改变的建筑普遍性特征，甚至包括中世纪吕贝克或者现代上海这样的城市侧影。

一个城市在历史的、经济的、科学的、文化的、社会的或者政治方面，在当地的、地方性的、国家的或者国际层面的意义是城市意象的第二个层面，这其中起到决定作用的不是大小而是重要性。其意义如同图宾根（Tuebingen）大学之于汽车城斯图加特，硅谷帕罗奥图（Palo Alto）之于旧金山，剑桥大学之于世界城伦敦。

最后，品牌的塑造是意象的第三个层面，它通过保护或者塑造像悉尼歌剧院、施特拉尔松海洋博物馆或者毕尔巴鄂博物馆这样的地标性建筑，通过有目的的长期的广告活动，或者通过世界博览会、奥林匹克运动会、双年展、节日汇演这样的特别活动，有意识地进行城市意象塑造。萨尔茨堡节日文艺会演、威尼斯艺术双年展、上海世界博览会和好莱坞电影节庆都属于此范畴。

图 03 施特拉尔松的城市意象：古老的城市侧影和新建的海洋博物馆整体塑造的城市品牌
Fig. 03 City Image Stralsund: branding through the ancient city skyline and new Maritime Museum

如果人们一开始并没有考虑城市意象的所有方面，在改善与发展的过程中，人们应该作出哪些必要的努力呢？以海湾国家和它们的城市为例——从迪拜到阿布扎比，最初只想借助令人印象极其深刻的现代城市景观来炫耀经济实力，但忽视了地方民众的"精神和灵魂"，现在规划部门则致力于在其建筑风格中植入教育和文化元素。所有这些不仅仅体现了意象对于现代生活和其多样性的重要意义，同时还说明了现在城市意象必须在哪些范围内被有意识地考虑在规划之中——我们认为城市意象的构想在将来会成为城市规划设计过程中的本质元素3)。

2. 城市景观——寻求美丽，拒绝荒芜的城市

在过去的30年中，中欧关于城市美丽的争论从思想家的檄文，如亚历山大·米切利希（Alexander Mitscherlich）和他的短小卓越的论著"我们城市的荒凉"4)延伸到充满质疑的联邦德国公民的求救——他们以"现在拯救我们的城市"为口号，在德国统一前手持蜡烛进行了大型游行示威。一开始是因为新的建筑都是冷漠的、非人性化的"混凝土块"，到后来是因为人们荒废了许多有价值的老城中心。新的城区和城市既缺乏特色，也不是家园。在逝去的旧城，市民丢失的不仅仅是家园，还有社会的平等、时间的连续和个人的识别性。

只有美丽可以拯救世界
但是一个城市的美丽体现了可见形式的一切：识别性、方向感、回忆和创造了所谓"家乡"及增强文化联系性的含义体系——每一个远离家乡的人，当他在外面遇到了家乡的人，就会体会到这点。然而，室内装饰、建筑和城市的美丽还代表着更多。一位拜占庭的智者说过："只有美丽可以拯救这个世界。"德国的哲学家鲁道夫·斯坦纳（Rudolf Steiner）也说："只有美丽，能够让我们再次成为有道义的人，成为不再撒谎、不再偷盗等的人。"5)

由此产生了一个似乎难以实现的要求：坚持每种形式的艺术，坚持所有东西的设计，坚持我们物质与精神环境的美丽。但是在今天的发展与实践中，越来越体现出设计这一行为本身在我们今天的现实生活中有多重要——我们周围的所有东西都是被设计的——我们看到的、手里拿的、使用的或者梦寐以求的所有东西。筷子或者刀叉，衣物或家居摆设，餐馆里美味的饭菜或者装饰的桌子，手机，电脑，电视或者汽车——所有我们看到的，所有我们使用的，所有我们的美食——都是以一种或者多种方式超越功能的必要性塑造出来的。

2. City Image – Beauty vs. Inhospitality

If all aspects concerning the city image are not taken into consideration from the very beginning, enormous efforts are necessary to complete the missing factors. This can be observed today in many Gulf States cities from Dubai to Abu Dhabi, that were primarily intended to develop as modern economic powers, but neglected the "spirit and soul" of the people. Today, great efforts are being taken to retrospectively insert educational and cultural values, in they civic form and content. All of the above examples reveal the significance of a city's image as it relates to the many facets of contemporary life, as well as the extent to which such image must be consciously designed. The conception of the city image design will become an increasingly fundamental element of the urban planning process in the future. 3)

2. City Image – Beauty vs. Inhospitality

For the last 30 years, a debate about the beauty of cities reaches from Alexander Mitscherlich's pamphlet: "the Inhospitality of our Cities" 4) to the outcry of desperate citizens: "Save our Cities Now," as the slogan for large candlelight demonstrations prior to the reunification of Germany. The reason: at first these new settlements turned into cold, inhuman "concrete castles." Many valuable ancient city centers were left to themselves, only to collapse. Neither identity nor a sense of home could develop from these new cities and city quarters. Citizens lost their habitat as well as their civic identities - continuity and selfhood in the dying historic city centres.

Only Beauty can save the World
But the beauty of a city contains all of this in a visible form: identity, orientation, memory, and meaning create a home and support a common identity. Anyone who has ever met another person from their own home city by chance in a far flung location knows this feeling well. But the beauty of interior design, architecture and urban design stands for even more than that. An old Byzantine proverb proclaims: "only beauty can save the world." And "only her," the German philosopher Rudolf Steiner proposes, "can turn us into moral beings again, human beings that do not lie nor steal etc., anymore." 5)

This may be an unrealistic demand, to instill beauty in every kind of form, to design all things, to make beauty in our environment. But experience shows more and more how important design has become for us in our everyday life. Everything surrounding us is designed, everything we see, hold in our hands, use or dream of: chopsticks, knives, forks, clothes, furniture, delicious food, a decorated table in a restaurant, mobile phones, computers, televisions and cars... Everything we see, everything we use, even everything we eat and drink is, in one way or another, designed beyond purely functional necessity.

图 04　设计无处不在：食物、材料、发型、浴室、汽车
Fig. 04　Everything is design: food, fabric, haircut, bathroom, car

The Power of Aesthetic Surplus

We could not live without the aesthetic surplus that refines necessity. To create this added value is the only purpose of design. It is the versatile, time-bound form of beauty and it satisfies a fundamental human need. Therefore design is a significant economic factor, as every entrepreneur producing consumer goods knows very well. It is the reason why nowadays everything must be "designed," from food packaging, to retail stores, to the exterior shape of a car and even its engine which remains almost entirely out of sight. Why? Does the object still work without the design?

People have not only material needs, but immaterial ones as well. The objects around us that we tend to favor, transport feelings, sentiments, and atmosphere. They amplify our own image. Brands like Nike, Dior or Boss are as important to us as the function, size and price of an object. The objects that surround us can even covertly, yet somehow thoroughly, even lead to changes in attitude, views and behavior. Objects of beauty refine us inside and out. Great statesmen have always known this, from the Chinese Emperors to European Kings. For all of these reasons, designing is not only an economic task, but also largely a social and cultural one. Design in urban planning - city design - is therefore a central task in urban planning. As experience shows; the beauty of a city's image is indispensable.[6]

Maintenance of City Image as an Economic Stimulus Program

This insight has lead to an historically unparalleled wave of urban redesign throughout Europe over the past few decades, especially in relation to economy, infrastructure and design. Predominantly funded by governments – for instance through the urban development subsidy law for states and federations – these funding programs support infrastructure renewal for historic city centers, business development, stimulation of private investments and the refinement of the cityscape. At the same time, the programs are very effective governmental economic stimulus programs. Today the effects of such past actions can be observed in large cities as well as in the renewal and refinement of small villages, where the standard of living has risen substantially. Furthermore, with the introduction of additional measures and modern communications technology, the rural exodus has been slowed down, if not stopped completely.

由于美学而造成的增值

在必要性以上，如果没有美学增值，我们已经无法想象我们的生活是什么样子——创造这个美学增值就是设计的任务。它是多样性的，当代性的，同样是人们一种天然的渴望与需求。由此，设计成为一种重要的经济含义要素，就像每一个生产消费品的企业所了解的一样。因此，现在所有的东西都必须是"被设计的"：从食品的包装到商店的外观，从汽车的外形款式到隐藏其中的发动机。为什么呢？没有经过设计的东西就不会起作用吗？

人类不仅有物质需求，也有非物质需求。人类周围促生我们渴望的物体，往往具有感觉、情绪和气氛上的特殊要素。它们各自的意象因此被大大加强——如同Nike、Dior或者Boss的品牌力量——而这些对人类来说往往与它的功能，大小或者价格一样重要。甚至这些物品，在不知不觉中，深深地改变了我们——它们导致了另一种立场，另一种视角，另一种行为。另外，美丽的东西还会从内部美化我们。伟大的政治学家往往都意识到了这一点——无论中国的皇帝还是欧洲的国王。出于所有这些原因，设计不仅仅是一种经济的，也是一种社会的和文化的任务。城市规划中的设计——城市设计，正如经验所表明的一样，是城市规划的核心任务，其中创造美丽的城市景观又是核心的工作。[6]

作为城市更新发展政策与计划案的城市景观改造

在过去几十年中，这种认识在欧洲引起了一场史无先例的以艺术创造性地进行城市更新的高潮，包括经济的、基础设施的和艺术创造的方面。主要靠政府的财政支持，如借助联邦和州的都市建设促进法，这个计划有利于旧城基础设施的更新、经济的发展，刺激私人投资，改善城市景观。这样，它同时也成了政府成功的发展促进计划。现在，这些措施的成效从大城市到小城市的更新和改善工作中都还可以看到，这些地方的人民生活质量也得到了本质的提高。更多的是，这个计划——与其他的措施和现代信息技术一起，有效遏制了农村人口向城市外流的现象，虽然这个流动还没有完全停止。

图 05　维尔茨堡的城市景观：法制的和经济的城市建设下生活质量的提高
Fig. 05　City Image Wuerzburg-living quality by judicial and economic urban development stimulation

内容空洞的城市景观就像冷漠的面孔

城市景观不能仅仅是装饰品——在某些城市，城市景观要比实际城市生活品质更美好，在这些城市中通常缺乏的是市民的内心责任感、财政措施，或者简单地说，是一种意识，真正美丽的地方必须反映一种活跃的经济的、社会的和文化的内心生活，而不仅仅是美丽的外壳。实施工作中，更新过的城市景观必须每天维护——就像每天早上洗脸一样——从道路的拐弯处到一个门脸的修葺，并且特别是在经济危机的时期，尤其显示了城市居民对城市景观的关切程度，对其认同感的实际效果。正如仅靠一张美丽脸庞无法塑造一个真正美好的人一样，只有内心的价值让一个人的脸庞美丽，这种没有反映一个活跃的经济、社会和文化社会的美好城市景观，就像是一个没有感情的模特的美好脸庞，极易衰坏。仅仅美丽的城市景观是远远不够的——要想持续地对人们造成影响，它还必须是一个吸引人的、活跃的、充满文化的城市有机体的反映。

3. 公共空间——城市景观的重要载体

在过去的三十多年里，欧洲地区的城市公共空间变得越来越重要。城市改变了自己的面貌，那些或多或少由行驶车辆占据的街道和广场，变成了带有人行道的行人优先的空间和步行街，随后又转化为包括行人专用区、步行街和人行道的多元化的步行网络区域。

这样，改变的不仅仅是公共空间的利用情况，其利用的时间长度和街道景观效果也发生了很大的变化。人们的生活习惯也因此发生了很大程度的变化，大城市和现在的小城市的公共空间已经发展为商业空间的拓展，特别是在目前的餐饮和零售业中，还有一些公共空间变成了居住空间的拓展。在这里，人们和朋友见面，建立联系，谈生意，并一块玩这个"看和被看"的游戏。公共空间改造成功的关键是它的利用率、形态，以及由此产生的新改造街道或广场空间的情调和氛围。

A City Image without content is like a Soulless Face

Sometimes, a city's picture is prettier than the reality. In many of these cases, the necessary commitment level from citizens or the financial resources are lacking. Often it is simply a lack of awareness that truly beautiful places must also reflect a lively economic, social and cultural heart and not just a pretty face. Add to the mix the fact that every redesigned cityscape requires daily maintenance, much like one's own face every morning. A city's daily regimen can be anything from street cleaning, to the renewal of facades, including and perhaps especially in times of economic crisis. A pretty face on its own does not a beautiful person make - only interior virtues can truly do that. And, in much the same way, an attractive cityscape that does not reflect a lively economic, social and cultural community, is nothing more than the enamoring face of a pretty, but soulless doll. A cityscape must therefore reflect an engaging, lively and culturally prosperous "cvic-organism" if it is intended to have a lasting effect on people.

3. Public Space as a Carrier of City Image?

Throughout the last thirty years, the significance of urban space in Europe has increased. Cities have changed their appearances substantially. Streets and squares once dominated by vehicular traffic have transformed their character step by step, first by introducing pedestrian friendly spaces with walkways, followed by pedestrian streets and later entire pedestrian areas based on a network of pedestrian squares, streets and alleyways.

But not only have the use and time-span changed considerably, the everyday habits of the people have changed as well in a quite sensitive way. In large and small cities alike, public space has become an expanded workspace – especially in gastronomy and private commerce – as well as an expanded living space in which we meet friends, establish contacts, conduct business and play at the game of "see and be seen." While usability and appearance are certainly crucial to the success of redesigned public spaces like pedestrian streets and squares, so are their spirit and atmosphere.

图 06 埃斯林根和埃尔旺根：通过街道空间改造实现的生活质量的提高
Fig. 06 Esslingen and Ellwangen: living quality by urban street design

Conditions for a Successful Urban Street and Square Design

Cities and villages therefore dedicate not only considerable financial resources – supported by their respective governments – but also great amounts of time and effort from municipal administrations, municipal councils and involved planners, to the redevelopment design and realization of such streets and squares. A memorable design concept that is easy for everyone to understand from the very beginning is crucial. Sometimes quite a bit of endurance is required to discuss an intervention with everyone involved - a process that can indeed take many years. The care involved in taking everyone's needs into account in a street or square – from placing a trash can in the right spot, to setting aside a space for a pleasant but modest bench, to just the right lighting day or night – is paramount. Furthermore all technical, practical and traffic-related demands need to be deliberately thought through; as does the planning of maintenance, upkeep and repair of the public space and its components, including watering trees, cleaning fountains, lighting, etc. If all of these aspects are taken into consideration during the planning phase, the newly designed streets and square spaces can be recognized and accepted as ideal places for street festivals, events and markets from the get-go. Without proper positioning of street lamps, market stands, as well as well placed connections for water, electricity, street cleaning, etc. events like these are stunted.

Design as Economic Factor

The significance of the urban space nowadays has become an important element for the quality of life within a city. And the experience of the last thirty years shows very plainly that city design - in this case, the design of streets and squares - can influence the behaviour and lifestyle of the people who use these spaces. It also shows how the street and cityscapes can have an effect, not only on the body – e.g. resting on a bench in the city – but can also influence the spirit as well, for example, by creating an atmosphere that is bright and warm. Spaces like these can even have an effect on the human mind, by being surprising, stimulating, encouraging etc. As tourists, we often spend a lot of money to feel this triple layered experience: physical, mental and spiritual; as inhabitants of one city, we often don't even have the idea that something similar could be possible.

成功的街道和广场改造的前提

城市和村镇对街道和广场空间改造进行的规划工作与财政支持代价不菲，不仅包括巨大的财政资源调用，也包括以及市政、乡镇议会和规划人员贡献的大量时间与精力。其核心往往是一个明确彰显的规划理念，它作为目标——一个具有可实现意义的理想愿景从一开始就很容易让市民理解。这其中包含着宽容性的空间，允许所有的参与者长期讨论与修正。同时具有综合性的考虑，满足一个人从早到晚在一个街道或者一个广场上的所有需求，包括合适位置的垃圾箱、舒服美好的座位以及适意的灯光。所有的技术的、交通的和实践的要求，其综合考虑是不可忽视的，如对实施、维护和修理进行规划，包括灌溉树木、清洗喷泉、照明等。新改造的街道和广场空间到处都可以成为街头庆祝活动、大事件和市场的理想地点——如果它的种种细节在规划的时候就已经被考虑进去的话：路灯位置经过准确规划，能够摆设市场摊位；提供了电和水的连接管道，方便清洗街道，等等。

作为经济要素的城市公共空间改造

公共空间的意义现在已经成为了一个城市生活质量的重要因素。近30年的经验清楚表明，城市改造——这里指街道和广场的改造——能够本质性地影响人们的行为方式甚至生活方式。城市景观，包括街道景观，不仅仅影响了身体——如在一个长凳上休息，也影响了心理氛围——如感受到清新温暖的氛围，同时还对思想有影响——如沉思或启迪新思路。作为游客，我们常常费金不菲来体验这三重经历——身体的、心理的和精神的；作为某个城市的居民，我们常常不会想到，同样的事情可能也会发生在这里。

图 07 斯图加特荷尔德林广场：公共空间中作为文化载体的技术个体

Fig. 07 Hoelderlinplatz in Stuttgart: technology as culture carrier in the public space

Fig. 08 City Ecology: from landscape to building

The resulting fact is that an attractive public space can be a significant contributing factor to the quality of life within a city, which in turn is an important economic factor. This is a fundamental motivation for the combined commitment of the municipal government and the citizens, when it comes to the planning and realization of redevelopment in public spaces. Minor disputes notwithstanding, citizens, local politicians and municipal administrations can work together in principal agreement on the opening, reclamation and refinement of public spaces intended for the use of the people. This can be accomplished through workshops, competitions, substantial planning and often elaborate and costly realizations.

4. City Ecology – Added Individual Actions or Integral Organism?

Ecological aspects are becoming more and more important in urban planning. Today they include all forms of sustainable water, lighting, ventilation and heating technologies, as well as a constantly growing specialized expertise alongside increasingly efficient specialized technology. There is architecture to build zero energy houses, as well as various approaches to a sustainable urban development; everything from planning to realization to maintenance of a settlement or a whole new city. The planning strategies span categories from building position, e.g. orientation of roof and facade, to organizing traffic infrastructure in order to maximize short distances, to considering seasonal airstreams as natural form of urban "air-conditioning." In the realization process that is the construction of settlements, city quarters or new towns, possible measures can range from construction site organization – for example humus removal, groundwater protection, etc. – to the usage of locally excavated soil instead of disposing it off site. For the maintenance of a new settlement, factors like onsite water treatment, preservation of fresh air corridors for a pleasant microclimate in the city, the supervision and cultivation of vegetation in the public spaces, etc., are all vital.

Practical Aspects of Urban Ecology

In many of the urban planning projects mentioned above, the fundamental ecological conditions have been analyzed and ecological urban concepts have been developed that include the natural location, the microclimatic conditions, the local vegetation and their influence on the overall environmental quality. This might affect, for example, the height of future developments. From project to project, the differing new adaptations of urban ecological aspects reflect the ever-increasing experience of the planning group, as well as the varying urban ecological depth of that planning and the often dramatically different microclimatic situations. In China alone, microclimates can range from the snowy north of Harbin to the subtropical south on Hainan Island.

An excellent example of this can be found in the ecological development planning of the eastern part of the Chongming Island in Shanghai, in a regional planning project with a small new town as its center. There, the urban and regional ecological aspects have been analyzed and planned. Included in this was the protection of ecological biotopes and maintenance of a bird sanctuary of global significance which created an important stop for migratory birds. Additionally, water supply balance, wind protection and airstream corridors, as well as combined ecological energy systems based on garbage incineration, solar energy, geothermal energy and wind energy were included as part of the plan.

Development of a Model for an Integral Urban Ecology
These urban ecological aspects are being deepened, evolved and broadened, in the matters of planning, construction and maintenance of urban facilities. Today, even in general terms, ecological knowledge and technology are rapidly increasing. But the approaches that dominate practices the of today are still, from the view of integral urban development planning, mostly an urban ecological patchwork. Missing is a model of integral urban ecology, in which all singular parameters are not only viewed together, but also in terms of their cumulative effectiveness, recognized as the result of synthesis, so they can be specifically applied.

More than ever, Descartes' old adage, "the whole is more than the sum of its parts," is especially true in the field of urban ecology. The planning group is working with the basic elements of a model of integral urban ecology and is refining it project by project. But what must become the actual content of all urban ecological efforts is the comprehension of the urban ecological organism that is already present and will be transformed in case of urban renewal or development, or will be created with the construction of a new settlement or town. An extended urban development model is necessary, in which urban ecology would be considered an equal factor to economic, infrastructural, or sociocultural parameters, not underestimated, or unilaterally ignored as it all too often is at present.

5. Civic Participation – Passive Performance of Duties or Active Partnership?

Civic participation in its various forms is an integral component of urban planning in Europe today. It reaches from the statutory minimal participation of citizens in regular planning procedures, like in the preparation of development plans, up to local referendums in general planning decisions. All dimensions of urban planning, everything from street design to city development, are being presented and discussed in citizen councils. The results of this type of passive participation, in which citizens discuss the plans of the municipality or investors, but do not attempt to create a plan of their own, are being taken more and more seriously by municipal governments and administrations. However, because they have almost no actual legal effect, citizen councils of this nature can, but need not be taken into account at all by the municipal government.

Civic Participation – the Emergence of a Third Power
Today a third power is materializing in the local political decision-making processes when it comes to urban planning. In addition to the elected committee, the municipal council, and the city administration with its various departments led by the mayor, cross-party topic-oriented citizen participation can sometimes support or dissolve resolutions put forth by the city administration and/or municipal council.

早在2003年进行的上海崇明陈家镇候鸟保护区区域规划及城镇总体规划项目中，就已经对区域和城市生态的各个方面进行了深入分析和规划，其中包括对生态群落的保护，形成了具有全球意义的候鸟保护区——作为候鸟迁徙途中的重要落脚点；还包括生态水体系统、生态风系统以及基于垃圾焚烧能、太阳能、地热能、风能的复合型生态能源系统等方面。

整体性的城市生态模式的发展
城市生态的因素在项目中就像这里所展示的一样处于不断的深化、发展和拓宽之中，在对城市的规划、建筑和维护等方面。整体来看，现在的生态知识和生态技术在快速增长。但是这些因素——从整体的城市发展规划来看——大多数情况下是城市生态的零碎要素，在实践中成为主导。所缺少的是一种整体的城市生态模式，在这种模式下，不仅所有的单个参数应该整体考虑，它们的整体影响、相互作用后的放大性效果也应该能够被理解并有目的地被应用。

而且这正好适用于笛卡儿的一句名言"整体要大于它们各部分的总和"。设计团队以这样一种整体的城市生态模式的基本因素工作，并从一个项目到下一个项目不断进行完善。事实上，必须成为城市生态努力本质内容的应该是来自于对城市生态有机体的理解，这个当前存在于城市修葺或者城市发展中的有机体，正在被改变或者在新住宅区与新城市的建造过程中被创造。除此之外，一种整体性的城市发展模式也是必要的，在这种模式中，城市生态处于与经济、基础设施或者——比如说——社会文化参数一样重要的地位——没有被低估，或者——就像现在经常发生的一样——也不会单方面高估。

5. 市民参与——被动的参与者还是主动的合作伙伴？

在今天，不同形式的市民参与是欧洲城市规划的整体组成部分。它涉及从正常规划程序中法律规定的最低参与——被动地参与，如建筑方案的确立——到总体规划等领域的市民决策。不同规模的规划——从街道改造到城市发展规划，都要通过市民大会提议和讨论。这些被动性市民参与的结果——市民为市政府与投资者的规划提出建议，但是自己不参与规划工作——常常没有法律效力：也就是说，他们的意见能够，但不是必须被那些承担责任的城镇议会考虑。但是目前却越来越多地纳入负责者的考虑之中。

市民参与——第三种力量的形成
原因在于，城市规划的地方政治决议程序中，市民这第三种力量变得越来越重要——现在，在党派选举形成的决策机构和城镇议会之外，市政府和市政当局之外，跨党派的专题性市民大会有时候会非常成功地支持市政府和乡镇代表大会的决议或者使其失败。

图 09 市民参与：从被动到主动的参与
Fig. 09 Civic participation: from passive to active participation

随着时间推进，主动型市民参与数量在不断增长。积极的市民、地产所有者或者商人组成小团体，着力推动、资助公共空间的规划和实施，甚至自己参与规划和实施——市政府的态度对此非常重要。这种主动的市民参与可以以很多不同的形式出现。

通过个人力量汇合而实现的城市规划
在埃斯林根的班霍夫大街购物街的改造项目中，一个由商人和房屋所有者组成的积极市民团队促成了这一工作，他们由一位经济界人士倡导，成功地促使市政府，发展一个独特奇妙的但具有可实现性的方案，以提高街道商业、生活的价值。这个市民协会随即开始自费打印吸引人的海报，将透视图作为街道的意象，在所有的居民、所有者和商人当中加以宣传。在每周的市民协调会议中，在建筑资金极其困难的情况下，逐步推进其实施，最终得到一笔可观的资金作为财政支持，竣工之际举办了盛大的市民活动庆祝。

通过市民力量的城市设计
由3个商人组成的团队，以其积极性和坚持的精神为基础，利用计划中的市铁重建机会，启动了斯图加特荷尔德林广场的重新设计。为了提高设计质量，他们自发捐款，提前为这个区域内广场和街道可能的重新设计的初步计划融资，因为城市既不愿意又不能立时启动。区咨询委员会和公众对这个计划的设想，以及建筑委员会和乡镇代表大会全体会议之后的其他设想引起了跨党派的支持和后面这些积极商人核心方案本质部分的实施。这个方案从一开始就一方面伴随着团结的改革者的支持，另一方面是面临城区议会到市长阶层的反对。设计前期工作所形成的优美形象，看起来童话般的规划理念，让市民、商人、地方政治家和冷静的消防行政部门、市铁工程师和土木工程专家了解了这个目标，并逐步转变为支持这个计划。正因如此才产生了弗里德里希荷尔德林（Friedrich Hoelderlin）的现代文化纪念柱，它改变了斯图加特市这一地铁站点单一无序的外貌。

But recently, the amount of active civic participation is also growing. Small groups of interested citizens, homeowners and business people organize to initiate plans in the public sphere, provide input on recent planning, and support or even develop and realize their own plans. And in most cases, they do so hand in hand with a grateful city administration. This type of active civic participation can occur in many forms..

City Design by the Power of One Personality
In the case of the redesign of the significant shopping street, Bahnhofstrasse in Esslingen, it was a group initiative of business people and homeowners, led by an influential segment of the commerce sector, that successfully pushed the city administration to develop an original, but feasible idea for the renovation of the street. Following that, the group made the idea known to all the residents, proprietors and business people through posters showing a perspective rendering of one possible future design. In weekly acclimation talks, the group accompanied the implementation in a difficult construction time, subsidized the financing with a large figure, and celebrated the opening with a large scale citizen festival.

City Design through the Power of Citizen Teams
The redesign of the Hoelderlinplatz in Stuttgart, due to the construction of the terminal station of the city tram line, is based completely on the initiative and endurance of three business people. They took the planned reconstruction of the city tram line as an opportunity to fund a possible redesign of the square and neighboring streets at their own cost, because the city could neither afford to, nor cared to do so. The presentation of these development plans in the district advisory board informing the public, as well as the later presentations in the construction committee and general assembly of the municipal council led to cross party support and implementation of major parts of the concept. The project was lead, from its very beginnings, by these determined business people on one side, as well as by its opponents, reaching from the municipal council to the highest level of city executive - the mayor himself. But in this case as well, a clear vision, an almost fabulous planning idea was indispensable in convincing citizens, business people, local politicians, fire fighters, city-tram engineers, and experts in the city's civil engineering department, to back the plan. It was not least because of this, that a modern memorial statue was erected for Friedrich Hoelderlin around one of the city line's catenaries masts.

City Design through the Creative Design Ability of Home Owners

It was In an area well known for the critical resistance of its citizens, in one of the most feared city quarters, where a program for improved living environment led to the most successful form of civic participation. The greatest form of active civic participation was the redesign of a small residential and business street in Esslingen Zell, by the homeowners and the residents themselves. Suggested by the residents and supported by the city administration and professional planners, the participants of a design workshop were able to develop detailed, realistic design alternatives through the aid of sketches and working models. The professional planners that took over the project afterwards were quite surprised by the quality of the plans created. In the following stages, it became their task to translate the ideas of the workshop as well as the results of the subsequent discussions into a design that would summarize the intentions of the citizens into a licensable draft. Ultimately, the Backhaeusleplatz plan was realized through the work and the funds of the citizens.

What insights can be gathered from the decades of involvement in these passive and active forms of civic participation? For one thing, passive civic participation can become a regular part of the planning process and while it may sometimes serve to complicate and prolong the process, it also enriches and makes it more successful. Passive civic participation that arises from protests often reflects unmet needs and if not responded to sufficiently, can even put an end to otherwise useful projects. There are many examples of this phenomenon. This underlines the insight that active civic participation can become a most useful, practicable and effective supplement to future urban planning and development. Even if in isolated cases the projects of the ISA Group were prolonged or cancelled, the experiences are extraordinarily positive overall, showcasing opportunities for real Public-Private-Partnership that could lead to new models of sustainable urban development in the future.

6. City Planning – Synthesis of Free Market Economy and Social Society?

Every urban planning task is a quest to strike a healthy balance between individual interests and the needs of the community. That is to say, individual economic interests on the one hand, and the collective interest of an urban community on the other, must be carefully weighed and balanced out in each case, with the interests of the collective community given priority within the legal framework.

Cityscape: Reflection on a Marriage of Architecture and Urban Planning

As history and experience teaches, this balance is reflected in what distinguishes genuine city character. The secret of beautiful cities always lies in the conscious linkage of architectural individuality and urban collectiveness. In the past as well as today, the shared design objectives always affect the common cityscape and lead to joint design regulations, within which the building owner and the architect have been left the largest possible freedom for their individual design inclinations. This is still valid in many present-day European cities like Venice, Florence, Paris, Luebeck, Stockholm or St. Petersburg. Whether it is the maximum height, width and depth that is mandatory, or even, depending on the city, the composition of the facade, proportion of the window openings, material and color, it is basically always the same. It is essential for authentic city character for there to exist a certain, distinctive, intentional relationship between individual freedom and collective commitment that naturally varies from city to city and era to era. Thus it follows: every individual, unique cityscape is always the child of a marriage between individual solitary architecture and the collective cityscape.

通过产权所有者具有创造性的设计能力而实现的城市设计
市民运动的焦点是那些城市环境恶化的城区，它们更应该实施改善居住环境的行动，从而成为积极的市民参与最成功的形式。我们主持过的范围最大、最成功的市民主动参与的活动之一是完全由房屋所有者和居民进行的埃斯林根-采尔(Esslingen-Zell)的居住区和小企业区街道改造计划工作坊(workshop)。由居民发起，市政府和专业规划师支持，工作坊(workshop)的参与者用草图和模型的方式进行了详细考虑，去发展一些现实的规划选择，连专业的规划师都惊异于其质量。然后，在接下来的讨论中将其结果转换成了初步规划，它总结了市民的意图，在一个法规性的初步规划中得以体现：其中"Backhaeusleplatz"广场方式甚至是通过市民自己的金钱和劳动实现的。

在被动和主动的市民参与方面，从数十年的参与经验中我们能得出什么结论呢？一方面，被动的市民参与已经成为规划过程的正常部分，不仅可以消极性地将其复杂化、加长，而且能够积极性地加深并使其更成功。被动的居民参与从抗议中产生，反映了一种没有满足的需求，如果没有对其作出充分反应的话，优秀的项目也可能受挫——这也有大量实例证明。这同时也进一步加强了我们一直以来的论点：积极的市民参与是将来城市规划最有意义、最切实可行、最有效率的补充。即便在个别情况下它导致了城市与ISA集团某些项目的延迟甚至失败，但是这些力量本身，展示了一种真正的公私合作的机会，可能引领可持续城市发展的新模式。

6. 城市规划——自由市场经济和福利社会的合成？

每一个城市规划的任务都是在寻找一种个人利益和集体需求之间的平衡。这意味着一方面需要重视的是经济方面的个人利益——企业的或者投资者的，另一方面是城市整体的集体利益——个别情况下两者必须重新进行内部衡量。在法律范围内，集体利益占据明显的优势地位。

城市景观：建筑和城市建设的联姻
历史和经验告诉我们，由这种联姻诞生了真正的城市个性——美丽城市的秘密总是存在于建筑学个性和城市建造共性的有意识结合之中。共同的规划目标——无论在历史与今天——总是涉及共同的城市景观、要素之间共同的规划联系，在这之内，又给了建设者和建筑师的个人规划意愿充分的自由。直到今天，大量欧洲的城市中心都是依照这一原则塑造并形成其魅力形象，如威尼斯、佛罗伦萨、巴黎、吕贝克、斯德哥尔摩和圣彼得堡。控制原则涉及最大高度、宽度和深度允许范围，甚至根据地点，规定房屋立面的布局、开窗比例、材料和颜色，整体体现了同样的原则。重要的是，在真正的城市个性中，个人自由和集体责任中存在着一种特定的、有意识寻求的联系，这种联系在不同地点不同时期当然是不同的。由此形成：每一个单独的、唯一的城市景观都是特别的单个建筑师所形成的组群和共同的城市规划结合的合法产物。

自由和限制之间存在矛盾？

当然，这个城市景观是个人自由和集体限制性长久争论的结果。每个城市都反映了当地的、受时间制约的矛盾，即个人对自己建筑自由设计的需求和公众对其限制框架之间的矛盾，从邻居开始，到共同街道上的建筑形态及其设计原则方面的争论。几乎每一个建设项目——从改造到新建——都会对集体空间产生影响，这种矛盾当然是正常持久的普遍性问题，只能通过两种极端相反力量——个性和共性——有意识的融合才能解决。但这又是城市规划中一项极其有创造性的任务：通过这两种极端特性的融合才会产生第三种力量——城市的个性与文化特征，世界文化遗产城市最清楚地表明了这一点。

个人和集体间的合作关系

就像城市景观所反映的一样——个性和共性的合成——城市的经济、社会和文化集体也遵循这一原则。在欧洲，这种"我"和"我们"间合成的基本思想——每种人类间合作关系的基础——发展成为多样化的方式，"二战"后欧洲以高社会保障的市场经济作为经济体系发展的基础，市场经济是社会主义和资本主义的一种结合，联合了两种体制的优点：为所有人以及每个个体，其最可能的经济发展机会提供社会保障。

经济上的考虑，必须从它的单独要求——股东需求，转化为整体性的要求中的合理位置，作为人的所有需求的一部分加以对待，而不是相反。另一方面，福利社会思想也不允许强制性地实施泛滥性的平均主义，在不给社会造成损失的情况下应当给个人发展留出足够的空间。建筑的每种形式，从单体房屋的小范围扩建到一个新城的规划建造，都应该去寻找这两种合法需求的合理平衡。这样，城市规划才能真正代表我们的发展理念，不论是小的局部还是大的整体，永远力图达到合法的个人利益和必要的集体需求之间的一种平衡。

Dichotomy between Freedom and Necessity

Naturally, the cityscape is the result of a constant debate between individual freedom and collective necessity. Every city reflects the local, time related debate between the individual right of free design of one's own building, and the reasonable demands of the public, starting with neighbors, for shared design regulations of property developments on a common street. Since almost every construction task, from redevelopment to new construction, has effects on the public space, this dichotomy is normal, constantly present and can only be resolved through the conscious synthesis of the opposing poles of individuality and community. And this is the eminent creative task of urban planning: to create a new third quality from the synthesis of those opposing forces, as shown most clearly in world heritage site cities.

Partnership between Individual and Community

What is reflected in a city's image – a synthesis of individuality and collectiveness – is also applicable for the economic, social and cultural community of a city. In Europe, the fundamental idea of a synthesis between "me" and "we," the foundation of every human partnership, has developed in many forms. After the Second World War, it evolved into a social market economic system that describes a marriage between socialism and capitalism: the basic social security to support the livelihood of everyone and the largest possible economic development potential for each individual. But its maintenance does have consequences.

Economic thinking on the one hand must be diverted from its exclusive claim of shareholder value, to its rightful place as a component part of human needs. Capitalism must not, as has happened so many times in the past, dominate everything. Social thinking, on the other hand, must not lead to a forced egalitarianism of everyone, but must leave as much space for individual development as is possible without harming the community. In every method of building, from a small annex, to a house, all the way up to the construction of a new city, a suitable balance between both legitimate ideals should be the ultimate goal. Urban planning therefore means to develop future visions in detail as well as in large scale that constitute a perpetually renewed balance between the legitimate interests of the individual and the necessities of the community.

图 10 城市团体：个人和集体间的合作关系
Fig. 10 City community: partnership between individual and community

7. The Conception of Man as the Basis of Future Urban Planning Theory

The central experience of the presented works in Europe, Asia and South America, is that any approach to urban planning that does not have at least a simple but integral conception of man to rely on, will always remain an unbalanced patchwork. Every urban concept, whether an idea for street furniture in a single town square or a total urban development plan for a metropolis of 10 million, should be based on a simple attempt to imagine what man is, and what his legitimate needs will be in the settlement, regardless of its scale. Such a concept can be simple, incomplete, differing with regard to time and culture, differing in its content and free of ideology or religion. The point is that an idea, a vision of human beings, a conception of man should be laid as foundation for the urban planning task in large scale as well as in detail.

This sounds obvious but would appear not to be, as evidenced through all levels of urban planning. The lack of a conception of man is the real cause of the dominant "curse of urban planning" plaguing cities throughout the world. It is the repetitively unbalanced overexposure of legitimate but overvalued urban decision parameters – i.e. taking into account the entire urban organism: representation of power, profitability or share holder value, transport, ecology, cityscape and city image – that often determines urban planning decisions. In the same way that human beings must satisfy their various needs from food to education, a city must meet the various demands placed upon it in the most harmonious way possible; from ecological waste water disposal, to creating residential neighborhoods, all the way to maintaining a unique city image.

7. 人性个体作为将来城市建设理论的基础

在欧洲、亚洲和南美洲的核心经验是：以简单的而不是整体的人物形象为基础的城市规划都会成为单方面的不完整的工作。每一次城市建设的规划，不管是一个广场的城市设施配备还是一个上千万人口大都市的城市发展规划，都应该至少去尝试这种设想，人到底是什么？在一个人类的居住区内他会有哪些合法需求？不管是在乡村还是城市。这样的一种人性个体是简单的，受时间和文化限制的，不同内容、思想形态和宗教自由均不同的。重要的是，尝试形成一种理想，一种人类的设想，一种城市建设工作所服务的抽象个体——整体上与细节上，作为我们工作的核心理论基础与实践服务对象。

这听起来是理所当然的，但是在今天，像所有城市建造的规划层面所展示的，既没有在理论上也没有在实践中加以应用。人性个体背景的缺失是"城市规划灾难"的本质原因，这种灾难统治着世界，让我们的城市变成病态。这就是一直不断变幻中的城市建设决定参数，它具有单方面的合法性，但是相对于城市有机体这个整体却被高估，比如说权力展示、经济价值、交通、生态、城市景观或者城市形象，它们如今太多地决定了城市建设的决定。人们有极其不同的需求必须满足——从饮食到教育，一个城市也必须以尽可能和谐的方式满足这些最不同的要求——从生态的废水处理、具有家园氛围的邻里关系到不可混淆的城市形象。

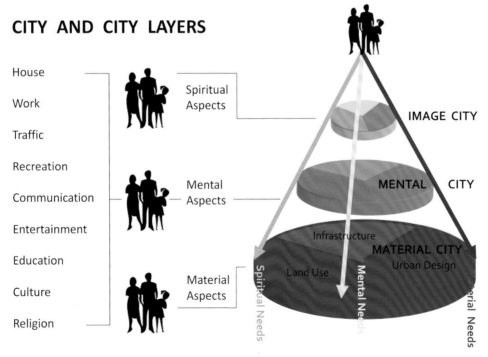

图 11　城市经历是身体的、心理的和精神需求的反映
Fig. 11 City experience as a reflection of physical, mental, intellectual and spiritual needs

From the Physical City, to the Mental City, to the Image of the City
A simple observation of human beings shows that they have physical, psychological, spiritual and intellectual needs that are defined by dichotomous characteristics. Material or corporeal needs would be, for example, eating, drinking, moving and defecating, as well as relaxing and sleeping. Psychological or emotional needs would be, for example, security, home, friendship and love, as well as independence, freedom and individuality. Examples of Intellectual, cultural and spiritual needs are, amongst others, education, culture and philosophy, as well as art, knowledge and religion. It is well known thanks to modern medical science that all of these factors influence each other.[7]

Necessities of Human Urban Experience
Foundations of a conception of man can therefore even develop from pure empiricism. Humans, no matter of what sex, race, age or cultural background, exhibit three essential sets of needs from themselves and their environment - in many cases a city: physical needs, psychological needs and intellectual or spiritual needs. These are the three major driving forces on which human existence relies.

Layers of the Human Urban Experience
These three sets of human needs are reflected in the three layers of urban experience: the physical city, visible in the cityscape; the mental city, tangible in the city appearance; and the imaginative city, apparent in the city's image. The physical city mainly represents the aspect of function; the mental city the individual visible and invisible perception of the city; and the imaginative city the individual and collective meaning of the city to people, or in other words, its image.[8]

Model of Human Urban Experience
A city therefore is experienced by every human in different layers, as a physical, mental and imagined city, and then assessed on the quality of those three most important criteria. The more divergent individual needs and layers of experience become depending on factors such as sex, race, age and cultural background, the more certain it becomes that it is exactly this inter-subjective superposition which should be the foundation of all urban planning.[9]

This conception of man should consider the three kinds of essential human needs, but should also take into account the functional, perceptual and imagined aspect of the city for the human. Practical experience has shown that such a conception of man can be used as a foundation for the simplest and most complex urban planning tasks - from the design of a street bench to the urban development of a metropolis – and thus establishes it as a universal, globally valid guide for future integral urban planning theory.

Experiences: Future Tasks for Research and Practice
A number of insights can be derived from these multi-layered experiences that may prove helpful for future urban planning tasks. For practical experience, it shows that people's immaterial complaints about their cities are growing constantly, but can't yet be satisfied sufficiently. It appears that there is still much scientific research to be done in all the following areas of experience on which ISA group is working.

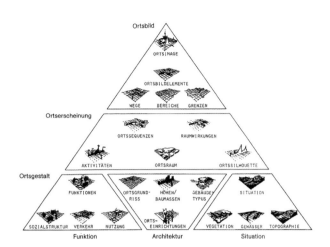
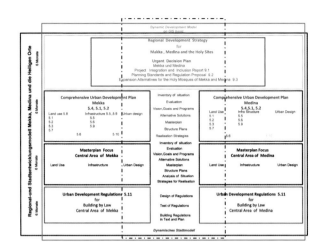

图 12 现代城市建设理论的层面：麦加和麦地那城市发展规划模拟中的人类需求的典范融合
Fig. 12 Aspects of a modern urban planning theory: exemplary integration of human needs in the simulation of an urban development strategy for Mecca and Medina

City Image: Reflection of the lived City

The city image is the reflection of the city, experienced subjectively and inter-subjectively by the human being. It is based on the level of satisfaction or dissatisfaction of physical, psychological and spiritual human needs towards the city. Consequentially, the city image is the actual, profound basis of all actions and reactions of the people in a city. To create a high quality city image, therefore is the central task of every kind of urban planning.

Beauty: a cultural Task of the Cityscape

The significance a cityscape has with regard to the quality of life in a city can never be overestimated, given that it is not an end in and of itself, but the expression of the real, existing or planned material and immaterial qualities of that city. The beauty of a cityscape has within it the cultural mission of "teacher" as well as being a "power source" for residents.

Quality of Life: a Question of Public Space

Providing urban living quality is a central task of urban space. The layout of streets and squares therefore is obtaining a growing significance throughout the world. The characteristics of the street and square space determine the spirit, the atmosphere, and the psychological sensation. These factors can also determine the extent of possible behaviors, e.g. walking, standing, sitting, etc., the satisfaction of physical needs, as well as the design and its underlying meaning, or the spiritual needs.

Ecology: Foundation of an integral City Organism

Ecological strategies and courses of action for cities are becoming more and more numerous. Reaching from energetic modernization of historic city centre buildings to the integral, sustainable ecological urban concepts in the planning of new towns. The ecological hardware of singular aspects, e.g. alternative forms of energy, and their systematic linkage are growing rapidly. The future task will be to integrate all parameters such as soil, vegetation, water, light and air into the planning, implementation and maintenance processes, in order to develop an integral, sustainable organism.

Civic Participation: from Hearing to Partnership

Civic participation can range from its passive form of a non-binding hearing, over to the active form of binding participation, e.g. a referendum, to real Public-Private-Partnerships. The active partnership has, as recent experience shows, a strong potential for development and can enable the short-to-middle-term implementation of projects that could not be realized on their own, either through public or private means. In times of decreasing governmental financial capacity, reveals the large hidden potential of alternative forces for urban renewal and development.

城市意象：城市生活体系的反映

城市意象是人们——主体的和主体间的——对所生活的城市的心理与文化的映像。它与城市对人们的身体的、心理的和精神的需求的满足或者不满足的程度密切相关。对于城市个体的行为和反馈，城市意象同时是其本质的和进一步加深的基础。塑造有质量的城市意象也就成了每种形式的城市规划的核心任务。

美丽：城市景观的文化任务

城市景观对于一个城市生活质量的重大意义几乎是无法超越的，前提是它没有被看做目的本身，而是一种真正的、现实的或者计划中的城市的物质和非物质质量的表现。城市景观的美丽既有作为"教育者"的文化任务，也有作为人类"动力源泉"的心理任务。

生活质量：公共空间的任务

提高城市的生活质量是公共空间的核心任务。街道和广场的规划在全世界范围内有越来越重要的意义。街道或者广场空间的特征与品质决定了情调、氛围和心理感觉，促进了不同可能的行为方式，包括其物质需求的满足，最终形成了精神需求的满足——意象和含义体系。

生态：一个完整城市有机体的基础

在城市规划中，城市生态的可能性越来越多。它们在旧城市建筑能源的现代化和新城市规划的总体可持续城市生态理念都有所涉及。其内容包括每一方面的生态硬件——比如说替代能源，和它们间的系统联系与整体性的快速增长。将来的任务将是把所有的参数——土地、植被、水、光、空气，在规划、实施和运营中发展成为一个整体的、可持续的有机体。

市民参与：从倾听到合作

市民参与从不承担义务的被动倾听形式，延伸到有责任参与的主动形式（公民投票），直至真正意义上的公众合作。就像上述最新的经验所表明的，主动的合作关系，在未来的城市发展中有很大的发展潜力，包括让那些既不能单独由公众、也不能单独由私人解决的项目在短期或中期的实施成为可能。不管是城市重建还是城市发展，这些耗资巨大的项目均受到国家财政前景的较大影响，经常表现出发展力量不足的问题。

Urban Planning: Synthesis of Social Society and Free Market Economy

Urban planning does not only plan the city, but also regulates the structural cohabitation of its citizens. Its principal task is to grant the individual citizen, from homeowner to investor, as much freedom as possible, and provide the community at large and the city as a representative of all residents, as much binding strength as possible. This is a balancing act between the interests of the individual and the interests of the community. In cases where balance is lacking, there is no question that cities as a whole suffer. The conscious synthesis of individual freedoms and social bonding is consequently a permanent task of urban planning.

Conception of Man: The Foundations of Future Urban Planning Theory

But the most important experience is that without a realistic conception of man, there cannot be truly integral, sustainable urban planning. Cities not only exist to satisfy individual demands, like traffic needs, Share Holder Value or representation of power, but for the satisfaction of all material and immaterial needs that a human being develops in a city. Only a conception of man that has the general physical, psychological and spiritual needs of a human being as the foundation for the thought and action in its urban planning, can lead to a future with an integral, sustainable city.

References

1) Mitscherlich, A. Inhospitality of the cities, Frankfurt a.M. 1965
2) Steiner, R. Way to a new architectural style, Stuttgart 1957
3) Trieb, M. Urban aesthetics as social tast: in winter; J.Mack,J (ed.): urban challenge - aspects of human ecology, Frankfurt a. M. and Berlin 1988
4) Trieb, M. Urban design and design regulations. In: Hanseatic City Luebeck – City image analysis and design regulations for the center of Luebeck, Stuttgart/ Luebeck 1977
5) Trieb, M. Aspects of city image; Trieb, M.: Urban design – Theory and Practice, 2. Edition, Braunschweig 1977
6) Helmy, M. Urban Branding Strategies and the Emerging Arab Cityscape – The Image of the Gulf City, Stuttgart 2008, Doctoral Dissertation University of Stuttgart
7) Eijk, Ph. Body, soul, spirit - views about psychosomatic interactions, ed. University Trier, lecture, Trier 2007
8) Trieb, M. Urban design – Theory and Practice, 2. Edition, Braunschweig 1977
9) Maslow, A. Toward a Psychology of Being, 1968

我们应以何种风格建造？
——21世纪初建筑学定位的探讨

In Which Style should We Build?
Architecture at the Beginning of the 21st Century
-A Positioning Approach

Phillipp Dechow

1. Every architectural era probably has it's own crucial question.

The question of the 19th century seems to be matched best by the title of a book written by Heinrich Huebsch in 1928: "In which style shall we build?" This question will accompany the entire 19th century.

In this century numerous architectural styles of the past were reintroduced, and used for a variety of construction projects. Build a factory in the gothic style, or maybe in the style of the Neo-Renaissance, a Doric, a Corinthian or a Gothic church?

1. 每一个时期的建筑学都会遇到各自的棘手问题

Heinrich Huebsch于1928年所撰书籍的题目恰如其分地表现19世纪的问题："我们应以何种风格建造？"这个问题贯穿了整个19世纪。

在这个世纪里，过去各类建筑风格再次流行，并被应用到不同的项目中。一个工厂应该建成哥特风格还是新文艺复兴风格呢？教堂应该是多立克式、科林斯式还是哥特式风格呢？

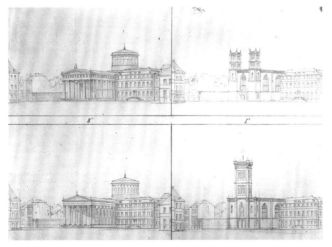

图 01　Schinkel的四个设计中，出现了四种不同的风格，这明确体现了19世纪建筑风格的不定性

Fig. 01　Four drafts by Schinkel, in four different styles, showing clearly the exchangeability of architectural styles in the 19th century

图 02　1900年前后的立面以及风格的随意性和过饰（罗马的Giustizia广场，1889~1910年），由此就很容易理解那个时代建筑师对"真实性"的诉求了

Fig. 02　In view of the randomness and the stylistic overload of the facades around 1900 (in this picture the Palazzo della Giustizia, Rome, 1889-1910), the longing for "truth" in architecture is very understandable

理论文章 ARTICLES

在这个时期，过去的各种建筑风格都是可能的，人们根据喜好加以采用，但这也给建筑师带来了烦恼。人们多次尝试去寻找一些规则，来确定哪种风格适合哪些建筑，但每次尝试都反而说明了各种历史主义和折衷主义风格的主观性和可替换性。这不断地质疑着建筑学的所有规则和约束，使这个时代的建筑看起来像"粉饰建筑"。

不仅仅是风格按照喜好被照搬，建筑材料也被模仿和冒充：例如原本厚重的石墙和石柱变成了只有薄石板遮盖的钢结构，甚至有些薄石板还是假的。很多原本应由昂贵石材构成的立面被便宜的工业预制石膏装饰，而且往往仅是街道立面被装饰得很漂亮，后院却采光通风不足。恶劣的居住条件与街道立面形成鲜明对比。荷兰建筑师Hendrik Berlage是这种状况激烈的批评者之一，他简明扼要地总结道："谎言成了规则，真相则成例外"。

于是人们希望发展出一种风格，它能剔除所有的随意性和伪装，使建筑能够呈现自身的纯净和本质。这种追求在那个时代是不难理解的，在当时的许多文章中，我们可以清晰地感受到这种诉求，如Hendrik Berlage所写："我们追求建筑的本质，那就是真实性，除此之外还是真实性"。人们对建筑也相应提出了改建的要求，尤其是要求针对立面进行改造，去除"遮盖形体的表层装饰"，正如Semper在书中所解释的：应该"冷酷地剔除所有没用的东西……甚至最后一个装饰物和遮盖物也应清除，因为我们想要的真实应是完全赤裸的"。

All previous architectural styles were available and were applied at will, but it was this approach that started to worry the architects. There were many attempts to find objective principles for which style to use for which type of building, but in the end every attempt at an answer only reinforced the arbitrariness and replaceability of the different historicizing and eclectic styles. This situation seemed to question all rules and commitments of architecture, until it could only be regarded as "mock architecture" by its contemporaries.

Not only were architectural styles copied at will, materiality was also imitated or faked: what looked like massive stone columns and walls actually consisted of concealed iron structures, covered by stone-plates that weren't necessarily authentic themselves. Many facades were covered with cheap industrially prefabricated plaster, which was used to simulate expensive stone decorations. Thus created, pretty street fronts stood in blatant contrast to the badly lit and ventilated backyards. The Dutch Architect Hendrik Berlage, one of the great critics of this time, stated it like this: "The lie has become the standard, and the truth has become the exception."

The longing for a style that does away with all the arbitrariness and hypocrisy and shows the building in its purity and its character, is – in this context – only understandable. This can be felt in many articles of this time, including when Berlage states: "We want the nature of architecture, that means the truth and the truth again." As a logical consequence the new architecture became in conflict with the facade, "the cloth hiding the body," as Semper had tried to prove in his writings. "All that is useless" must "recklessly be disposed of [...]. Every last cover, even the Fig's Leaf must be removed, because the truth we want is totally naked."

图 03　柏林的"白城"（1929～1931年）是人们追求建筑的真实性以及去除装饰的结果

Fig. 03　The "White City" in Berlin (1929-1931) is a consequential realization of the quest for truth and for the dismissal of masquerade in architecture

图 04　乏味平庸的战后现代主义风格占据了德国的大量街区（慕尼黑的Mitterhofer大街）
Fig. 04　The often uninspiring and trivial architecture of post-war Modernism characterizes many urban quarters in Germany (Mitterhoferstrasse, Munich)

And so it happened. The modern architecture, which became more and more popular in the first decades of the 20th century, seems to embody the ideals demanded by Berlage in the most consequential way. The slick white walls of classical Modernism do not cover nor hide anything, the buildings are pure bodies, undressed and undisguised.

But the renewal did not only take place on a formal, stylistic level. Much more than that, it was the expression of an internal reorientation concerning different areas, based on a fundamental criticism of the industrialized city. The disastrous hygienic conditions, not to mention the generally poor quality of life, were to be improved by this new manner of building. Less density and an orientation towards the sun were to bring light, air and green into the settlements. The separation of living and working was to solve the emission problems, and with the help of standardization and industrialization of the construction-process, more health and comfort would be accessible for a broad sector of the population.

2. The inhospitality of Modernism

But with the modern style, a new question soon emerged that was to perpetually accompany 20th century architecture. Arbitrariness and hypocrisy had been banned, yet a subtle coldness was surrounding the new type of modern architecture, since it had been stripped to the naked body. The Architect Josef Frank, who designed one of the houses of the "Weissenhofsiedlung," one of the most important architectural exhibitions of Modernism, already wrote back in 1931: "The modern German architecture maybe objective, practical, generally correct, often appealing, but it remains lifeless."

这种建筑之后出现了。所谓的"新建筑"于1910年前后启动了它的胜利号列车。Hendrik Berlage所要求的建筑学理念似乎获得了实现。现代主义光滑的白色墙壁没有遮盖或掩饰任何东西，建筑就是纯粹的形体，没有覆层和装饰。

这个演化并非仅仅发生在形式和风格层面，更是多个领域内在的革新要求的结果，其根基是对工业时期城市的批判。新建筑要求改善当时工业城市灾难性的卫生状况，保证生活质量；利用较低的建筑密度和合理的建筑朝向给社区带来光照、新鲜空气和绿化；将住宅区和工业区分离以解决工业污染带来的问题，并通过建设过程的标准化和工业化，使更多的社会阶层可以低成本地享用新建筑。

2. 现代主义的冷漠

伴随着新建筑又出现了一个棘手的问题，这个问题从20世纪一直延续到今天。因为当从建筑中剔除任意性和装饰后，只剩下赤裸裸的建筑形体，这样就会有一种可疑的冷漠感环绕着新诞生的现代建筑。Weissenhofsiedlung住宅区是现代主义建筑最重要的代表作之一，其中一座住宅的设计者Josef Frank在1931年写到："德国的现代建筑是真实客观的，符合实用性和原则性要求，甚至常常是吸引人的，但是缺乏生命力"。

"二战"后，对现代主义的批判主要集中在城市和城市空间方面。简·雅各布斯1961年所写的富有影响力的批判文献《美国大城市的死与生》是对现代城市建设批判的开始，现代城市建设被指为常常导致城市空间质量的下降和城市生活的缺失。在德国，作家和心理分析学家Alexander Mitscherlich在1965年出版的著作《我们城市中的冷漠》中讨论了这个话题。正如Heinrich Huebsch的书题一样，这本书的题目也切中了所在时代的问题，Alexander Mitscherlich围绕"冷漠"这个词汇展开了讨论，这是一个一直沿用到今天且富有概括力的概念。

在接下来的几年里，这些文献引发了对城市空间冷漠化的大量讨论，也发现和解决了许多欧洲城市建设理论和实践中存在的具体问题。与此同时，现代主义也体现出了其惊人的灵活性：最初的功能分离原则反被功能混合原则代替，交通混合原则代替了人车分离原则，密度城市代替了疏散城市。考虑日照朝向、与街道分离的行列式布局也渐渐被沿街的建筑布局取代。最后还放弃了工业化和标准化的理念，取而代之的小地块开发模式更利于实现个性和多样性。几十年中，这种影响深远的范式转变得到了贯彻实施，现代主义反都市性的城市模型被放弃，人们又开始推崇都市性，局部又回归到了前现代的城市理念。

毫无疑问，这些措施促进了公共空间质量的持续提升，在这种新精神的影响下，产生了一些新城市和新城区，我们可以对它们加以评定，例如柏林附近的滨水城市Spandau、慕尼黑的展览城Riem、汉堡的港口新城。但这并不是讨论的终点，更多的是下一轮讨论的开始。

After the Second World War, criticism of Modernism soon arose around the topic of the city and its urban spaces as a key subject of the discussion. The influential article, "The death and life of great American Cities," written by Jane Jacobs in 1961, was the beginning of a fundamental criticism of Modern Urbanism that resulted much too often in a loss of urban quality and in the disappearance of urban life in the city. In Germany, the author and psychoanalyst Alexander Mitscherlich took up the subject in his book, "The Inhospitality of our Cities," published in 1965. As previously done by Huebsch, once more the title of a book states the question of the time. Mitscherlich delivers the notion of "inhospitality," which is an incisive and until now central expression to the debate.

Since then, there had been a lively discussion about the "inhospitality" of the cities, and many problems of urban planning theory and practice were identified and corrected in Europe. Modernism proved to be surprisingly flexible in the implementation of these changes: a mixture of functions replaced the separation of functions, and coexistence replaced the separation of pedestrians and motorized traffic. The ideal of the low-density, green city was replaced by the demand for urbanity and density. The typical row-buildings oriented towards the sun but without any link to the street and its direction, were slowly substituted by buildings creating urban space. In the end, even the idea of industrialization and standardization was abandoned in favour of a small-scale plot concept, promising individuality and diversity. Over the course of decades a broad paradigm shift had been completed abandoning the anti-urban model of the Modern City, and returning to urban, sometimes pre-modern urban concepts.

All of these actions have undoubtedly contributed positively to the quality of urban spaces. Today several new cities and quarters have been built in the modern spirit, such as the Water City Spandau near Berlin, the Trade Fair City Munich Riem, the Hafen City Hamburg. The discussion, however, is by no means coming to an end, but merely entering the next round.

图 05　两条街道的对比：一条是19世纪的风格，一条是战后现代主义风格，其对比表明了立面的简化如何降低了城市空间的质量
Fig. 05　The comparison of two streets: One in the style of the 19th century, one in the style of Post War Modernism, shows the lack of quality of urban spaces resulting from the reduction of the facades

Roman Hollenstein for example writes in the "Neue Zuericher Zeitung": "many city expansions and urban redevelopments still [...] appear soulless." One can also find this judgment in many critiques of recent urban projects. Joachim Schirrmacher writes in the NZZ about the HafenCity Hamburg:

"Whilst the massive construction volume of the neighboring "Speicherstadt" is generally seen as balanced, the atmosphere here – despite singular successful buildings, despite many green areas, cafés and restaurants – seems strangely ambivalent. One cannot feel comfortable here. [...] An accumulation of self-centered solitary designs, that will not form an entity, has evolved here."

Other articles about the same project call the buildings "strangely lost and relationless," and talk about "cold and heartless architecture." The architects on the other hand don't seem to be able to defend their style of building against this criticism, as shown in the following interview by Hanno Rauterberg with the famous German architects and urban planners Meinhardt von Gerkan and Volkwin Marg, for the German weekly newspaper "Die Zeit":

"ZEIT: [...] Most people seem to prefer traditional urban squares to modern ones.
von Gerkan: Reasonably.
ZEIT: Do you feel the same way?
von Gerkan: Of course I feel the same way, even if I don't like to accept it. [...]
ZEIT: Is there even one modern urban square that seems as well designed as some squares of previous times?
Marg: (silence) None comes to my mind spontaneously."

Roman Hollenstein在《新苏黎世报》中写道:"很多城市扩张和城市改建显得……缺少灵魂",一些针对城市建设项目的批判也反映出了这点,如Joachim Schirrmacher在NZZ中是这样评价汉堡港口新城的:

"相邻的仓储城区虽然有着巨大的建筑体量但是也给人以和谐的感觉,但在港口新城,即使有单个成功的建筑,有很多绿化、咖啡厅和餐馆,氛围还是没有被充分营造出来。还是没有舒服的感觉……出现了一系列以自我为中心的单体设计,却缺乏整体感。"

对同一项目进行评论的其他文章写道,房子像是"一个个落了单,变得互相没有关系",是一种来自"冷漠和没有灵魂的建筑学"的做法。建筑师也并不反对这种批判,就像Hanno Rauterberg对建筑师和规划师Meinhardt von Gerkan和Volkwin Marg为周报DIE ZEIT中所作的采访:

"ZEIT:……大多数人都更喜欢传统而不是现代的广场。
von Gerkan:对的。
ZEIT:您也这样认为吗?
von Gerkan:我当然这样认为,就算我不愿承认这是真的……
ZEIT:您有没有发现一个现代的城市广场,让您感觉它像过去的广场一样成功?
Marg:(沉默)暂时想不起有哪个很成功。"

图 06 清晰的建筑美学常常会导致建筑物互相分离,没有整体感,这里是慕尼黑的展览城Riem(1998~2009年)。在批判的声音中也越来越多地听到"冷漠"或者"冷酷"这样的词汇

Fig. 06 The pure aesthetics of architecture, shown here in the Trade Fair City Munich Riem (1998-2009), often result in isolated buildings, unable to form an entity. Expressions like "soulless" and "heartless" are used more and more frequently in critiques

图 07 慕尼黑展览城Riem:围合人权广场的建筑立面之一。随之产生的问题是,这样的立面能够构成漂亮的城市空间吗?

Fig. 07 One of the facades making up "the human rights square" in Munich Riem. The question is, if these reduced facades can create good quality urban spaces?

有趣的是，这些批判很少涉及建筑自身的美学质量(有时甚至被认为是成功的，例如Schirrmacher就这样认为)，更多的是涉及建筑之间的城市空间，建筑之间的关系或观者与建筑间的关系，这些常常被判断为"冰冷或没有灵魂"或者"舒适"。换而言之，批判不是针对建筑，而是针对城市，以及其中的空间和空间环境质量。建筑师Christoph Maeckler的发言切中了要害：

"过去的一百年中没有一个广场，能让到那里的市民说：'我在这里感觉很舒服。'"

3. 互相矛盾的建筑单体和城市空间

究竟是为什么呢？现在的城市中，这些问题的缘由是什么呢？其中一种矛盾的存在引起人的注意，这种矛盾不管是在对新项目的批评中还是在很多建筑师的思考中都可以发现。Christoph Maeckler的一句话对此进行了清楚的阐释：

"我们必须注意到前现代时期的城市品质，不是从建筑风格的层面，而是城市设计的层面。"

这句话说出了我们这个时代建筑的深刻内部矛盾，因为Maeckler在这里所批评的，是城市设计和建筑形式风格的分离。从已经建设的实例中可以看出：来自19世纪的城市街区结构被沿用到今天，但是却套着现代主义建筑的外套。毫无疑问，这样的分离将会引发问题。这样的状况反映了两个系统及其各自的内部逻辑，这两个系统应该统一起来：

在Maeckler 所说的"前现代"城市规划和设计中，单个的建筑隶属于一个连续的建筑组团，这样，建筑只有朝向城市空间的二维立面可以被看到。处于这种语境下的历史建筑风格将很大的造型精力放在了建筑立面上，并且实现了很好的造型效果，由此将立面从二维局限中解放出来，成为集体营造城市空间的一个层面。这些立面还有一种令人赞赏的能力——一方面让自己隶属于一个公共系统，另一方面通过单体的变化体现自己的个性。

与之相反，现代主义建筑有着完全不同的前提。如前所言，对工业化时期城市及其卫生和社会状况的拒绝是现代建筑从旧式布局中脱离的功能性原因，即拒绝临街式的连续建筑布局和所谓的"巴洛克协会"。对这种建筑布局方式的拒绝在勒·柯布西耶著名的口号"我们必须消灭巷子！(Il faut tuer la rue corridor!)"中达到了顶峰。但最实用和艺术解放成为最终动因，正如在他另外一段著名的文字中所表现出来的：

"建筑艺术是针对阳光下的建筑体块进行的一种明智的、准确的、伟大的游戏。上帝创造了眼睛，让我们可以看到阳光下的形式：光和影揭示了形式：立方体、圆锥体、球体、圆柱体或棱锥体是伟大的基本形式，太阳更容易凸显出它们的存在：它们的影像让我们感觉到纯粹、实在和明确。"

Interestingly, the criticism rarely relates to the aesthetic quality of the buildings themselves – these can even be called successful, as in Schirrmacher's article – except for the space between the buildings, the relation of the buildings to each other, or the viewer's relation to the building, as implied by the expressions "cold and heartless," or "comfortable." In other words: the critics do not relate to the architecture but to the city, its spaces and its sojourn qualities. The architect Christoph Maeckler states it like this:

"There is not one modern urban square of the last 100 years, of which a citizen could say: I feel comfortable here."

3. Architectural bodies and urban space in contradiction

What might be the difficulties of our present cities? In this context, the contradiction that can be found in criticized new plans, as well as in the thoughts of many architects, is striking:

"We have to turn back to a kind of Pre-Modernism, not in a stylistic, but in an urban planning sense."

One can already see a profound inner disruption of the contemporary way of building in this quote, because what Maeckler demands is nothing less than the separation of urban content and architectural style. This idea can already be seen in several existing examples: in many places, the urban block of the 19th century is adopted in different variations, but it occurs in the shape of Modernism. It's no surprise that this contrast causes difficulties. The reasons for these difficulties lie within the inner logic when both systems are combined.

The urban planning of "Pre-Modernism," addressed by Maeckler, subordinates the individual building to the usually closed alignment of buildings so that in the end, the two-dimensional plane of the outer wall directed towards the city is the only part of the entire building which is visible. Previous styles used to cope with this fact mainly by focusing on the craftsmanship of the facade and thus produced enormous artistic accomplishments that would free the facade of its two-dimensionality, transforming it to a space-occupying and space-defining element. These facades have the astonishing capability of integrating themselves in a common system on the one hand, while still reaching a certain individuality on the other, through the variation of small details.

The architecture of Modernism, however, is based on opposing conditions. The previously mentioned rejection of the industrialized city, with its hygienic and social problems, generates the functional justification for a detachment of the building from the continuous, street-following building alignment, the so-called "baroque unit." The rejection of this architectural norm came to its climax with Le Corbusier's famous exclamation, "Il faut tuer la rue corridor!" But in the end, it was not only motivated by practical reasons but also creative liberation, as can be seen in another popular quote:

"The art of architecture is the wise, accurate, grandiose game of buildings, gathering in the sun. Our eyes are created to see shapes in the sun; lights and shadows unveil the shapes; the cubes, the cones, the balls, the cylinders, or the pyramids are the great primary shapes, which are unveiled by the sun: their image seems pure, tangible, unique to us."

Le Corbusier describes precisely the appeal of modern architecture, the joy of the pure geometric bodies, playing with volumes, its aesthetics of clarity and the precision of autonomous architectural objects. For this architectural style, created for rows, slices, point houses or other detached buildings within a green environment, the facade, while so important for the old town, became unnecessary, even disruptive.

The liberation of the structural body from the unit, and the new aesthetics of proportional interplay created conditions to fulfill the previously mentioned demand for "truth" in architecture, to avoid any "overload," to recover "the naked wall in all its simple beauty."

But what happens if – as proposed by Maeckler – the basic urban planning principles of one system are combined with the stylistic principles of another? What remains of the proud and autonomous architecture of Modernism, that wants to liberate itself, and for which the structural body of a building is a central design feature, when reduced to a wall in a continuous alignment of buildings? Following construction, can this architecture – criticized for its "denial of urban space" and "inability of forming a unit" by architect Kaspar Kraemer – now simply be used as public space that is supposed to define and create urban areas like avenues and squares? Or is it not obvious that these buildings appear "strangely lost and relationless," as a "sequence of self-centered individual designs, unable to create a whole"? The already constructed examples demonstrate that urban space cannot be created in this way.

And besides, what remains of the Modernism that introduced the architectural form as a result of the urban modernization movement? Did Modernism simply become a "style," an "architectural illusion" and a formal gimmick? How much Modernity is left in this much reduced idea of Modernism?

4. New Historicism or Retro-Look: a Fata Morgana

In this context, Historicism, which has lately come into fashion in urban planning under the name of "New Urbanism," seems to have a more consequential outlook. If urban planning turns back to pre-modern principles, why not use the matching architecture, as practiced in the US by the representatives of New Urbanism, in England by a movement close to Prince Charles, and in Germany by the domain of the architect Hans Kollhoff? After all, these styles still offer the repertoire for designing facades as urban walls, so important for the creation of urban spaces, and which had been thrown overboard by Modernism.

In Germany, Hans Kollhoff was one of the first architects to approach historical archetypes without the ironic distance of Post-Modernism, and he justifies this step with a revised urban planning condition:

"Kollhoff: Until 1989, we architects were only dreaming of the city, but in reality mainly built subsidized apartments for the periphery. Even the residential areas of the International Building Exhibition were located on the city margins, because of the (Berlin) Wall. It was a big reality shock, after the crumbling of the Wall, to be planning for the inner city, with its entire functional range, and not only for the periphery anymore. This has led us to a different architecture.

柯布西耶这里对现代建筑的吸引力进行了简洁的诠释，即对纯粹的几何形体的兴趣，对体块的游戏，以及建筑单体清晰性和鲜明性的美感。对处于绿地中的行列式、板式、点式或者其他独立的建筑布局和建筑风格来说，对于老城很重要的立面就显得多余，甚至会破坏气氛。

建筑形体从整体中的脱离以及针对体块游戏的新式美学为满足前面所提的建筑"真实性"提供了前提，避免了任何的"过饰"，以便可以"展示赤裸墙壁的简约之美"。

如果像Maeckler所建议的那样，将城市设计和建筑设计的风格结合起来，将会发生什么呢？现代主义建筑一向希望自由发挥，把体块游戏作为主要造型手段，骄傲且独立，一旦被置入连续的临街式布局，被缩减为只剩一面墙的时候将会发生什么呢？这些被建筑师Kaspar Kraemer认为"毁坏城市空间"以及"不能形成整体"的建筑可以纳入临街式布局，并成为构成街道空间和广场空间的一部分么？建筑看上去是否像是"一个个落了单和互相没有了联系"，或者"没有联系的单体的罗列"？参照已经建成的类似实例就可以看到：这样确实不能产生正确的城市空间。

在这种情况下，作为现代主义运动重要成果的现代建筑还会保留有什么呢？在其当初所反对的城市布局下，现代建筑会不会再次变成其所反对的形式游戏、空间布景和"粉饰建筑"的集合？这样的现代主义建筑还能代表现代主义吗？

4. 新历史主义或者复古样式：海市蜃楼

在过去的几年中，历史主义风格在新城市主义的旗帜下再次成为潮流，在此背景下，历史主义看上去顺理成章。既然城市规划和城市设计再次推崇前现代的原则，为什么建筑学不应对此作出反应呢——就像美国新城市主义运动、英国的查理王子运动以及德国建筑师Hans Kollhoff所实践的那样？这些建筑风格仍将立面同时作为城市公共空间界面来设计城市空间构造的重要手段，但是被现代主义认为是多余的。

Hans Kollhoff已经摒弃了后现代主义富有讽刺性的做法，彻底重新回到前现代建筑学中，在德国他是第一批这样做的建筑师之一，他利用城市建设语境的变化来解释这一转变的必要性。

"Kollhoff：到1989年我们建筑师才对城市有了憧憬，但是实际上建造的都是城市边缘的补贴性社会住宅。国际建筑展的住宅区也因柏林墙的存在而建设在城市边缘。因此柏林墙倒了以后，人们不再为城市边缘设计，而是为功能丰富的内城进行设计。这是来自现实的巨大冲击，将我们引向了另一种建筑风格。

SPIEGEL: But why has it led to the architecture of the 19th century?
Kollhoff: Because it has worked especially well."

At second glance, however, Historicism quickly loses its logical consistency. Nowadays, various archetypes are used for different construction tasks: Classicism for a mansion, the style of the 1930s for skyscrapers. Predictably this leads straight back to the debates of the 19th century. The traditionalists have to ask themselves how to make the discussion go another way, so as not to end up with the same criticism as before: that their architecture is "only the degeneration of true styles, only the product of past times," nothing but "illusionary architecture" and "tasteless shapes."

Fig. 08 In the "Playa Vista" quarter of Los Angeles, many residential blocks are constructed in historicizing European styles. While the quality of urban spaces seems to improve, the question remains if these examples can be models for the future

Additionally, buildings have changed drastically since the 19th century. New typologies have arisen, shopping malls for example, and the historic design guidelines are not suited, not flexible enough for new technologies and materials. Especially in housing, which still makes up the majority of construction, far-ranging changes have taken place. Even if contemporary urbanism has largely distanced itself from the ideals of Modernism, it still led to a profound change, which no one wishes to completely reverse. Despite all demands for mixing and density, some principles are not being abandoned because they are too useful and too profoundly linked with the construction legislation, e.g. the separation of disturbing industry from housing areas, or the determination of maximum permitted density and minimum distances. These achievements of Modernism, which prevent us sliding back to the inhuman conditions of the cities of the Industrialization, have become a part of contemporary building that is taken for granted. A building block from the 21st century has only become acceptable because of Modernism, and it cannot be compared with a building block from the 19th century.

All this may present Historicism with unsolvable problems. With this background, it doesn't seem to be the natural answer anymore, but rather a style born out of embarrassment and which nurtures on the weaknesses of Modernism and the lack of alternatives. This is exactly from where new concepts for dealing with the city will have to depart.

5. In which style should we build?

Does that mean the 21st century faces the same question as the 19th century? The discussion about style is certainly not caused by the diversification of architectural methodologies (e.g. Deconstructionism, Blob-architecture, Second-Modernism, Regionalism and the different historicizing styles), since the idea of one dominant style does not fit into the pluralism of present times.

Still it seems that a discussion about style is inevitable, if the paradoxical situation of separating urban principles and architectural style is to be resolved without winding up in the dead-end of Historicism.

Style is to be understood as a shared fundamental orientation of all the different trends towards one main goal. This is possible because, unlike in the 19th century, the discussion about style nowadays is linked to a detailed analysis and confrontation of the aspects that affect our interaction with urban space. In Europe, a common goal of almost all architectural trends is the "desire for a spatially functioning city," for an architecture whose "subject is not the singular building, but the urban ensemble," in other words, for an urban style of building.

In the search for such a style, the development of Urbanism towards the end of the 20th century could set the example: important and useful principles of Pre-Modernism combined with the necessary changes of Modernism. Contemporary urbanism consists of a synthesis of Modernism and Pre-Modernism, in a new, independent shape.

This strategy could work for architecture, as long as there is a successful acknowledgement of the qualities of Pre-Modernism. But if so, the task becomes more complex. The understanding of architectural principles of pre-modern times brings up difficulties: how can these buildings create such pleasant squares and streets, whereas those of Modernism appear so unappealing? How can these buildings come together so

19世纪以来，建筑发生了本质性的转变，产生了许多新的建筑类型，例如购物中心。对于新技术和新材料来说，历史主义风格的形式规则显得不够灵活。特别是占建筑总量很大比重的住宅建筑，发生了更多变化。即使当下城市设计在很大程度上离开了现代主义的理念，现代主义还是带来了很大的转变，这种转变是没有办法逆转的。尽管今天人们强调功能混合和城市密度，但是有些现代主义的基本原则仍然没有被放弃，其以无可置疑的合理性写入了建筑和规划法规中，例如污染工业与住宅区的分离，或者最高密度上限和最小建筑间距的规定。这些现代主义的成就避免城市出现工业化时期非人性的状况。21世纪的Block街区首先要满足现代的要求，这和19世纪的Block街区已大相径庭。

这并不意味着历史主义无法复苏，但在这种背景下，它又确实更像是一个随手拈来的答案，更像是一种困境中催生的风格，这种风格来自于现代主义弊端甚众且缺少替代方案的背景。在这种情况下我们更应该找寻新的方案，给城市带来新的道路。

5. 我们应以何种风格建造？

21世纪还面临着19世纪这个棘手的问题吗？对风格的讨论必须注意到过去几年解构主义、流体建筑、第二次现代主义、地方主义和各种历史主义风格中建筑风格的不断分化。建筑风格应该具有多样性，将其统一为共同的风格也与我们多元化的时代不相符。

但是，为了克服城市设计和建筑风格分离的悖论性处境，同时不陷入历史主义的死胡同，对风格的讨论是必要的。

这里风格应更多被理解为所有建筑流派的一种共同方向，这是可能的。因为不像19世纪，这里讨论的新风格强调的是将不同的内容联系起来，重新获得城市空间。至少在欧洲，几乎不同流派的所有代表，现代主义的和历史主义的，都有一个共同的目标："对空间上良好运营的城市的向往"，对"不局限于单体的设计，而是塑造出城市整体"的建筑学的向往，或简而言之，对营造都市化品质建造方式的向往。

在对这一目标的追求过程中，20世纪末城市设计的进展成为了典范：重新采用很重要、具有意义的前现代规则，同时现代主义必要的新主张也融入其中，所以我们今天的城市设计是现代主义和前现代的融合体，是一种新的、独立的形式。

如果当代建筑学能够具备类似前现代建筑学的品质，那么这个尝试应该可以成功。但在这一方面，任务比预想的复杂。对前现代建筑设计原则的理解提出了一系列的问题：为什么这些建筑可以创造舒适的广场和街道，而现代主义却遭人拒绝呢？为什么源自完全不同时期的旧建筑，例如巴洛克和哥特式建筑，可以轻易地融合在一起，而现代主义建筑常常看起来疏远、毫无联系，不能成为一个整体呢？

我们正在慢慢地理解这些问题。这背后似乎有一个对应着一些知觉心理学现象的复杂系统，只能逐步被破译，唤醒对过往成就的高度重视。在之前的讨论中，已经产生了第一个答案：城市建筑学需要立面！

从这个角度来说，立面不再是多余的装饰，不能被削减为功能性的围护，必须认识到其在整体中的重要作用。立面不仅仅属于建筑，同时也是城市空间的组成部分，它不仅仅是建筑物的围护，同时也是围合城市空间的墙面。
鉴于其对城市空间的意义，立面应成为独立的建筑艺术设计内容，可以并且必须遵循自身的设计原则。这可能会导致立面与建筑的内部逻辑陷入矛盾，但如历史所表明的，对这种内部断层的处理和解决可以释放出巨大的艺术能量。

作为"城市空间界面"的立面应该被如何设计，还不能最终定论。过去对立面的研究无疑是找到答案的关键，从中可以得到两种可能的方案——当然将其结合起来更好。一种是通过设计实践中的操作和实验，以历史立面为基础，通过变形来获得新形式，再对其效果进行验证。另一种方案是根据知觉心理学的规范对立面进行分析，找出其影响我们城市空间的规律性。从而从历史的典范中解放出来，将现代主义和这些风格融合起来。
规划师和建筑师因此面临着一个艰巨的任务。当然，这其中也存在着很好的革新契机，它会引领一种新的建筑风格，这种风格的奠定应同时结合相应的城市设计。这种风格以城市空间为对象进行思考和设计，而不只是建筑的体块。而且，这种风格能够重新实现城市设计和建筑设计的统一，并由此为建筑艺术和建筑文化的联合奠定基础，而这些正是我们在过去几十年中常常缺失的品质。

effortlessly to form an ensemble, even when they are from different architectural eras – Baroque and Gothic, for example – whereas those of Modernism appear distant and incoherent and cannot form a unit?

We are just starting to understand these questions. It seems like there must be a complex system of cognitive-psychological phenomena that can only be decoded little by little, and which raises great respect for the achievements of the past. One answer however seems to be obvious after the thoughts exposed so far: urban architecture needs facades!

From this point of view, the facade can no longer be seen as unnecessary decoration, reduced to the single role of a cover, but it must be recognized for its general significance to the greater whole. This construction element is not only consequential to the architectural part of a building, but is also equally critical to the urban development of a city. It is not just a building shell, but a space-forming component of its urban environment.
The facade gains its justification as independent structural element from its importance to the urban space, which must pursue its own purpose and design principles. This can result in a contradiction of the facade to the inner logic of the building, but the act of solving problems like that can often evoke enormous creative potential.

What principles the facade should follow in order to fulfill its importance as urban wall, cannot be said conclusively. Studying the facades of the past is certainly the key to a solution, and there are two possible approaches that when combined may provide this conclusion. One is working and experimenting with historic facades in the design process, generating and verifying new forms and metamorphoses. The other approach is the analysis of facades with cognitive-psychological criteria to detect the principles that determine our perception of urban space, with the facade as medium. Only this enables the emancipation from historic archetypes and carries the aspect of Modernism into a new style.
Urban planners and architects face a new great challenge, which includes an equally great chance for development that could lead to a new architectural style. A style which once again finds its reasoning in urban needs, that allows thinking and designing spatiality instead of building volumes. Consequently this could lead to a reunion of Urbanism and Architecture, and thus create the basis for a comprehensive art of architecture and culture of building, which we seem to have missed so routinely in the last decades.

从两种文化中学习　LEARNING FROM TWO CULTURES

从数量到质量：
韩国城市规划控制的经验
From Quantity to Quality:
Experiences from City Supervision in South Korea

Seog-Jeong Lee

Since becoming an urban planning and design professor in Seoul, Korea, I have in some ways seen a large difference between the German and Korean planning culture. Many things have been painful for me to see, especially the careless handling of old things, from a house to an entire city district, that are usually replaced by something new as quickly as possible. Even if they look grungy and underdeveloped at first glance, they are still valuable because each of them carries a life story.

Korean architecture and urban planning have been strongly influenced by American planning theories of the 1950's. During that time, urban planning was only understood as a matter of traffic, social and economical importance and only quantitatively rated. This way of thinking, combined with the rapid growth and urbanization process, caused an absurd discrepancy between the existing, obsolete small city structure, and the "modern" large-scale structure. High-rises and oversized streets became the symbols of prosperity and development.

However, much has recently changed in Korea. Currently there is an 'upheaval' in the planning and construction fields, which stems from when the elevated road in the heart of Seoul's old city was torn up and the formal Cheonggyae stream (Cheonggye Project) was uncovered. This is a change from quantitative to qualitative thinking. Many things that used to be self-evident are now being questioned, especially how to handle existing cities with existing buildings. In the recent past, it was simply demolition to build new city districts and buildings. However, many new perspectives are making the headlines: from quantitative expansion to qualitative development, urban renewal instead of new towns, upkeep and rejuvenation of historical city districts, green cities, smart cities, and the beauty of cities.

自从成为韩国首尔的大学教授以来，德国与韩国规划设计理念的一些差别开始愈加显著。韩国的许多情况都令人困惑，例如冷漠地对待老建筑及老城区，使得它们经常轻易地被拆除而为新建筑所取代。这些老建筑尽管可能第一眼看上去有些破旧和过时，但它们依然是珍贵的，因为它们承载了城市生活的历史。

韩国的建筑和规划自19世纪50年代以来受到美国的强烈影响，开始过于注重交通、社会和经济等技术层面，单一依靠各种数据作为规划的基础。这种思维方式，伴随着快速增长和城市化进程，造成了老的、小尺度的传统城市街区与新的、大尺度的现代城市街区之间的荒谬性的疏离关系。高层建筑和尺度巨大的街道一度成为城市发展和经济富足的标志。

今天韩国的情况正在发生改变。拆除首尔老城区的快速路并恢复被覆盖的河流Cheonggye项目，表明规划和建设领域的思想已经开始从注重数量到追求质量的巨大转变。许多以前被看做是理所当然的事情今天都被打上了问号，特别是在对待老城区的态度上。很多新的观点都表明：城市发展应从量的膨胀转变为质的发展，以城市更新取代新城建设，维护并复兴历史旧城区，建设绿色、智能、美丽的城市。

图 01　城墙内的历史城市中心，首尔2011年
Fig. 01　Historic city core within the city wall, Seoul 2011

图 02　历史城市中心"老的"小尺度的地块和建筑结构，首尔2011年
Fig. 02　"Obsolete" and small plot building structure in historic city core, Seoul 2011

图 03　高层建筑林立的"辉煌"现代街道，首尔2011年
Fig. 03　"Splendid" and modern streets with high-rise buildings, Seoul 2011

理论文章 ARTICLES

图 04 重新恢复的Cheonggyae溪流成为了公共空间，首尔2011年
Fig. 04 Re-opened Cheonggyae stream as public space, Seoul 2011

图 05 交通干道间的"绿岛"，首尔2011年
Fig. 05 A "Green Island" between traffic roads, Seoul 2011

当前在韩国出现的这种转变可以同19世纪70年代在欧洲各国，包括德国发生的城市规划思想变革进行比较。当时的欧洲经历了极富戏剧性的政权更迭，城市中老城区的品质被重新评判，首先就是街道空间被重新作为各种社团及活动的社会空间来看待。许多建筑师和城市规划师，如Colin Rowe、Edmund N. Bacon、Rob和Leon Krier等，批判现代主义——将建筑视作"自由存在的物体"而与街道空间相对立和隔离开来考虑的观念。最晚至80年代初，欧洲的专业人士就开始策动"城市规划的进化"，旨在维护历史建筑，将旧城修复和现状居住环境的改善作为城市规划的第一要务。当前韩国及首尔的很多政策性目标及示范项目都会让人想到那时的欧洲规划政策。

尽管许多良好愿望看似都指向了高质量城市规划与建设这一目标，但要想实现它还有很长一段距离。根据我们的观察，问题主要基于以下几方面：

• 规划方法太过强调分析，而没有提出具体的指导思想；
• 规划的初始目标往往过于追求特殊性和轰动效应；
• 将街道看作社会空间的意识还很薄弱；
• 当今被偏爱的高层住宅区(Danji)形态过于单调，并缺乏与城市街道空间的关联。

在承接的每一个韩国项目中，例如城市更新项目"Gangseogu"和"Bangbaedong"，我们都有意识地深入研究了上述问题，并尝试寻找创新和现实的解决方案。以下是ISA德国意厦针对上述问题的一些规划设想。

This change in Korea is comparable to that in European and German urban planning in the 1970's. During this time, Europe experienced a dramatic paradigm shift. The quality of medieval cities was re-evaluated. Street space, especially, was rediscovered as social space where diverse activities and communication could take place. Many architects and urban planners like Collin Rowe, Edmund N. Bacon, Rob and Leon Krier criticized Modernism, which considered buildings unaffected and isolated from the street space and evoked 'free standing objects.' In the beginning of the 1980's, European experts started to pay attention to the evolutionary phenomenon of cities. They began to see it as an important task of urban planning to preserve historical fabrics and improve the living condition of old cities along with existing buildings. Many current political goals and pilot projects in Korea and Seoul remind one of the former urban planning politics in Europe.

Despite good intensions, there seem to be many convictions that conflict with the actual goals of qualitative urban planning. From my observation, this is mainly based on the following basic problematic attitudes:

• The planning method emphasizes analysis of the existing situation too much and works without any concrete 'vision'
• The planning approach aims too much for events and a spectacular effect
• The awareness of the street as social space is poorly developed
• The preferred housing developments, high rises (Danji), are monotone and without any urban spatial context

With every Korean project, the urban renewal programs in "Gangseogu" or "Bangbaedong" for example, we have consciously confronted such problems and searched for a creative and realistic solution. The following ideas are some planning approaches that our office and I have used in order to change the above-mentioned basic problematic attitudes.

1. Planning approach against a one-sided analytical planning method without a concrete "vision"

"Playful planning" should be implemented at the very beginning: first develop a 'dream picture', then 'analysis and evaluate'

A strong emphasis is still placed on the analysis of existing conditions. Many times everything is analyzed in meaningless detail. But even if one knows exactly what the problems are and how they came about, one stands helplessly in front of 'a high wall' after comprehensive analysis. Now, what should the future look like? One knows that it should become different, but one doesn't know how. The philosopher Karl Popper has already warned us about this problem, "Because the future is open. There is no historical law for progress... It is wrong to even attempt to extrapolate from history, also when we try to find out what will happen tomorrow through current trends... This is why our basic approach should not be controlled by the question, 'What will come?' but by the question, 'What should we do?' Do anything that will make the world a little better." (Alles Leben ist Problemloesen: 272-275, Piper, 2008)

The qualities that fulfill people's needs should first be questioned, and the approach should derive from those results. If people generally have the same needs, does it consequently mean that all cities should be based on the same vision? Maybe political approaches could look the same but not concepts for the physical spatial qualities of each city because the spatial configurations and conditions of varying cities are just as different as their social and cultural backgrounds. The 'dream picture' of three-dimensional urban physical elements is all the more important. Such a 'dream picture,' a vision, serves as rating criteria for the analysis results and at the same time designates a guideline for planning whereupon this process is not a linear, but an iterative process.

1. 规划之初应当避免自下而上的工作方法——没有指导原则而单纯依靠分析

首先形成"规划愿景",而后进行"分析与评估"

现状分析是规划工作的重要基础环节之一,仍然是需要重视的,但是对现状的细枝末节进行面面俱到的分析却往往毫无意义。即使人们经过了事无巨细的分析后知道了存在哪些现状问题以及问题是如何产生的,但当他们面对如何描绘未来蓝图这一问题时,依然会面临一堵无法逾越的高墙。尽管人们知道它应当变得不一样,但是却不知道如何能实现这种改变。如同哲学家Karl Popper的观点:因为未来是开放的。历史的进步没有法则……哪怕仅仅是尝试从历史进行推演——比如从当今的趋势来判定未来的格局——都是错误的……因此我们的基本观点不应囿于"未来会发生什么",而是要去思考"我们能为未来做些什么",要尽可能地做一些事情来使我们这个世界变得更美好。(《生即为问题的解决》,272～275页,Piper出版社,2008年)

提供能够满足人们需求的品质是首要问题。应该以此为出发点来确定未来发展的指导思想。那么是否由于人们的需求相同就会使所有城市的发展目标都相同?也许各个城市发展的政策性目标可以是相同的,但是它们的空间品质的发展愿景却不尽相同——因为各个城市的特性,特别是空间条件,如它们的社会文化背景一般存在着各种差异。所以三维形态化的城市空间发展指导目标之确立是非常重要的,一方面可以作为现状分析结果的评价标准,另一方面也作为城市发展过程中重要的城市管理控制原则——因为城市发展的进程不可能是单一线性的模式,而是一个不断更新发展的过程。

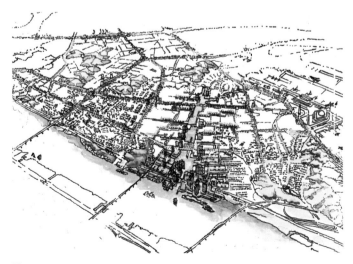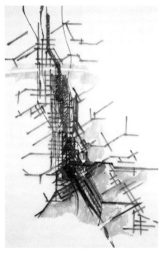

图 06 具有开放性的头脑风暴
Fig. 06 Playful planning through brainstorming

通过以提升生活质量为目标的高品质"城市设计"，来校核确定量化的指标要求

只要城市规划还只是单一地从技术、经济和法规的视角去考量，高品质的城市建设几乎就是不可能实现的。对于高品质的城市建设而言，城市景观塑造是所有规划过程中必不可少的前提。城市景观塑造及城市景观规划应该被作为一项重要的社会任务来理解，这项工作能够满足人们的灵魂、精神和心灵上的需求。

在城市规划实践中，定量的分析和规划也是必要的。借助这些分析的数据结果，可以明确我们的物质需求，也就是经济、技术、文化和社会方面的需求，并对总量进行预测。然而这些量化数据并不是"最终的"成果，而是应该通过城市景观规划的过程将之体现到三维空间中。

市民们对一个公园的确切规模没有兴趣。他们关心的是：这是个大公园还是小公园，是漂亮的还是丑陋的公园。而这种感受并不单取决于这个公园的面积。如果使用者能在公园获得丰富多样的体验，那么一个面积不大的公园也能让他们从心理上觉得这是个大公园。通过不同的空间划分和塑造，同样的面积能够让人获得或大、或小的不同感受。由此可见，城市规划更加关注的应该是品质的提升，而不仅仅停留在满足数量的需求。

意厦展开每个项目都会从这两方面入手。项目之初首先进行头脑风暴，以得到有创意的设计及指导思想，而后通过分析、评价规划区内的日常生活品质，在进一步的规划过程中验证这些构想。

2. 反对以追求轰动效应或特殊效果为目的的规划

面向日常生活的城市规划

作为全球化的现象之一，许多城市都希望吸引著名企业集团以及其他特殊的企业。不仅仅是为了能提供更多的工作岗位，也是为了它们带来的影响力。与之相关的，旅游也成了一个重要的城市功能。几乎所有的韩国城市都在追求突出自身的闪光点和美丽之处以作为吸引游客的手段。这些城市凭借一些宏伟的高层建筑而自称为IT城、BT城或者媒体城。各地纷纷修建林荫大道、博物馆和剧院。活动策划和市场营销被看得比其他事务都重要。但是人们不应忘记，世界著名的旅游城市，如巴黎、伦敦、威尼斯、佛罗伦萨、纽约和旧金山，都不是为了某些企业或游客而建设的。它们直至今天仍然首先是普通市民日常生活的地方。

从我们的观点而言，城市建设应该以两方面的改变为前提，一是应该更为全面地理解历史建筑文化遗产的意义，二是应该强调以现状为基础进行建设。

Verifying the quantitative demands through a qualitative urban design for everyday living quality

Qualitative urban planning will never be possible as long as urban planning is only regarded in technical, economical and legal terms. Qualitative urban planning sets the design of the city as a primary objective of the entire planning process from the beginning. In doing so, the design of the city or the city image should be understood as a social function that fulfills the psychological, spiritual and physical needs of the people.

In the realization of urban planning, a quantitative analysis and plan is also needed. These results are conveyed in numbers and square meters so that we can grasp our material needs and the economical, technical, cultural and social needs, as well as estimate our overall space needs. Such numbers, however, should not be understood as the final solution but should be transformed into three-dimensional physical elements to be allocated in adequate spaces through the process of urban design.

The exact size of a park doesn't really bother the user or citizen. They understand it as either a 'large' or 'small' park, or a 'beautiful' or 'ugly' park. But they don't exactly know 'how large or small' it is in terms of square meters. For example, users can find a small park large, when they can experience diversity there. The same quantity can be perceived as either smaller or larger through the arrangement and design of the space. Because of this, the quality of the urban spaces should be inquired about and searched for, instead of the quantity.

These two basic approaches are placed in the foreground of every project in our office. At the beginning of every project there is a brainstorming session to find unbiased creative ideas or 'dream pictures.' Afterwards they are tested further through the planning process. Likewise, the everyday qualities of planning areas are also analyzed and rated.

2. Planning approach against designing for events and spectacular effects

Creating cities for everyday

The desire for many cities to attract famous corporations and other specialized businesses and industry branches is a phenomenon of globalization. This is not just about creating more jobs, but also the city's presence through its strength and culture. In connection to this, tourism also becomes an important function of the city. Almost every city in Korea tries to rise above the others, to reach more tourists through shine and beauty. These cities, with their impressive high rises, are called IT-cities, BT-cities or media cities. Boulevards with new museums and theatres are built. Event and marketing strategies become more important than anything else. But one must not forget that the world's famous cities like Paris, London, Venice, Florence, New York and San Francisco were not built for specialized manufacturing or tourism. They were and are firstly built for living and daily spaces for the regular citizen.

In my opinion, urban planning for daily life calls for changes to two other aspects. First, the term, "inherited architectural culture," should be interpreted more thoroughly. Second, building within existing urban structures should be strengthened.

Comprehensive interpretation of the inherited architectural culture

In the past few years, experts and citizens have recognized that not only should monumental buildings be kept as architectural heritage, but so should historical houses constructed during or before the 19th century. Recently, some downtown areas have been allocated for this type of building, which should exist in a design language of 19th century historical reconstructions. Such areas form 'islands' in the middle of high rises. In terms of urban space, style and dimension, they look strange and disruptive. Such an obsession is a reaction against the neglect of traditional architecture and the culture of the past. But a demolished building or a wiped-out district cannot be easily 'replaced' like a stolen 14th-century porcelain that was found again and brought back to a museum.

Keeping a city's lively architectural culture means that existing buildings like courthouses could be preserved, but should not be reproduced. They should be metamorphosed and built in contemporary design languages. Only through such a way can the vivid building culture be continually cultivated from the past to the present. Today's buildings will be tomorrow's heritage.

Step by step development, respecting the existing urban fabrics

Another new school of thought for daily urban planning and design is to turn away from complete demolition. Even the 'everyday' buildings should be respected, and new structures should be developed step by step. Districts with smaller structures that have grown over the decades, especially, should be improved in a way that doesn't require them to be completely torn down. Residential districts make up more than 70% of the total area of all cities in the world. The occupant in the residential district is playing the main role, and the city is their living stage to consume and to create a culture. Residential districts are not only important because of their size, but also as a qualitative element for urban planning for everyday life.

全面理解历史建筑文化遗产的意义

近年来，除纪念性建筑外，19世纪前建设的历史民居的保留价值也得到了专家和市民的认可。历史庭院、南山韩屋村甚至得到重建——虽然被保留或重建的建筑往往成为现代高层建筑包围的岛屿，高层建筑过于都市化、风格化以及超常的尺度，严重影响了城市形象。这种创意可以理解为对过去不注重保护历史建筑及历史建筑文化遗产的反击。但是，被拆除的建筑或者街区是很难恢复的，正如被偷走的14世纪瓷器很难被找到并归还给博物馆一样。

为了创造生动的城市建筑文化，应该保护现状的庭院住宅，而不是简单地模仿复制。并应当由"传统庭院住宅"发展出一种使用现代建筑语言表达的新的建筑形态。只有这样才能保持从过去到现在乃至未来的延续性，因为当下的建筑就是未来潜在的建筑文化遗产。

在现状基础上逐步发展

另一个城市建设中的必要思想转变在于放弃完全拆除的做法。"日常建筑"也应当得到相当的重视，并且新的城市空间结构应该逐步发展形成。首先应该使那些已经建成几十年的小尺度居住区的价值在不被拆除的情况下得到提升。居住区在全世界的城市中都占有超过70%的面积。正是居住区中的居民以城市为舞台塑造着城市的文化。因此居住区无论是在面积还是品质上，都是城市建设的一个重要元素。

图 07　20世纪初期重建的"Hanok"庭院住宅，首尔2011年
Fig. 07　Rehabilitated 'Hanok' courthouses, built at the beginning of 20th century, Seoul 2011

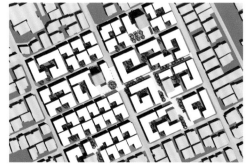

图 08　ISA设计的当代庭院住宅，首尔Hwagokdong，2011年
Fig. 08　Contemporary courthouses, designed by ISA-Stadtbauatelier, Hwagokdong, Seoul 2011

理论文章 ARTICLES

Many city districts in Korea are missing an ordered street structure, as well as green spaces. This is why building new streets and parks is regarded as the most important task in the rehabilitation of the city. Usually, when large boulevards and oversized parks are built, like that of Central Park in New York, buildings on the old streets have to be removed. A so-called 'Central Park' will normally be built on the city edge because the huge space required can be found cheaper and more easily there than in downtown, which is usually totally built up and with a land price that is too expensive. But such a park cannot be used as a daily establishment. It might only be visited once a year because the distance is too far and because of daily stress. Instead, many smaller parks and green spaces, or even simply trees along the streets, should be built to raise the quality of daily life. Such things in the citizen's normal environment are easily experienced and enjoyed. Only in this way can the social and cultural worth be preserved – as well as people's personal memories or the city history. Building amongst the existing conditions is the shortest way to healthy and lively urban planning.

This kind of a planning process requires a very slow development over a long period of time. On one hand, there is a long-term approach and on the other, a short-term goal is required. Clear visions for each step of development can serve as guidelines and steer every step for the whole city. Some short-term goals can be formulated as pilot projects that produce inspiration and have a short-term effect on the whole city.

3. Planning approach against insufficient awareness of streets as social spaces

Reclaiming the streets as social spaces

An important qualitative aspect of urban planning is to reclaim the streets as social spaces. Streets were originally not only seen as a space for traffic to reach a destination, but also as 'a place to meet and play.' Until the automobile arrived, this was a general phenomenon all over the world, independent of culture. It isn't enough to build one gorgeous pedestrian street while the entire street system consists of cars. There must also be a good network of pedestrian walkways.

在韩国，很多城市缺乏清晰有序的街道结构以及充足的绿化面积，因而建设新的街道和公园被视为城市更新的首要任务，而这些新的街道和公园通常都会被规划为大型林荫道和像纽约"中央公园"那样的超大尺度的公园。这样的话那些原有街道两侧的建筑就势必被拆除。由于市区合适的土地有限且地价过高，通常规划的"中央公园"都被设置在城市边缘，然而这样的公园根本无法服务于居民的日常生活。由于距离过远、日常压力过大，居民也许一年才会去游赏一次这些城市边缘的大型公园。为提高日常人居品质，应该在市区布置小型绿地、公园或栽植行道树，取代城市边缘的大型园林，使居民在日常生活中可以方便地享受这些绿色。只有这样才能使城市的社会、文化价值在居民的记忆中或城市的历史中保存下来。基于现状的建设是健康而富于活力的城市发展的捷径。

上述这样的规划方案是一个渐进的过程，需时长久。一方面需要树立长远目标，另一方面也需要设置短期目标。明确的长远目标可以引导各个阶段的发展，并且调控全局。而短期目标可以作为示范性项目，起到有力的带动作用。

3. 反对对于街道的社会空间属性认识不足的规划

恢复街道的社会空间职能

关于城市设计的一个重要观点就是应该使街道重新作为社会空间。街道的最初不仅仅只是作为通往某个目的地的交通通道，同时也是"一个愉快交往的场所"。这在汽车出现前一直是一种极其常见的场景，无论是何种文化背景。当整个城市的道路系统都由机动车道路构成时，仅仅只建造一条华丽的步行街是不够的，应该同时建立一个完整的步行系统。

 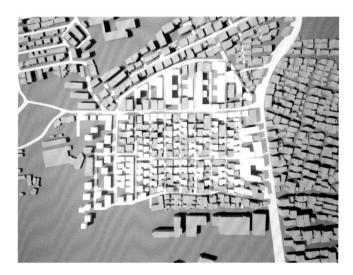

图 09 通过遵循设计导则的小尺度和单独建筑项目，进行逐步开发的示范设计
Fig. 09 Test design of piecemeal development by small and individual building projects, regarding design guidelines

Since the 60's and 70's, experts like Jane Jacobs, Jan Gehl and Donald D. Appleyard have demonstrated that well-functioning neighborhoods require different social activities and contact, where again emerges the importance of pedestrian-friendly street spaces with diverse uses.

The question is: What does pedestrian-friendly mean? Prior to redesigning and reconstructing streets with gorgeous materials and wide sidewalks, the Genius Loci of the existing streets should be respected. Most of the streets in old districts in Korea and Asia are market streets, where different shops are easily mixed together. A passerby is drawn from one store to the next and from one building to the next. One comes across acquaintances and neighbors, stopping briefly to have a chat, gather memories and materials, satisfying immaterial needs at the same time.

早在20世纪六七十年代，就有简·雅各布斯，Jan Gehl和Donald D. Appleyard 等专家证明，以不同社会活动及交往形成的良好友邻关系为基础，能再次构建出具有多样化功能的行人友好型街道空间。

问题在于，什么是行人友好型？在街道铺上绚丽的面砖和修建宽敞的人行道之前，应该首先尊重街道的现状和特色。在韩国和亚洲其他国家，旧居住区内的大多数街道都是"商业街"，这里各种商店鳞次栉比，人们会在各个商店之间流连忘返，偶遇熟人和邻居，停下来短暂寒暄一番。这样就会使人们的记忆在这里沉淀下来，使人们的物质和非物质需求都得到满足。

图 10 一个现存的邻里街道
Fig. 10 An existing neighborhood street

图 11 带有绿色广场的街道，首尔愿景2011年
Fig. 11 Rearrangement of the street with a green plaza designed by ISA Stadtbauatelier

图 12 不受欢迎的小地块住宅区
Fig. 12 Unpopular housing area with small plots

图 13 通过人性化尺度的单独建筑项目逐步改善现有建筑和空间
Fig. 13 Block by block improvement of the existing buildings and spaces through individual building projects in human scale

由此可知，封闭的街道空间界面以及沿街建筑底层的功能混合是建设行人友好型街道的重要前提。

这就为以狭窄路网为特性的现状居住区提供了新的发展契机。这类街区可以被划分为适宜的邻里单元，其中的街道可以被改造成行人、汽车拥有同等路权的慢速交通区，而且大部分机动车将被拦截在邻里单元的外部。这样可以使现状街区就整体而言得以保留及更新，即便它们在视觉中略为破旧。

人性化的尺度
在这里，人性化的尺度扮演着重要角色。人性化的尺度不仅仅使城市舒适宜人，而且赋予其生动有趣的视觉外观。小尺度的空间、分隔的结构和生动的立面传递更多信息，并且满足了人们的精神和心理需求。

行人通常可以步行1公里，如果道路的功能和景观吸引人的话，行人可以步行2公里。与此相符，欧洲城市老城区主要步行轴线的平均长度大约为1公里。从这个角度出发，我们在规划中，将整个Gangseugu区划分成相互联系的13个步行邻里街区作为公共空间组织结构。

4. 反对"Danji居住区"设计中没有考虑城市空间关联性的规划方式

住房建设的多样性
影响规划品质的最后一个方面就是住宅建设以及住宅类型的多样化。韩国被认为是高层住宅的国度。其中南北向的高层住宅一直是最受欢迎的住宅，大约占韩国住宅总量的75%。所有的这类南北向高层住宅都有相同的平面图，外观看上去毫无识别性可言。它们也不能形成多样化的外部公共空间，形成的仅仅是楼房与楼房之间的剩余空间。在过去十年扩张的城区中充斥着这种由高层住宅构成的新居住区（Danji）。旧的街道和道路被一个个居住区的围墙阻隔而形成了大量的尽端路。

情况已经开始发生改变。在过去的2～3年中对于住房的多样性问题，规划界已经展开了激烈讨论，由"Danji"高层居住区建设引发的城市建设和社会问题正在思辩中。

典范案例
首尔市政府自2009年4月起启动了一个示范项目——"市民参与的邻里项目"。该项目的目标是，将现状小尺度的居住区，尤其以1～4层小体量建筑为主的居住区保护起来，避免高层居住区造成的压抑感，并被确定为备选的住宅形式之一——最终目的是在保护的基础上更新居住环境。首尔市政府在4个城区选出4个不同的居住区作为试点。选择的过程类似于一次竞赛，有意向参加的各区政府都必须申请。其中Gangseugu区政府在一年以前就在意厦所作的总体规划的框架之下确定了一些备选项目，因而准备得最为充分，所以Gaehwadong被纳入首尔的这个示范项目，意厦也受邀继续指导改造项目的推进。

Old streets like that teach us that continuous street walls and different uses next to each other on the ground floor, are important requirements for a pedestrian friendly street.

In this way, existing city districts with narrow streets and alleys do have a chance. These types of districts organize themselves into appropriate neighborhoods and change the streets to slow traffic zones so cars and pedestrians have the same rights. This leads to most traffic being kept on the district edge, giving us an idea on how existing neighborhoods can be vividly kept and renewed, even when they look shabby at first.

Human scale
The human scale plays an important role. A human scale not only makes the city more 'comfortable' and pedestrian-friendly, but also lively and visually interesting. Smaller spaces, articulated building masses and facades with plasticity offer more information and gratify people's emotional and spiritual needs.

It has become evident that pedestrians usually walk about one kilometer, but if the streets have an interesting design and function, people will walk up to two kilometers. Correspondingly, the average length of a pedestrian axis in historical European cities is about one kilometer. In respect to this, the entire Gangseogu district was divided into 13 pedestrian neighborhoods, which were again linked with each other.

4. Planning concept against 'Danji residential areas' without any urban space context

Variety in residential building
The last aspect of qualitative urban planning is the design of housing, or rather, housing forms with a variety of types. Korea is often called 'the republic of residential slabs.' Residential slabs oriented north to south are accepted as the best housing form and make up about 75% of all residential buildings in Korea. All these types of buildings have the same floor plan and look the same from the outside. They do not develop any differentiated exterior public spaces, but leave only the leftover space between buildings. City districts that have grown over decades were replaced with new residential districts made of these high rises (Danji). Their old streets and paths are cut off and become dead-ends because of the bordering walls. This is how new districts form immense foreign elements in their surroundings.

However, the situation is changing. In the last two-to-three years there have been many discussions about the variety of residential buildings, as well as the urban and social problems cause by the 'Danji.'

Model project: Gaehwadong
Starting April 2009, the city government of Seoul started a pilot program called the 'Neighborhood Project with Citizen's Participation.' The goal of this project is to protect a city district's existing small structures, especially the 1-to-4 story buildings, from being replaced with high-rise residential buildings and to preserve them as an alternative residence type. Ultimately, the program would serve to rehabilitate and preserve these residential areas. The city government of Seoul chose a total of four different residential districts in four different areas. This transpired in a manner similar to that of a competition, wherein different local authorities had to apply. The local authority for the Gangseogu district was exceptionally prepared because a year earlier they had defined a framework for pilot projects within their master plans. This is how the Gaehwadong area was accepted into the program, and why our office will continue to direct the master plan.

从两种文化中学习 LEARNING FROM TWO CULTURES

Subsidiary goals of this project were to improve the quality of public spaces through public means, as well as to make existing streets more pedestrian friendly and thus more appreciated. Through these objectives, social contact and communication in public spaces would be cultivated. Public facilities would be strengthened or renovated, especially small neighborhood parks, collective parking lots, and facilities for small children, etc. Most importantly, the coverage- and floor space ratios will not allowed to be increased above that of the zoning plan, and must be kept at the current percentages. The entire district should develop a better image, and likewise the property worth should increase through improvements to the public spaces. Until now it was unimaginable to have an urban renewal or urban improvement project without increasing the coverage ratio or floor space ratio ("up zoning").

Because of both the goals and the planning process, this program is a milestone for the future of urban renewal and redevelopment. At the start, residents will attend a planning workshop and bring their written or drawn ideas. This is a totally new way of public participation, since the public would normally only be informed of plans after their creation and could only use their right to veto.

该项目的下一级目标是由政府主导改善公共空间的品质，设计行人友好型街道，提升街道空间的价值。这就需要加强社会交往及交流。公共设施，特别是小型邻里公园、聚会场地、儿童游乐设施等得到了补充和更新。并且首先要保证由土地使用规划确定的法定最大建筑容积率和建筑密度不得提高，而应该维持现状。提升公共空间品质可以为整个街区塑造更好的形象，并实现住宅价值的提升。一个不提高建筑容积率和建筑密度的城市改造或更新项目直到今天仍然让人觉得难以置信。

无论从规划目标上，还是从规划程序上，该项目都称得上是城市改造及更新项目的里程碑。参加规划进程的市民从一开始就通过研讨会积极参与，并且以书面或图纸的形式表达观点、献计献策。这种市民参与规划的程度与迄今为止的市民参与是大不相同的。此前通常情况下市民只是被通知来行使一下否决权。

图 14　低层高密度住宅的设计
Fig. 14　Design of a housing type, high density but low-rise

图 15　Eunpyeong新城的高层住宅
Fig. 15　Residential high-rises in Eunpyeong New Town

图 16　低层城镇住房是Paju的另一种住宅类型
Fig. 16　Low-rise town houses as an alternative housing typology in Paju

这种新形式的市民深度参与尽管费时又费力，然而却是非常重要的，可以通过这个过程过滤出市民的真正需求。比如一个人说"请给我吃饭"，这句话的意思只是说，这个人饿了想要吃点什么，而并不一定就是指他想要煮米饭，也许是他不知道还有别的选择。专业人士应该设身处地为民众考虑，提供多种选择方案，并且指引公众作出正确选择。

无论从哪个角度看，该项目都是一次有意义的勇敢尝试。我们所有的参与者都深知这个项目的实施将给当前的规划手段带来必要的改变和完善。如当政府希望某栋住宅的更新修复得到赞助，然而却缺乏相应的政策时，这与之前通过高层住宅区的建设来完成住宅区的更新是大不相同的。这种不合理的现象表明，迄今为止，住房建设政策还把注意力集中在"高层住宅"这一类型上。

尽管存在许多问题，所有参与者还是踌躇满志。首尔市政府已经开始着手适应和补充现状实施手段的研究项目。衷心希望，这个项目只是个开始，更多积极的变化还将相继出现。

This new form of public participation requires a lot of strength and time because it is essential to filter through the actual needs of the people. It's like when someone says, "please give me rice," it could really mean that he is still hungry but doesn't know anything else other than rice. Perhaps he might enjoy spaghetti or steak? The experts must put themselves in the place of the people so that they can find alternative solutions and help with important decisions.

It is most definitely a very brave and ambitious project. We, as participants, are very conscious that the implementation of this program would bring about needed changes and additions to the existing planning tools. As an example, the public might financially support the renovation of individual residential units. However, the instruments for this are missing, other than simply renovating residential districts with high-rises (Danji). This absurd fact shows how residential building politics have only concentrated on the typology of residential high-rises.

In spite of many problems, all the participants are very confident and hopeful. The city government of Seoul has already announced a research project for the adaptation and amendment to the existing implementation tools. And I hope that this project is just the beginning and that many positive changes follow.

图 17　通过各项实施策略改善公共街巷环境
Fig. 17　Improvement of public alleys through small implementation strategies

图 18　通过提高新建筑的密度改善居住区环境
Fig. 18　Improvement of residential areas through new buildings in terms of redensification

图 19　保护现有建筑的总体规划
Fig. 19　Master plan in terms of preserving the existing buildings

新城：
有关城市的梦想
New Town:
Dream City

张亚津　Yajin Zhang

The construction of a new town is unlike the conventional development process of a city. It is not a result of a natural mutation, but the result of many "decisions." It begins with political, economical, military and cultural intentions. In other words, it is the result of city planning. From its history, we can see that new town constructions represent the dreams of the builders of the generation, a Utopia.

1. New City Utopia: A mirror of the society

Utopia is a mirror, a mirror of the reality. The Utopia of the eighteenth century represented the clouded real life of the industrial city.

The European industrial era was a crisis of human civilization. In the 19th century, the British population quadrupled. London's population reached 6 million and Berlin 7 million. The rapid urbanization at that time put tremendous pressure on the original medieval cities. The farmers lost their land and the adventurous moved into the city to live in sewage-filled slums and rely on poor daily wages. In Berlin, the density of workers' residential areas at that time reached seven people per unit, and the interior block space was completely filled with flats for rent. Over a hundred year span, the beautiful pastures and green fields had changed into an industrial city full of noise, dust and crowds. From 1760 to 1844, Parliament enacted more than 3,800 Enclosure Acts, occupying a landmass of more than 7,000,000 acres. Woodland was cleared to build factories and the steam trains roared across the fields puffing their black smoke.

Population growth and unemployed farmers caused a drop in workers income levels despite an increase in productivity. In 1810, the British textile worker's average wage was 42 shillings 6 pence a week, and that fell to 25 shillings 6 pence by 1825. So the easily oppressed women and children went to work for the lowest wages. In 1839, there were 419,560 workers in England, including 242,296 women, accounting for 57.75% of the total number. Children of six or seven years old went to work. In the lace industry, "... children of nine or ten years old were pulled up from their dirty beds at 2, 3, or 4 o'clock in the morning. In order to make a living, they had to work until 10, 11, or 12 o'clock at night. They had thin limbs and their bodies were withered away with atrophy. They looked slow-witted, numb as a stone-like people, which made people awkward." A survey in 1840 showed that in Liverpool, the workers' average life expectancy was only 15 years, and more than 57% of the working children in Manchester of less than 5 years old died. The

不同于城市的常规发展过程，新城建设的产生不是来自于自然的发展过程，而来自于一个突变，它是一个"决定"的结果，背景是政治、经济、军事甚至文化意图，换言之，也就是一个城市规划的结果。纵观新城建设的历史，可以说新城建设代表了各代城市建设者的一个梦想，一个乌托邦。

1. 新城乌托邦：社会的一面镜子

乌托邦是一面镜子，一面现实的镜子。18世纪乌托邦的甚嚣一时，代表了以工业城市为背景的现实人类生活的全面困惑。

欧洲工业时代的发展曾经是人类文明的一次危机。这一个世纪英国人口增长了4倍。20世纪60年代，英国的城市化程度已达到61.8%。1900年，伦敦人口达到600万，柏林700万。这种迅速的城市化过程必然为原有的中世纪城市带来巨大的压力。失去土地的农民与冒险者涌入城市，依靠每日工资在污水横流的贫民窟过活。柏林当时的工人住宅区容积率已经达到7，相当一部分仅靠中心天井采光。一百年间，温和的地中海气候呵护下的优美田园变成了工业城市——噪声、烟尘、拥挤不堪。与此同时，1760～1844年，议会颁布的圈地法案多达3800多个，圈占土地700多万英亩。林地被砍伐修建工厂，冒着黑烟的蒸汽机车轰鸣着穿过原野。

人口的增长、失业农民的流入造成了工人收入水平随着生产率的提高而下降。1810年英国纺织工人每周的平均工资为42先令6便士，1825年下降为25先令6便士。迫不得已，工资低廉又易受欺压的妇女、儿童都进入工作岗位（1839年，英国共有工人419560人，其中女工就有242296人，占工人总数的57.75%）。工人们的子女六七岁就得进厂做工。在花边行业，"……九岁到十岁的孩子，在大清早2、3、4点钟就从肮脏的床上被拉起来，为了勉强糊口，不得不一直干到夜里10、11、12点钟。他们四肢瘦弱，身躯萎缩，神态呆痴，麻木得像石头人一样，使人看一眼都感到不寒而栗。"1840年的调查，利物浦工人的平均寿命只有15岁，曼彻斯特工人的孩子57%以上不到5岁就死亡。"斯温运动"、"卢德运动"就是一场工人农民憎恨工业革命、机器效率的直接爆发。欧洲当时当地，工业革命摧毁的不仅仅是农业者的耕田与古老的城市，更摧毁了人的尊严和传统的价值体系。

2. 乌托邦绘制者们：思想者？社会革命者？

每个"桃源"的梦想都来自于对此"世"的厌恶，如同阴阳两极，彼此依存。托马斯·莫尔等社会革命者借助乌托邦小说这样一种文艺性手段，充分表达了他们对于工业革命时期种种弊端的一系列反思与改革的畅想。

各个文化背景中大约都有这样的梦想：现实煎熬着肌体，于是思想插上了翅膀，飞越煤矿的尘雾、肮脏的收容所，寻找一个"他在"。但在这一阶段，面对前所未有的经济、社会、文化变革，寻求新格局的任务不仅仅是少数精英代表国家政府提出方案并实施——如巴黎1852年的Baron Haussmann所领导的大巴黎规划，而是在知识分子层面形成广泛的回响，甚至以一己之力或私人团体的力量进行中小型城镇的建设——欧洲这一时期乌托邦的梦想者，更有身体力行的特性，梦想先行，随之是社会与城市的深入实践。这使乌托邦理想国这种理想性质的文学作品具有了迥然不同的意义。

事实上明确地提出通过城市形态的改造进行社会改造这一概念，起始于工业化革命，这种将社会与经济的弊端部分归结为城市结构的问题是当时大量政治思想家的普遍思路。经济力量的突飞猛进与民主政治的初步萌芽是其根基所在，这两者形成了改造城市、建设城市的多元化新力量，也基本是现代城市规划的开端。

workers and farmers hated the industrial revolution and the machine efficiency, directly leading to the "Swain movement" and "Luddite movement". At that time in Europe, the industrial revolution destroyed not only the agriculture fields and old cities, but the human dignity and traditional value system as well.

2. Utopia designers: Thinkers? Social revolutionaries?

Every dream of "Heaven and Peace" comes from battles in society, like the two sides of Yin and Yang, depending on each other. Social revolutionaries like Thomas Moore used the artistic means of Utopian novels to give full expression to a series of reflections and reforms due to the malpractice of the industrial revolution.

In the background of various cultures, there is such a dream: reality is troubling the body, so the thoughts take on wings, flying over the coal dust and the dirty houses to look for another space. But at this stage, in the face of unknown economic, social and cultural changes, only a few elites on behalf of the government sought the new duties to put forward a plan and implement it. An example of such an act was the plan of Great Paris in 1852 by Baron Haussmann. However, most intellectuals formed all-embracing conclusions, adding to the power of their own ability and private organization to take construction to medium or small cities. During that period, European utopian dreamers had more distinguishing physical efforts, first the dream, then the development practice of the society and city.

In fact, a large number of political thinkers from the industrial revolution clearly put forward this concept of social transformation through the transformation of city morphology. There was a general thought that social and economic problems arose from the structural problems of the city. Surely, such a thought was based on the rapid economic development and the seeds of democracy. In any case, that was the beginning of modern city planning.

图 01 描绘工人居住状况的版画
Fig. 01 The workers' living conditions

图 02 新和谐城
Fig. 02 New Harmonious City

These social innovators dedicated to the construction of the dream are divided into two categories. One is the industrialist who wanted to keep the large number of workers in relative human living conditions and thus built a large number of excellent quality worker homes in Europe. Another category is the rich mastermind, who spent their fortune to build a city as a social experiment. Erwin's famous "New Harmony" is a typical example of the second category. The former had a broad industrial background, which in fact formed a great number of outstanding works, but the latter often ended in failure with a lack of practical significance.

Taking the famous Sun Village as an example (Fig. 03 village of Port Sunlight, source: Wikimedia Commons), in 1887 William Hesketh Lever, a successful soap magnate and philanthropist, planned the expansion of his suburban soap factory. One year later, the village of Port Sunlight, as a workers residential area, was built on a swamp. He funded the building of the garden town with residential buildings, accompanied by a large number of welfare, educational and entertainment facilities. He encouraged the workers to take active part in musical, scientific and artistic activities. He also built other public facilities including hospitals, churches and technical schools, even an art school for women. The value was the merging of work and living spaces, strongly sealed from the outside, which allowed the Lever brothers to carry out a series of ambitious administrative measures.

Although there appeared to be a large number of the Utopian dreams being implemented in the late eighteenth century, the examples of large-scale cities that were actually built were not great. Whether successful or not, the size of these towns was for about three-thousand people, which could not change the workers' overall living conditions.

New town planning cannot depend only on a utopian's dream. Likewise, a new town cannot rely on a weak foundation. Construction of new town requires a series of tasks, such as the construction of infrastructure, the organization of an administration structure, the relocation of the population and jobs, all of which means large-scale investment with a high risk. That is a main reason why new town planning always needs a willing, common and broad consensus of administration, economy, politic and military to finance it in great amount.

这些致力于投身理想国建设的社会革新者，据其背景不同可分为两类。一是企业主，他们本身即具有建设大型工人住区的需要；结合这一阶段的人文思想，在欧洲形成了大量具有优良品质的工人住区。另外一类是这一时期出现的一些利用私人募集资金、倾家荡产进行城市与社会实验的先行者。欧文建设的著名"New Harmony"即属于其中的典型。前者产业背景明晰，事实上形成了大量的建成优秀作品；后者往往因为缺乏现实意义，以失败告终。

以著名的Port Sunlight为例，1887年，William Hesketh Lever作为一个成功的肥皂商人与慈善家，计划了郊区肥皂厂的扩建工作。一年后作为工人住宅区，在一片沼泽地上兴建了Port Sunlight村，作为一个花园新镇，他资助了居住房屋的修建，与此伴随的是大量的福利设施、教育设施、娱乐设施，积极鼓励工人从事音乐、科学、艺术等活动。为此建设了包括医院、教堂、技术学校甚至一个女子艺术学校等公共设施。其代价是工作岗位与居住空间的绑定，利用这一方式，Lever兄弟推行了一系列强迫性的管理措施，对外封闭性较强。

虽然在18世纪末期，突出涌现了大量乌托邦梦想的实施实例，但是真正形成的大规模城市建设作品总量并不大。这些城镇无论成功与否，事实上规模都在3000人左右，这对于影响改变整体工人居住情况显然是不够的。

新城规划的这一面不是乌托邦梦想者的一腔热血可以支撑得了的。新城的产生一方面无法依赖于薄弱的原有基础，一方面目标却是在短期内形成城市建设效果。其具体工作或多或少地包括城市基础设施的建设、管理系统的建立、人口的吸引与搬迁、产业的吸引与建设等一系列问题，这意味着在对城市基础设施前期性的大尺度整体投资与高风险，如果在区域发展中没有特殊的原因与强有力的后盾是难以想象的。新城规划者的规划意图必须与城市经济规划、政治规划、军事规划的意图、计划与实施过程相匹配，才可能赢得大量投资支持这一经济活动。

图 03 Port Sunlight村
Fig. 03 Village of Port Sunlight

3. 当乌托邦成为国家与居住者的梦想

芒福德的"城市发展史"将田园城市概念称为19世纪最重要的发明之一，"它让我们得以在地球上更好地生活"。霍华德的思想本质上影响了整个具有现代城市规划意义上的新城发展运动，其社会、经济、空间发展目标基本成为第一代新城的指导原则。这一规划理念切合当时政府对于工业人口居住品质提高、战后人口安置、产业调整的需求，成为欧洲"一战"至"二战"后城市发展的重要手段。

霍华德1850年生于伦敦，父亲是位小面包店店主，21岁赴美国，4年后重返英国并成为议会和宫廷的速记员。此间阅读了Edward Bellamy于1888年所著的乌托邦小说"Looking Backward"，为其中的社会化城市思想深深打动，稍后形成了"田园城市"的构想。这是一个极其完整的城市结构——不仅仅是城市结构，更是城市建设、经济、管理、文化的全面体系。

3. When Utopia becomes the dream of the state and residents.

In the "History of the City", Mumford says the concept of a Garden City should be celebrated as one of the most important inventions of this century. "It gives us a better life on earth." Howard's thoughts in essence have affected the whole movement of new city development and of modern city planning. The social, economic, and spatial development goals become the guiding principle of a new town's next generation. The planning concept became concerned with the improvement of living quality and the adjustment of industry, and therefore became one of the most important means of city development in Europe after World War I, all the way through World War II.

Howard was born in London in 1850. His father was a small bakery owner, and he went to the United States when he was 21 years old. 4 years later he returned to England and became the secretary of Parliament and Court. He read Edward Bellamy's 1888 utopian novel, "Looking Backward," in which the city deeply affected Howard's thoughts. He later defined the conception of a "Garden City." This idea reflected a complete city structure, not just the composition itself, but a comprehensive system of city construction, economy, management, and culture.

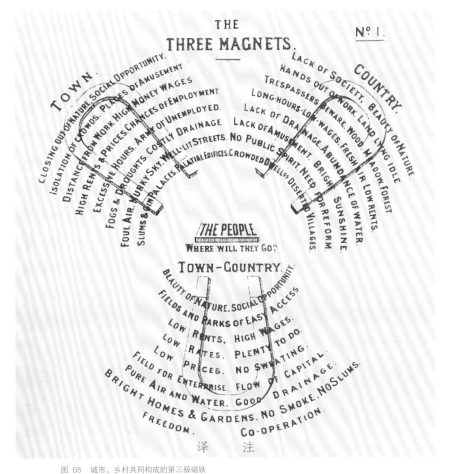

图 04　霍华德诞生地的纪念碑
Fig. 04　The monument of Howard's birthplace

图 05　城市、乡村共同构成的第三极磁铁
Fig. 05　City and countryside together constitute the magnets of the third

The basis of this theory is that human life is fusion of urban and rural qualities. At first, Howard put forward that the "garden city is a place for people to live and work healthily."That means a conscious combination of urban quality and rural quality: "All the advantages and aesthetics of the most vivid aspects of the city life and the pleasant rural environment in harmony together." This was a breakthrough in analysis of the city shape.

In addition, the garden city is the material part of a highly integrated social body, as includes far more than just the construction of the city itself. The city planners headed by Howard did not just pursue the city form, but the social reflections as well. The construction of the city is the material embodiment of a "Utopian"society, while the latter must have a series of economic management and legal support. This is Howard's garden city's most important quality for success. Howard conducted in his book a series of discussions about land expropriation, urban administration, rights of renter and tenant, and thus formed a complete system of social organization and economic systems beyond the city's planning system. "The different innovation lies in the establishment of the combination of balance and autonomy through a rational and orderly process. Despite the different circumstances, it can maintain a good order. Despite growth and development, it can maintain a sense of cohesion and harmony. This is a kind of thought and idea which has the ability to change, to play a role in reform."

More significantly, Howard and a group of rich, experienced entrepreneurs and celebrities established the Garden City Association in 1899, later renamed the "Town and Country Planning Association" (Association of city planning). This association recognized the garden city space and social models, and at the same time saw its economic potential. New town planning is not only a charity, but also a path for new business opportunities. So Howard and his theoretical followers, through careful analysis and private investment, built two examples in London: Letchworth, built in 1903 and with a population of 31,146 in 1991; and Welwyn Garden City, built from 1919 to 1920, and with population of 40,665 in 1991. These are two new cities that have a healthy economy and a good social structure with excellent construction qualities, as well as a beautiful and friendly rural lifestyle. In 1909, Germany's first garden city, Hellerau, came into being through the work of Hermann Muthesius and Bruno Taut. This influenced the modern city planning at that time. In the First Modern Housing Exhibition, the style of low-density garden cities became the dominant force. The national housing policy changed with it.

这一理论的最基本思想是通过城市与田园品质的结合形成人性化生活。开宗明义，霍华德提出"田园城市是一个为人们健康地生活和工作而规划的城市"——田园城市有意识结合城市与农村两者之间的优良品质："把一切最生动活泼的城市生活的优点和美丽、愉快的乡村环境和谐地组合在一起"这是一次对城市形态突破性的实验。

此外这一田园城市仅仅是一个高度综合的社会肌体的物质部分。乌托邦以邦为名，所包容的其实远远超过城市建设。以霍华德为首的城市规划者，进行的不仅仅是一个城市形态的追求，更多的是社会形态的反思。城市建设是一个"乌托邦"社会的物质体现，而后面还必须有一系列的经济管理法律支持——这是霍华德田园城市成功的重要品质。针对城市发展的土地制约、管理机制、空间机制，霍华德在书中进行了一系列社会城市、地主地租的消亡和新城行政机构的讨论，从而在城市规划体系之外形成了一个完整的社会组织体系与经济组织体系。"它与别的城市全然不同的创新之处在于它通过一个组合体对错综复杂的情况加以合理而有序的处理，这个组合体能建立平衡和自治，并且——尽管情况千差万别——维持得整齐有序，尽管需要生长发展，能维持内聚力、协调与和谐。这是一种有变革能力的、能起改革作用的思想和主意。"

更具意义的是，霍华德与一批富人、有经验的企业家和社会名流在1899年建立了Garden Cities Association(田园城市协会)，后更名为"Town and Country Planning Association(城市规划协会)"。这个协会认同田园城市空间与社会模式的同时，敏锐地看到了其经济潜力——新城规划不仅仅是一个慈善事业，而且是新的商机。借此霍华德及其理论追随者倚靠缜密的综合分析，依靠私人投资在伦敦周边建成了两个实例：Letchworth(建于1903年，1991年人口31,146)与Welwyn Garden City(建于1919年–1920年，1991年人口40,665)。 这是两座经济运转健康、社会结构优良、城市建设环境与田园风格相互依存、友善优美的新城。1909年通过对

田园城市的综合发展目标		
空间目标	社会目标	组织管理目标
• 城市控制在一定的规模，对建成区用地的扩张进行限制 • 几个田园城市围绕一个中心城市组成系统 • 用绿带和其他敞地将相对独立的居住区隔开 • 合理的居住，工作，基础设施功能布局 • 各功能间拥有良好的铁路(交通)联系 • 可以便捷地与自然景观接触	• 通过土地价格公共政策规定限制房客的房息压力 • 资助各种形式的合作社 • 土地出租的利息归公共所有 • 建设各种社会基础设施 • 创造各种就业岗位，包括自我创造就业岗位的专业户	• 具有约束力的城市建设规划 • 城市规划指导下的建筑方案审查制度 • 社会作为公共设施建设的承担者 • 把私人资本的借贷利息限制在3～4 范围之内； • 公营(国有)或共营(集体)的企业的建立

图 06　田园城市的空间、社会、组织管理目标
Fig. 06 Space, social, organizational goals of garden city

图 07　列契沃斯
Fig. 07 Letchworth

Hermann Muthesius 与 Bruno Taut的影响，推广形成了Hellerau——德国第一个田园城市。这一理论进而影响了整个现代主义城市规划。在第一届现代主义住宅展览中，低密度田园城市的风格明显成为主导。

与之相适应的是国家住宅政策的改变。F.J. Osborn——TCPA(Town and Country Planning Association)协会会长，在两座新城初见成效后，1930年向政府提出通过城市建设挽救经济危机的主张，随即在1937年与皇家特别协会共同调查英国工人人口分布与住宅状况，形成了著名的《Barlow Report》，为城市发展模式的变化与田园城市的概念推广播下了种子。

最关键的因素是"二战"后的住宅需求。"二战"结束后，伦敦被德国轰炸得遍体鳞伤，战争胜利，但大批市民面临着无家可归的局面，同时拥挤不堪的伦敦城在残砖瓦砾之间露出了阳光、广场，伦敦市民惊讶地看到城市废墟上的阳光，恍然发现原来城市不仅仅是浓烟尘雾蔽日的工作地点，他们对未来城市的发展满怀憧憬。伦敦市政府下定决心，借新的发展机会彻底排解工业革命对伦敦城市造成的危害，同时应对未来人口膨胀的需求进行扩张。伦敦市政府出台一系列产业政策，要求伦敦城市内的工业企业向周边新城置换迁移，否则在税收等方面会有针对性政策。大伦敦规划中，围绕伦敦设计了8个不同规模的新城，这是人类历史上第一次以新城为理论基础的大规模城市规划工作。

这是新城规划第一次在历史上系统地整体尝试，当时哈罗的总规划师形容说：建设初期所有的参与者对于未来发展的前景，犹如面对升上天空的云梯，不知上面是什么。TCPA为新城发展制定了《新城镇建设法》（New Town Act）。模仿Letchworth1902年第一家新城开发公司的模式，通过政府支持的新城开发公司进行商业运营。10年后这一批新城取得了经济社会与城市生活品质的多面成功：哈罗新城的总投资来自于60年的低息贷款（总额为7200万英镑），这一投资基本如期收回，每年有800万英镑的税收收入。今天已经形成了35400个工作位置，事实上已经超出了哈罗新城的工作岗位需求。这一模式的成功迅速在欧洲普遍传播，成为各个发展中国家的城市扩张模式。保守的英国民族在这一次城市规划试验中表现的不仅仅是勇气，还有有条不紊地进行社会规划、经济规划、城市规划等各项工作的研究、实践过程所体现出的理性。

第一代的新城代表的田园城市，与其说是城市规划者的梦想，不如说是城市中产阶级的梦想。哈罗新城在成立30年后，整个城市出现了超过一千个各种协会，新城的音乐厅与交响乐团在伦敦周边盛名卓著。无独有偶，"二战"结束同时期，美国出现了大规模统一建设的雷德朋(Radburn)郊区住宅区。结合工业技术发展，被复制的独立住宅一夜之间拔地而起。一种新的生活方式就此出现。

F. J. Osborn, director of the TCPA (Town and Country Planning Association) in England, after the success of the two new towns, put forward an idea to the government in 1930 to use city construction as a instrument to save the economic crisis. A survey of British workers' residential status by the Royal Special Association in 1937 resulted in the famous Barlow Report, which sewed the seeds for a change of city development and the extension of the concept of garden cities.

The key factor in the demand for housing in England after the Second World War was that London was thoroughly bombarded by Germany. They won the war, but there were a large number of people left homeless. At the same time, crowded London was finally exposed to the sun in the squares of leftover remnant brick rubble. The citizens of London were surprised to see the sunshine above the city ruins. They not only suddenly found that the original city was a workspace with smoke and dust, but this gave them a full vision of the city's future development. The city government was determined to thoroughly resolve the hazards of the industrial revolution and the new London would get a new development opportunity to meet the demands of the future expansion. London's municipal government issued a series of policies requiring industrial enterprises to move to the surrounding new towns, otherwise there would be targeted tax policies. The Great London Plan designed 8 new towns of different dimensions around London, which was the first time in human history that city planning was carried out on such a large scale.

This was the first time for an overall attempt at new town planning. Harrow's chief planner claimed that at the early stages of construction, all the participants' prospects for future development were blurred, as people knew nothing when faced with designing heaven. TCPA made the "New Town Act" for the development of new towns. Imitating the Letchworth model of the first new town development organization in 1902, the new towns development organization was supported commercially by the government. 10 years later there was a defined measurement to the success of the economic society and city life quality: the total investment of Harlow New Town was from low-interest loans of 60 years (a total of 7. 2 million pounds), which would be retracted on schedule, each year with a tax income of 8 million pounds. Today there are 35,400 work places, which have in fact already exceeded the job demand. The success of this model quickly spread across Europe, and became the city expansion model for developing countries. What the conservative British people showed in the planning test was not only the courage, but also the rational reflection in the research and practice of social, economic and urban planning in good order.

In this context, the Garden City was a representative of the first generation of new town as city planner's dream and the dream of the middle class. During the 30 years after that establishment, over one thousand other associations appeared in Harlow new city, and the town hall and Symphony Orchestra are in good reputation around London. Coincidentally during the same period at the end of World War II, the Radburn suburb, a residential neighborhood with large-scale unified construction, appeared in the United States. In combination with industrial technology development, similar independent residential areas were built overnight, and so the new way of life appeared.

图 08　低密度建设的哈罗新城
Fig. 08　Low-density Harlow New City

图 09　哈罗多姿多彩的文化与生活
Fig. 09　Colorful culture and life in Harlow

In fact, in the 1970s and 1980s, a lot of high-density new towns failed but the low-density garden city model was consistently stable in the real estate market. This type of residential development represents on a deep level the balance of quiet rural life and urban life. A symbolized beautiful life often reminds you of the dream you had. What you buy is not only the house, but a way of life, a philosophy. In short, the realization of a dream.

4. If Utopia became the country's dream but not the resident's dream

Excellent new town planning itself often provides inspiration for city planning. During the crystallization of city planning concepts, the outstanding works often correspond to the spirit of the time and have incisive cultural characteristics, becoming a historical model of city planning. In the late industrialization period, the situation was very clear in the development of modern new towns. The modern theory lead the practice of city planning, which has since become one of the motives for modern city development.

In the year of 1957, the International Building Exhibition (IBA) demonstrated how the main venue Hansaviertel had changed greatly. Not satisfied with an ordinary solution, modernist masters explained life with the functionalism theory. Human life was "simplified" or "optimized" to be "living, working and recreation," which are respectively configured in order to form an efficient, functional allocation, eventually incorporating Corbusier's theory of an "ideal city" presented in "Tomorrow's City"(1922) and "Sunshine City" (1933). The most important points were: 1. Various city functions are separated, and the functions of living, working and recreation are respectively set. 2. Modification of high-rise buildings to low-density residences with large areas of green including sports fields to form the ideal pure living environment. 3. City infrastructure and the general road system should be designed according to transport function and speed to meet a variety of needs, advocating a regular road network similar to a chessboard – using the multi-layer traffic system made up of overhead, underground and others – to obtain high transport efficiency and the separation of people and vehicles. 4. City construction should

事实上在高密度新城大量失败的20世纪70与80年代，低密度的田园城市模型始终在房地产市场上得到稳定平衡。这一住宅类型在深层次上代表了一种平静的田园生活与城市生活的平衡点，一种符号化的美好生活提醒你，你曾有的那个梦想。出售的不仅仅是屋子，更是生活方式、生活哲学。总而言之，一个梦想的实现。

4. 当乌托邦成为国家的梦想而没有成为居住者的梦想

优秀的新城规划本身为城市规划思想提供了一个广阔的天地。作为城市规划思想的结晶，新城中的优秀作品往往带有突出的时代哲学思想特征，因此具有强烈的景观与文化特色，从而成为城市规划史上的典例。这一情况在工业化的后期，在现代主义的新城规划这一理论的发展中有突出的表现。现代主义城市规划理论领导城市规划实践，自此成为现代城市发展的动因之一。

1957年国际建筑展览(International Building Exhibition，IBA)主会场汉莎居住区(Hansaviertel)所展现的生活景象大大发生了变化。不满足于一个中庸化的解决方案，现代主义大师们以功能主义的理论，重新诠释人的生活图景。人的生活被"简化"抑或"优化"为"生活、工作、游憩"，为了形成功能配置的效率，分别设置，最终形成了《明日之城市》(1922年)、《阳光城》(1933年)中提出的柯布西耶"理想城市"理论。对新城影响最大的理论点为：1)城市各功能分离，居住、工作、游憩分别设置；2)高层建筑低密度的设置与大面积公园绿地和运动场的并置，所形成的理想化纯粹居住环境；3)城市道路系统应根据运输功能和车行速度分类设计，以适应各种交通的需要，主张采用规整的棋盘式道路网，采用高架、地下等多层的交通系统，以获得较高运输效率；人流与车流分离；4)城市建设要用直线式的几何体形所体现的秩序和标准来反映工业生产的时代精神，以标准化构件的方式进行工业化快速建造，降低成本。

战后第一代婴儿潮也在此时初现端倪，对城市的压力之大，不能再通过常速的城市发展加以疏解。战后工业技术的全面提升造成了标准化住宅建设的可能，大大降低了住宅成本。规划者的梦想与政府的资本背景一拍即合。欧洲在这一时期迫不及待地建设了大量低密度高层住宅群为主体的大型住区。柯布西耶的梦想绘成的简单草图，直到今天看起来还是那么心旷神怡：阿尔卑斯山一般的连绵绿草如茵，森林起伏；其间的高层建筑如云塔般典雅地矗立在云朵之间，底层架空，任游人们信步徜徉。童稚的笔画比任何一个房地产的谄媚都让人深吸一口气。我们是向往那样的天地的，那么开阔的格局，让人心胸都似乎有清风拂过。名媛们在汉莎街区优美的高层建筑前摆出pose，一时这样的居住方式代表的就是时尚。

第二代新城因此而形成了规划史的崭新名词——"卧城"，以居住为主体，没有任何产业，商业服务普遍薄弱，配套设施不足，就业与商业都依靠中心城市，形成严重的钟摆交通。此外在空间上呈现出无结构而散漫的城市空间，空间等级不清，无公共空间与私人空间的界定。典型的是：新城本身只有服务核，作为交通与服务的中心，往往没有真正的中心区，极其缺乏都市环境气质与吸引力。这一批评可以说几乎或多或少适用于所有的新城，但是在第二代新城中尤为突出。大型散漫的公共空间缺乏社会监视功能，也造成了安全问题。在战后第一代著名住区"汉莎居住区"可以看到，一面是格罗皮乌斯近于艺术品一般的精美建筑立面设计，一面是尼·迈耶设计的公寓楼下架空空间中的潮湿阴冷、生气了无。

对外交通在欧洲仍然以快速公共交通联系，但受到空间尺度散漫的影响，汽车交通仍然是重要的交通类型，大面积的停车场占据了大量空间，而且以地上空间为主，将公共空间进一步打碎。

use the order and standards of linear geometric shapes to reflect the spirit of industrial times production, utilizing the standardized components for rapid construction to reduce cost.

The first generation of baby boomers began to appear, and pressure on the city could not be eased by the normal speed of urban development. After the war, industrial development made standard residential construction possible, greatly reducing the cost of housing. Planners' dreams conformed to the government's capital. In Europe during this period, they built many large residential areas of low-density high-rise buildings. The simple sketch of Corbusier's dream still looked so fresh: green grass and forests like the Alps; the high-rises standing like elegant cloud towers. The traveling people stroll at the bottom. Childish strokes are more beautiful than any real estate descriptions. We are longing for that world with an open pattern, as the wind flows over the mind. Women in the Lufthansa District pose before the beautiful high-rise, which symbolized the fashionable way of living at that time.

The second generation of new towns formed a new word, "living city." In the history of planning, living was the main body, without any industry, with generally weak commercial services. Employment and business relied on a central city, causing the pendulum traffic. Additionally the city space showed no structure, it was rambling, so the spatial hierarchy is not clear; there was no spatial definition of public or private areas. New towns didn't have a real urban center. There was a lack of characteristics and attractiveness as an urban environment. This criticism can be more or less applied to almost all the new cities, but definitely to the second generation. A large rambling public space produced a lack of social surveillance functions, causing a security problem. In the famous "Hansa" residential area of the first generation, you could see beautifully artistic Gropius facades, while on the other side sat the cold and wet grouping of residences downstairs, designed by Nymayee, with no vitality whatsoever.

As for the external traffic, rapid public transport links were the main topic in Europe, but had been affected by the undisciplined spatial scale. The automobile was still an important transport method. Though a large portion of the park was allotted for it, public space on the grounds was broken up.

图 10 柯布西耶光辉城市
Fig. 10 Corbusier Glorious City

图 11 汉莎小区
Fig. 11 Lufthansa area

图 12 法兰克福西部新城
Fig. 12 Western new city of Frankfurt

5. The third generation has the same problem

The new towns of that era built high-density residential areas, and by increasing the building height, created a scattered spatial structure with a lack of urban characteristics. As for functions, they were designed with a combination of central squares and walking spaces, forming service and entertainment areas that should strengthen the urban temperament of the city center. A city's high-density environment will inevitably depend on the high-speed public transport flow. A considerable amount of these new towns solved the parking problem by moving it to the first two underground layers. Vertical transportation should further strengthen the function of the city's horizontal and vertical layout. Large green space was to be set within the surrounding environment.

With respect to the urban space, planners of this generation consciously used the spatial sequence of the traditional city to shape the public spaces, combining them with the natural landscapes, which garnered a positive result. But in actual use, the high-rise will cast a large shadow on the public spaces. For the wet European climate, sunshine is a precious resource and in winter it is blocked by the high-rise on almost every side. There also exists a prominent safety problem in some of the huge underground parking areas at night.

The new towns of these two generations become more failed practice attempts, thus creating the German word "Stadtflucht" (Fleeing from the city), which indicates a mass move out from the larger cities, forming a serious vacancy. Which in turn, will cause the loss of the city's assets and maintenance costs.

In the relatively successful example in the northwest new town of Frankfurt, more than 53% of the second generation left the town to select a superior place to live. However, more than 25% of these eventually returned to the city. 1/3 of Germany's high-rise housing is located in East Germany. The government's countermeasure was to refurbish some of these residential areas, and put them on the market again. On the other side, it would remove some residential areas, where the economy was extremely slow, and transition new towns into green space or sporting venues. If a city loses its citizens, it will no longer be a city. It is a simple truth but something that is easy to forget.

5. 具有相同问题的还有第三代新城

第三代新城针对第二代新城空间结构散漫造成缺乏都市性的问题，将城市结构人为水平化与垂直化进行密度提升，进行了以高层高密度为主的住宅建设。在功能上，提高了城市服务功能在建筑底层的密集设置，与中心区域的城市广场与步行空间结合起来设计，形成服务区域与娱乐区域，强化了城市中心的都市化气质。这一高密度的城市环境，必然造成对高速度、大流量公共交通的依赖。相当部分的新城区通过架高地平，在底下两层解决停车问题。垂直交通核心强化性地将城市功能水平与垂直地在这一区域之中布局，而大面积绿地位于周边环境之中。

就城市空间而言，规划师们在这一代新城中，有意识地应用传统城市景观的空间序列来塑造城市外部公共空间，并结合地面的高低处理、人工自然景观的内部设计，起到了一定的积极效果。但是在实际使用效果中，高层建筑给这些狭小的公共空间投下大面积的阴影。对于欧洲海洋性气候而言，阳光是弥足珍贵的资源，但是冬季这里几乎被各个方面的高层建筑遮挡，一些阴角空间与巨大的地下停车区域在夜晚有突出的安全问题。

这两代新城在规划上成为比较失败的新城实践结果，以至形成一个名词，叫做"Stadtflucht"（城市逃亡），就是指从这样的大型城市区域中搬迁出来，形成严重的空置，造成了城市资产与维护费用的流失。

在法兰克福西北新城这样相对较为成功的实例中，超过53%的第二代人口离开新城，择地而居，超过整个城市平均25%的情况。全德国1/3的大型高层板式住宅区位于东德，政府的对策是一方面竭尽力量对这些住宅区加以改造，重新投放建筑市场，一方面将经济运行极端低下的部分住宅区，甚至若干新城拆除还原为绿地或运动场地。城市一旦失去自己的市民，就不再成为城市，这是一个简单但又容易忘记的真理。

图 13　德国海德堡Emmertsgrund新城
Fig. 13　Heidelberg New City in Germany Emmertsgrund

图 14　城市逃亡：Cottbus德国联邦环境协会以107500欧元研究资金发展的有效拆除大型板式居住区的体系与机器
Fig. 14　Fleeing from the city: Cottbus German Federal Environment Association effectively removed the large plate residential system and machine with the research fund of 107.500 euros

6. Dream: Yours or Mine?

We see a problem. Some wonderful dreams are just a vision, but the planner's dream, not the residents'. Dreams will bring a new world, but can also bring a nightmare.

"God creates man, people create the city." The city represents the outcome of human civilization. The occupation and propagation of culture and civilization is another feature of the city. It is the container to store the human culture, civilization, and the citizens. In the construction of city, one problem is how to attract people into the city to live and develop, forming a cultural and emotional environment, which is difficult to create in a short time.

Compared to the central city, the real estate in a new town generally has a standard quality for inexpensive features, which offers a kind of relative balance for the urban environment. But the balance is fragile. When the economic level increases, people require a higher living quality. On the contrary, in an economic decline, when the city cannot provide enough jobs, people often leave the new towns because it does not provide diverse employment opportunities.

In a primal period, residents of a city chose houses from their material needs to find shelter from the wind and rain. But with the passing of time, people now require the needs of emotion and thought. For a city's residents, the city has a far-reaching influence on all aspects beyond just their own time. City residents shape the city, and the city will shape the residents, even dozens of generations over several centuries. The positive effect of the urban environment on the social environment can make the city space a fair place: efficient, humane and also beautiful. But its negative effects may also evolve to create a negative place: violent and selfish with a lack of education. The residential areas of the third generation have a high rate of greenery, but with poor quality due to quick constructions. This in turn caused public activity spaces to not meet the social and psychological needs of occupants, which brings with it unrest and social problems. Many vulnerable social groups would gradually be replaced.

Fig. 15 Turbulent city: in 2007 New Town Riots in Paris

7. In any case, people always need a dream.

I want a yard in which there is a jujube tree. I get up in the morning, set the table and make a pot of tea. Two jujubes dropped from the tree, which can be washed for breakfast, ice-cold and crisp. When I talked to a famous architect, he had lost these wishes. He had designed courtyard-type housing, but the developers were firmly in denial. The developers clearly did not believe this dream still existed, the life of a farmer or of the urban petty bourgeois.

Unlike the European utopian dream, the Chinese Utopia always tends to pursue the typical patterns of "Taoyuan," where a vast natural space is occupied only by a few occupants. As a whole, they completely deviate from the city life and its dominant function, completely back to nature. The disadvantage of this solution is very obvious. In fact, very few Chinese cultures have any praise for the city and city life. "Shuihu" is a folk novel. There are corrupt officials and scoundrels in the market place, who shelter evil people and endorse evil practices. Besides the illegal system of "ShuiboLiangshan," the dream of an anarchist, there was never any social space or physical space in China. It was only a concept in the presence of artistic landscape paintings and poetry.

7. 无论如何，人需要有一个梦想

我一直想有个院子，院子里面有棵枣树，早上起来，摆张小桌，先沏壶茶，捡起树下昨晚凉席上掉的俩枣，拿来洗洗当早点，冰脆的。餐桌上讲给一个著名的建筑设计师听，他莫名感慨：原来有这样的人喜欢这样的房子，然后告诉我，他刚刚设计了一个院落式住宅，被开发商坚决地给否了。开发商显然不相信这种梦想的存在——农民或小市民化的生活嘛。

和欧洲的乌托邦梦想者不同的是：中国乌托邦追求的典型模式"桃源"是在一个无限广阔的自然空间中仅有极少数的居住者，没有阶层，没有时间，空间无限。整体而言是彻底地背离城市与其附属统治功能，完全地回归自然。这一解决方案缺陷十分明显。事实上，中国文化中很少对城市与城市生活有什么褒奖，纵看一部《水浒》，其中市井描述，除了顷刻被上元灯节的繁华打动之外，城市无异于藏污纳垢之所在——除了贪官污吏就是市井无赖。除了"水泊梁山"这种非法体系之外，"桃源"这一无政府主义者气氛浓厚的梦想，在中国也没有社会空间与物理空间存在的可能——只存在文人山水画与诗文中，仅仅是一个概念。

图 16 汉堡新城
Fig. 16 Hamburg New Town

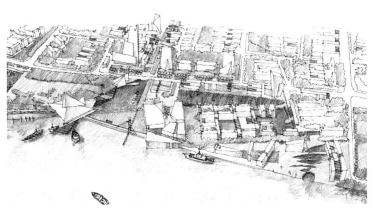

图 17 ISA意厦国际设计集团作品
Fig. 17 Works of ISA Internationales Stadtbauatelier

除此之外，我们并没有一个城市化的梦想，上海"一城九镇"移植九个欧洲城市的方法看似简单直接，实际上颇有童趣，为上海人民建设一个既可以工作，又可以生活，还可以娱乐的小镇；孩子们可以很得意地讲，它盖得和上海别的地方都不一样，跟童话似的。虽然这事实上又是某些人的一个梦想——罗店房地产商业上始终难以运作的是中间仿建的北欧小镇；除了照照婚纱照，住在里面想必是有点惶惑或者恍惚的，打开窗户，看着一窗的灰绿色松木墙瓦，一时间，是走入了别人的梦想，还是一个异国的城市闯入了我的梦想？

工作、生活与娱乐在一起——为了这个目标，无数小资还在前赴后继地前往丽江，买个小店，支个水边的露台，坐在树荫下面，拿台笔记本捎带做点网上生意。经过调研，我确定这样做经济上回报率实在有限。支持他们的是：原来人是可以这样生活的。汉堡新城和这个氛围有点像，坐落在码头上，防洪完全通过建筑间高架桥解决，低层与广场在高水位时就自然被淹没掉。唯一的目的就是这样离水更近些。你可以在楼上作设计，下楼来在广场上吃个意大利面，发会儿呆，海风吹过来很是惬意，晚上住在loft里面，能听到波涛荡漾的声音。柯桥那曲曲折折的廊子后面也是这样生活的——这不是西方人的发明。

霍华德在明日的花园城市中曾这样说："城市规划，作为一项与思想和计划相关的活动，是一件被遗忘了的艺术，至少在我们国家是这样。这里并不仅仅需要重新恢复其生命，而是要将此提升到至今为止所梦寐以求的、更崇高的理想境界。"

我们规划了一个玉龙新城，尝试了一下这个梦想。除此之外，我们还有些梦想，例如山里的城市在屋里可以看到山，出门拾阶上上下下，听得到鸟儿婉啭地叫；水边的城市，能让住宅也临着水，起居室外面水波荡漾，推窗跳到小船就去市场买菜；飞机场边的城市，就是一个超级交通枢纽，处处四通八达，上面的城市公寓日夜繁华，大购物中心酒店里四海宾客云集，脉搏自然快起来，透出些跃跃欲试的神气。除此之外，我们还在寻找一个空间，让这些梦想得以实现。

In addition, Chinese do not have urbanization dreams. Shanghai has a concept of "one city and nine towns," the transplantation of nine European cities, which appears to be simple but is actually quite absorbing. For the people of Shanghai, they are towns in which to work, live and entertain; the children can proudly say the new towns are not the same as other places in Shanghai, like in the fairy tales. Although it is a dream of certain people in reality, the Luodian Real Estate Company has always been ambitious in creating middle-class housing similar to Nordic towns. Besides taking an ideal wedding picture, people living in this area must be somewhat in a trance when they open the window and see the wall of lime green pine. For a time, people must live in it like in their dreams, as if a foreign city appeared in their dream?

Work, life and entertainment should be connected each other. For this, numerous citizens still travel to Lijiang to shop, to build a waterfront terrace, to sit in the shade, or to do business. By way of a survey, I determined that the rate of economic return is limited. Still, people like living there, where the atmosphere is a bit like the Hamburg new town. Located on a dock, where controlling floods can be completely solved by constructing a viaduct among buildings. The lower layer and the square would naturally be drowned at a high water level. The single purpose is to live much closer to the water. You can do your design work upstairs and eat authentic Italian noodles downstairs in the plaza, sitting there with the sea breeze blowing comfortably. At night, living inside the loft, you can hear the rippling sound of waves. Living behind a zigzag portico is really living, which is not the invention of westerners.

Howard, in tomorrow's Garden City, said "City planning, as an activity related to thought and plan, is a lost art, at least in our country. Here it needs to restore its life, even increase it to a more sublime ideal state which we can only dream of so far."

We have planned the Yulong new town trying for this dream. In addition, we also have some other dreams, such as a mountain city. From your house, you can see the mountains. And leaving your house, you will step up and down, hearing the birdsong. In the waterfront area, the house faces the water; and outside, the water is rippling, jumping out the window into the boat at the market. The city near the airport is a super traffic hub, where the traffic leads in all directions, downtown apartments are bustling day and night, with massive shopping hotels gathered with guests, so the pulse is naturally quick, with little stressful air in it. In addition, we are still looking for a space to let the dream come true.

项目实践
PROJECTS

城市发展
URBAN DEVELOPMENT

01 世界文化遗产城市波茨坦城市总体规划
The Urban Development Planning of World Cultural Heritage City of Potsdam

02 丽江玉龙新城规划
Masterplan of Yulong New Town, Yunnan

03 泉州市城市新区建筑天际轮廓控制规划方法研究
Planning of the Skyline of the City of Quanzhou, Fujian

04 泉州市区和环湾地区空间发展战略研究
Strategic Study on Development of the Urban Districts and General Area Around Quanzhou Bay, Fujian

05 广州市南沙光谷地区城市发展规划
Urban Development of Nansha Guanggu in Guangzhou

06 阳宗海旅游度假区西北部概念规划设计
Conceptual Planning of the Northwestern Part of Yangzonghai Resort, Yunnan

07 东山蝶岛发展战略规划
Dongshan Butterfly Island in Fujian: the Strategic Planning of Urban Development

08 十堰市东部新城概念规划及城市设计
Conceptual Planning of the East New Town in Shiyan, Hubei

09 福州市东部新城中心城市设计
City Planning for the Center of the Eastern New Town of Fuzhou, Fujian

10 杭州天堂鱼生活：杭州市运河新城概念规划
A Fishing Life Paradise in Hangzhou: Concept Planning for the Canal New Town in Hangzhou

11 唐山市南湖生态城概念性总体规划设计及起步区城市设计
Conceptual Planning and Urban Design for an Ecological Zone in Tangshan, Hebei

12 大同市御东新区概念性总体规划及核心区概念性城市设计
Conceptual Planning of Yudong District and Urban Design of the Core Area, Datong

13 文昌市抱虎角概念规划
Conceptual Planning for Baohujiao Area in Wenchang, Hainan

14 昆明市绿地系统概念规划
Concept Planning for Kunming Green Space System

城市更新
URBAN RENEWAL

01 伦茨堡控制性规划
Framework Development Planning of Rendsburg

02 "霍费尔天空"：提升城市品质的范例项目
"Heaven of Hof": an Initiative-project for Upgrading the City

03 埃尔旺根步行区
Pedestrian Zone of Ellwangen

04 埃斯林根总体规划和公共空间城市设计导则
Master Plan and Design Guidelines for the Public Spaces of Esslingen am Neckar

05 "粉红岛"：韩国首尔方背洞街区广场
"Pink Islands": a Street Square in Bangbae-dong, Seoul, South Korea

06 **Gangseogu** 城市更新规划
Urban Renewal Gangseogu

07 北京市昌平区奥运自行车训练馆及周边地区城市设计
Urban Design of the Area Surrounding Changping Olympic Cycling Training Center, Beijing

城市保护及管理
URBAN PRESERVATION AND MANAGEMENT

01 世界文化遗产城市施特拉尔松：城市景观规划
World Cultural Heritage City of Stralsund: Townscape Planning

02 新勃兰登堡城市景观规划
Townscape Planning Neubrandenburg

03 世界文化遗产城市波茨坦瑙恩郊区及耶格尔郊区的设计导则
Design Code for the Nauener Vorstadt and the Jaegervorstadt in World Cultural Heritage City of Potsdam

04 泉州市法石街区保护与整治规划及环境设计
Urban and Landscape Design for the Fashi District in Quanzhou, Fujian

05 泉州市城市新区公共空间控制规划方法研究
Research of the Planning Methods for Open Space in Quanzhou, Fujian

城市发展
URBAN DEVELOPMENT

01 世界文化遗产城市波茨坦城市总体规划
The Urban Development Planning of
World Cultural Heritage City of Potsdam

02 丽江玉龙新城规划
Masterplan of Yulong New Town, Yunnan

03 泉州市城市新区建筑天际轮廓控制规划方法研究
Planning of the Skyline of the City of Quanzhou, Fujian

04 泉州市区和环湾地区空间发展战略研究
Strategic Study on Development of the Urban Districts
and General Area around Quanzhou Bay, Fujian

05 广州市南沙光谷地区城市发展规划
Urban Development of Nansha Guanggu in Guangzhou

06 阳宗海旅游度假区西北部概念规划设计
Conceptual Planning of the Northwestern Part of
Yangzonghai Resort, Yunnan

07 东山蝶岛发展战略规划
Dongshan Butterfly Island in Fujian,
the Strategic Planning of Urban Development

08 十堰市东部新城概念规划及城市设计
Conceptual Planning of the East New Town
in Shiyan, Hubei

09 福州市东部新城中心城市设计
City Planning for the Center of
the Eastern New Town of Fuzhou, Fujian

10 杭州天堂鱼生活：杭州市运河新城概念规划
A Fishing Life Paradise in Hangzhou:
Concept Planning for the Canal New Town
in Hangzhou

11 唐山市南湖生态城概念性总体规划设计
及起步区城市设计
Conceptual Planning and Urban Design
for an Ecological Zone in Tangshan, Hebei

12 大同市御东新区概念性总体规划
及核心区概念性城市设计
Conceptual Planning of Yudong District and
Urban Design of the Core Area, Datong

13 文昌市抱虎角概念规划
Conceptual Planning for Baohujiao Area
in Wenchang, Hainan

14 昆明市绿地系统概念规划
Concept Planning for Kunming Green Space System

01 世界文化遗产城市
波茨坦城市总体规划
The Urban Development Planning of World Cultural Heritage City of Potsdam

规划面积 AREA		190 km²
人口规模 POPULATION		150000
完成时间 PROJECT DATE		1992.03

引言 Introduction

Potsdam is an extraordinary city. From its beginnings as the seat of a Slavic prince, to a noble margrave's castle in medieval times, and then into a "New Town" as a residence for electors, kings and emperors, the city gradually grew into today's state capital of Brandenburg. Throughout history it has been the cradle of the Prussian state, the seat of Germany's "Hollywood," and home to a European cultural monument of international repute.

Not only does Potsdam's history encompass world famous personalities and events like Frederick the Great, Voltaire, Alexander von Humboldt, and Einstein, but also the last German emperor, Wilhelm II and the dictator Hitler. This is where Einstein's theory of relativity was developed, and later Babelsberg Studios became the dream factory of the German film industry. In addition, U.S. President Truman gave instructions to drop the first atomic bomb on Hiroshima here and this is the place where Churchill, Truman and Stalin, divided the world into two parts. (Potsdam Conference, 1945)

Today it is a world heritage city (Sansouci Palace, Babelsberg Castle, Parks), a media city (Babelsberg film studios), a science town (universities and research institutes), a military town (foreign command of the German Federal Armed Forces), a recreation and sports city (rowing, sailing etc.) and also a part of the metropolitan region of Berlin. In many ways, Potsdam is also a "Central European think-tank," which now ranges in subjects from international earthquake monitoring and European biotech research to fashion and automotive design and development.

波茨坦是一个不同凡响的城市，从最初的斯拉夫王侯贵族的城堡，到中世纪边疆伯爵的城堡区，再到选帝侯、国王、皇帝官邸聚集于此的"新城"，到今已经发展成勃兰登堡州的首府。在历史发展的过程中它曾经是普鲁士王国的摇篮、德国好莱坞的所在地和世界级欧洲文化遗产。

波茨坦总是和世界历史名人和历史大事联系在一起，如弗里德里希大帝、伏尔泰、亚历山大·洪堡和爱因斯坦，也包括德国最后一位皇帝——威廉姆二世和独裁头子希特勒。爱因斯坦的相对论在这里成熟，巴本斯伯克的小摄影棚在这里成长为德国电影的梦工厂，美国总统杜鲁门在这里下令向日本的广岛投下第一颗原子弹，丘吉尔、杜鲁门和斯大林在这里把世界一分为二（1945年波茨坦会议）。

它现在是世界文化遗产城市（无忧宫、巴贝斯堡宫和公园）、媒体城市（巴贝斯堡的影视基地）、科学城市（大学和科研机构）、军事城市（联邦海外防御司令部）、休养和体育城市（划桨和帆船运动等），同时也是柏林大都会区的一部分。从多个方面来看，波茨坦都是中欧的"智囊"——包括国际地震监测、欧洲生物技术研究以及时装和汽车设计的发展。

图 01　在过去与未来之间的波茨坦
Fig. 01 Potsdam, between yesterday and tomorrow

城市发展 URBAN DEVELOPMENT

图 02 新镇：波茨坦的基尔希斯泰格菲尔德
Fig. 02 New Town Potsdam: Kirchsteigfeld

图 03 波茨坦的影视城：巴贝斯堡
Fig. 03 Film town Potsdam: Babelsberg

任务 The task

自东西德合并以来，社会主义经济制度逐渐被资本主义经济制度所取代，投资者担心的是，新联邦州的城市暂时会不受保护地移交给西德，社会市场经济需要经过数十年的检验，新联邦州(前东德)还没有掌握德国经济奇迹成功的根本秘诀——资本主义和社会主义的有意识的融合。

这促使联邦政府、州政府和城市一起，在各地进行了一系列包括城市发展、城市更新及城市形象规划在内的规划工作，以便把城市的发展纳入有序的轨道，在个人利益（资本价值）和集体利益（公共福利）之间找到一种平衡。

波茨坦市政府和新联邦州勃兰登堡的州政府也委托ISA集团开展了州首府波茨坦的城市发展规划。这项任务的定位很明确——为波茨坦的未来发展制定详细的、可操作的方案。但是如何完成这项任务则是不明朗的，因为这将是一种全新的城市发展规划方法。社会和经济制度的改变使前东德的所有数据及评价参数失去意义，这意味着它们不能被用于对新经济制度下的波茨坦未来的规划。而另一方面前西德的数据系统及评估参数由于时间和内容方面的原因也不能直接运用，如很多标准规范（例如单个居民人均居住面积指标）和经济预测方法（由于所有制的区别）都很难采用。[1]

这就把城市发展规划带入了一个新的模式，它首先以对城市内每个人的基本需求——工作、居住、教育、休养、健康、运动、休闲及文化娱乐等因素的分析为基础，再来研究在哪里、以什么样的方式、在什么样的环境中来满足这些需求。

After German reunification, the socialist economy was immediately replaced by the capitalist system and cities were initially at the mercy of investors from the new federal states, without any protection. The principle of the social market economy, which was tested for decades, was an important secret in the success of the German economic miracle – a conscious synthesis of capitalism and socialism – that had not yet been transferred to the new federal states (the former GDR).

This prompted the federal and provincial governments to work alongside cities. Their goal was to develop a universal plan for urban development, urban renewal, and – last but not least – the townscape, in order to guide the coming revisions and establish a balance between the individual single interest (shareholder value) and the common interests of the public at large.

Thus, the city government of Potsdam and the national government of Brandenburg commissioned the urban development planning of the state capital of Potsdam. The task was clear: to flesh out a feasible vision for the future development of Potsdam. But the path to that goal was unclear, as it would require entirely new methods of urban development planning. By changing the social and economic system, all data sets and parameters of the GDR became obsolete, i.e. no longer suitable for planning the future of Potsdam in a social market economy. The data management system and evaluation parameters of the FRG were not directly transferable either, due to constraints on time and content. There was a lack of standards, for example, as in the number of square meters required per inhabitant in living spaces. Nor, for that matter, could any new economic models be created that changed the state sector to the private. [1]

This led to the development of a new model for urban development planning, which is based primarily on an analysis of the basic needs of the population – for jobs, housing, education, recreation, health, sport, leisure, culture and entertainment – and it examines where, how and to what extent those needs can or should be met in the future.

社会和经济制度的改变使前东德的所有数据及评价参数失去意义，而另一方面前西德的数据系统及评估参数由于时间和内容方面的原因，也不能直接运用，因此城市发展规划必须引入一个新的模式。

By changing the social and economic system, all data sets and parameters of the GDR became obsolete. The data management system and evaluation parameters of the FRG were also not directly transferable because of time and content reasons. The development of a new model of urban development planning was necessary.

87

日常生活对城市的多方位需求必须在一个全面的、可持续的城市发展规划中，作为城市机体的一部分来被评价和规划。

Urban development planning is the two-and three-dimensional vision of the future development of a city. This means that the many diverse demands of daily city life have entered into a sustainable urban development plan, realized individually, but evaluated and designed as part of a whole organism.

Urban development planning is responsible for the countless functions that a city must meet; such as housing, employment, education, leisure, sociability, recreation, arts and culture, transport, supply and waste disposal, and ecological training. The urban development plan is the two-and three-dimensional vision of the desired future development of the whole organism of a city. This means that the many diverse demands of daily municipal life have entered into a holistic and sustainable program, realized individually, but evaluated and planned comprehensively as part of that whole organism.

An integrated urban development plan requires special consideration when it comes to compatibility and sustainability with regard to the type and scope of land use, transport, as well as social and technical infrastructure. Careful attention should also be paid to social factors, urban engineering, urban economy and urban ecology. This also implies design aspects of the city can exist as an identity for a community (image), a medium for citizen's education and training (cultivation by the environment) and as an economic factor (good location for business and quality of life). Such qualified development – supported by both expert review and multiple design proposals, as well as architectural competition between investors – may lead to a new culture of construction.

城市发展规划的任务之一是：将组成城市的众多功能，像居住、工作、教育、休闲、交际、休养、文化艺术、交通、供给和回收以生态的方式达到一种相互和谐的状态。城市发展规划是整个城市机体未来发展在二维和三维尺度上的体现。这意味着，日常生活对城市的多方位需求必须在一个全面的、可持续的城市发展规划中，作为城市机体的一部分来被评价和规划。

一个兼顾用地方式、用地规模、交通、社会和技术基础设施的城市整体发展需要特别重视承载力、社会因素、城市技术、城市经济和城市生态。同时城市作为社会形态的体现(形象)、教育和培训载体(环境培育)、重要经济要素(位置和生活质量)，其形象塑造也是整体发展的重要方面。这样一个高水平的城市发展可以通过专家的意见、建筑和投资竞赛来获得支撑，并且可以引领新的建筑文化潮流。

一种新的城市规划模式
A new urban planning model

发展道路：从城市发展的历史到新生活的蓝图
The route: from the history of urban development to the vision of new life scenarios

At the beginning of the planning work there was an extensive analysis of previous urban development in Potsdam. The City of Potsdam is the result of a complex historical process. It is based on the unique landscape, a composition of harmonic meadows and hills, greens and forests, rivers and many large lakes. This is how it became the former residence of Prussian kings, and why it developed a large state vision over the centuries. The Prussian state formed a new government and an administrative, educational and cultural system, which not only made strong efforts to "cultivate" citizens but also became known for its tolerance (the first mosque in central Europe) and apparent openness to foreign influences (Italian, Dutch and Russian districts). This political, social, and cultural destination embodied the idea of a city as a total work of art, comprising the urban and landscape area, which has been treasured for centuries and found its biggest attractors in the various parks and palaces that are now a magnet for visitors from all over the world.

在着手城市发展规划初期，ISA对波茨坦过去的城市发展作了广泛的分析。波茨坦城是多样化的历史发展进程的结果。从无与伦比的自然景观开始，在由草地和丘陵、绿地和森林、河流和湖泊组成的和谐景致中，出现了普鲁士国王的都城，它在数百年的发展过程中形成了承载多元职能的国家形态——具有新的执政、管理和文化教育体制的普鲁士王国，一方面对公民进行严格的教育和"培养"，同时对外国文化也是包容（中欧第一个清真寺）和开放（意大利、荷兰、俄罗斯城区）的。这种政治、社会和文化宗旨被植入了将都市作为由城市和景观空间共同构成的艺术品的理念中，数百年间作为社会行为准则，其形形色色的公园和宫殿充分代表了文化的多元性，它现在是一个吸引全世界游客的旅游胜地。

城市发展 URBAN DEVELOPMENT

图 04 城市整体艺术的组成部分
Fig. 04 Components of an urban all-embracing art form

在20世纪，城市空间逐渐向外扩展，随之产生了由不同区域组成的多样化的居住形态。一系列相互矛盾的发展目标导致了城市形态的混杂无序。在郊区，高质量住区及高层住宅区相邻；在内城，战后重建建筑及改造的历史建筑混杂。由城市和自然景观空间产生的城市整体艺术性也随之遗失。

城市发展

基于这种情况，首先对城市发展的中心要素——自然空间情况、历史联系、整体城市和单独区域、发展轴以及土地利用的重要因素——进行了SWOT分析，也就是一方面关于问题和威胁，另一方面关于机遇和目标的广泛的总结分析和评估。城市发展规划的第一步是确定每个要素的城市发展目标：如对于"自然空间情况"，目标就是让自然空间更好地被感受、保持和加强城区的绿化带或者联系内城与水系；目标设定为"历史联系"、"加强与历史相关的文化意向"、"让历史城市空间结构更好地被感知"等。其他要素也按照这种模式提出相应的目标设想，作为城市发展规划原则的基础，最终作为波茨坦市议会的决议草案。

During the 20th century, the city became more widely spread out and there was a heterogeneous settlement structure with highly divergent individual areas. A series of conflicting urban development visions led to the fact that high-quality residential areas and high-rise developments on the city outskirts, replaceable reconstruction architecture after the Second World War, and recreations of historical buildings in the inner city are now all standing side by side. In this way, some essential parts of the city's identity and landscape space as an all-embracing art form were lost.

Urban development

This was the reason why a SWOT analysis was required regarding the central parameters of urban development. The environmental situation, historical references, the whole city as compared to individual areas, development axes and key elements of the land use – i.e. a comprehensive inventory and assessment concerning potential problems and dangers on the one hand, and the opportunities and aims on the other – was carried out. This first section of planning was completed with the creation of urban development aims for each parameter. The goal of the "Nature Area Environments" parameter was to "improve the accessibility and overall experience of the landscape space". For the "Historical Reference Points," parameter, there were objectives such as reinforcing "historically relevant cultural identity" and "making historic urban spaces more engaging," etc. These tangible goals had been formulated as the basic principles for the urban development plan, which served as a template for Potsdam City Council meetings.

对城市发展的中心要素——自然空间情况、历史联系、整体城市和单独区域、发展轴以及土地利用的重要因素——进行了SWOT分析。

A SWOT analysis was required to determine the central parameters of urban development: the environmental situation, historical references, the whole city versus individual areas, development axes and key elements of land use. This step was completed with the creation of urban development goals for each parameter.

图 05　内城：多文化影响的拼贴
Fig. 05 Downtown: collage of multicultural influences

Urban development history and scenarios

Potsdam was a Slavic establishment in the 13th century, but for the next four hundred years developed only very little. The modern city of Potsdam actually began in the 17th century. Therefore, urban development was first investigated at that time and was illustrated in the city's layout plans, particularly the various phases of city planning. Then, different models of future development were investigated, with careful regard given to character, interior and exterior development, park or urban city centers, structure and function, along with integration into the metropolitan region of Berlin. Guidelines for future development were derived from these models.

Life scenarios as the basis of urban development planning

As previously described, the new model of urban development planning is, at its core, based on the essential needs of the people - their right to work, housing, education, recreation, health, sport, leisure, culture and entertainment. The citizens shape the character and uniqueness of their city. Their potential, activities, needs, feelings and ideas are reflected in the appearance of their urban environment and their complex interactions within the urban organism.

Human needs were integrated into the notion of urban development and divided into three prospective groups according to their varying demands on the city: land use, urban design and urban infrastructure.

In order to be able to analyze and plan these complex municipal organisms, these human needs were integrated into the concept of urban development and divided into three prospective groups according to their varying demands on the city: land use, urban design and urban infrastructure. The concept of land use for Potsdam includes primarily material demands: a city to live in, the city of jobs, the city of education and science, governors and government town, a shopping city as well as recreation and sports. The concept of urban design is to create Potsdam as a city of gardens and parks, history and culture, tourism, water gardens and neighborhoods, which primarily implies immaterial demands. The concept of urban infrastructure involves themes such as city engineering and utilities, social organizations, ecology, urban climate and a sustainable traffic system founded on public transportation.

城市发展的历史和场景

波茨坦是13世纪由斯拉夫人建立的城市，到17世纪为止发展相对缓慢，当今波茨坦城的发展实际上开始于17世纪，因此对该城市发展的研究也开始于这个时间点。首先描绘城市发展不同时期的城市总图，接下来对未来城市发展的重要特征、内部和外部发展、公园绿地与城市中心以及考虑到与大都市柏林相联系和协调的功能分区及结构进行研究，以便得出未来城市发展的指导方针。

生活场景作为城市发展规划的基础

正如前面所述，新的城市发展规划的模式，核心是以在城市中生活的人的基本需求——他们对工作、居住、教育、休息、健康、体育、休闲和文化娱乐的需求——为基础。一个城市的居民能反映这个城市的特点和特性。他们的可能性、活动、需求、感情以及观点都会在城市环境中以及综合协作的城市机体中反映出来。

为了对这个综合的城市机体进行分析和规划，人们的各种需求被纳入城市发展规划的指导原则中，并被分为三种类型的指导原则，即功能指导原则、形象指导原则和专业指导原则。功能指导原则提出波茨坦承载了居住、提供工作岗位、教育和科研、州首府与行政、军事、购物和餐饮以及休养、体育等城市功能。形象指导原则提出，波茨坦是一个公园和花园城市，是历史文化、旅游、水上花园城市，是城区的一部分。专业指导原则包括城市技术和供应能力、社会设施、城市生态、城市气候、可承载的交通发展和公共短途交通等方面。其中功能指导原则主要侧重物质方面的需求，而形象指导原则侧重非物质方面。

图 06 波茨坦作为教育和科研城市、州首府、购物与美食城市
Fig. 06 Concepts for education and science, government, state capital as well as a shopping city

图 07 居住城市：指导原则和生活场景
Fig. 07 Living city: vision and scenarios

According to the concepts of land use and urban design, the target groups have been determined, the existing situation described and evaluated, general life scenarios simulated, concepts for future urban development defined, along with planning and building projects derived from the examination of bearing capacity and feasibility. All these contents have been summarized and mapped as "theme plans."

The concept of land use, "Potsdam – living city," takes into account the people who will be affected by the space, e.g. whether they are children, adolescents, adults, or senior citizens? Singles, couples or families? It then examines the area, i.e. where to live in Potsdam, where and what are the necessary facilities, from kindergartens, schools, youth centers to culture and club houses, to retirement homes and cemeteries. Then it develops scenarios as to what living actually means in respect to sleeping, eating, talking, drinking, watching TV, cleaning, educating, but also learning to love, argue, dream, have hobbies etc. This is done in order to keep track of the demands of living at home, the flat, the neighborhood, districts and the city. From these parameters a vision was then developed, which included goals such as providing a sustainable sense of "home" by increasing the identity of settlements, designing public space as a kind of "living room" for residents, thereby increasing the overall quality of life. Then, the planning draft followed and described what could be renovated, made denser by filling the vacant lots, extended or built anew. The "theme plan – living city" stated that the inner city area should be developed before outer areas, and defined the type and location of existing and future residential areas with regard to preservation, reduction, extension and new development. This "theme plan" continues to serve as a part of the zoning plan, and those zoning plan laws instruct its presentation form and depth.

The policies, and consequently the representation of them, consisted of a thematic master plan, which was then combined into an overlay of the desired image of Potsdam's future profile.

一般情况下，每个功能和形象指导原则都会把参与其中的人群考虑在内：描述和评估现状，模拟一般的生活场景，表达未来城市发展的目标构想，提出规划和建设任务，并在承载能力和容量研究的基础上，通过相应位置和内容的主题性专业规划总结形成规划成果。

以波茨坦功能指导原则中的居住城市功能为例，相关人群划分的一种方式是分为儿童、青年、成年人、老年人，另一种方式是分为单身、伴侣和家庭。然后要对现状进行研究，波茨坦现有与未来可支配居住用地，还有与之相关的公共设施，从幼儿园、学校、青年中心、文化馆、俱乐部到敬老院和墓地。通过概括地叙述出居住的真正含义，包括生活、休息、吃喝、谈话、看电视和打扫，还有学习、教育、恋爱、争吵、做梦及业余爱好等，形成了居住功能对住房、居住环境、居住区和城市的要求。在这些参数基础上形成目标构想，如把居住区的特性提升为可持续的"故乡"，把公共空间塑造成居住者的生活空间，提高生活质量。规划确定波茨坦可以更新、加密、扩展或新建的区域及内容，此外要对青年之家、敬老院等后续设施作出表述。在总结性的居住城市规划专题中，提出内城先于外城和郊区发展的目标，以及可以保留、加密、扩展或新建的现有和未来居住区的位置，它的表现形式和深度相当于居住用地专题规划。

这些指导原则和其规划表达——专项指导规划，叠加起来共同组成波茨坦未来城市发展的蓝图。

图 08 波茨坦：巴贝斯堡北部的一个历史别墅区
Fig. 08 Potsdam: "Babelsberg" North - an historic, villa area

城市发展 URBAN DEVELOPMENT

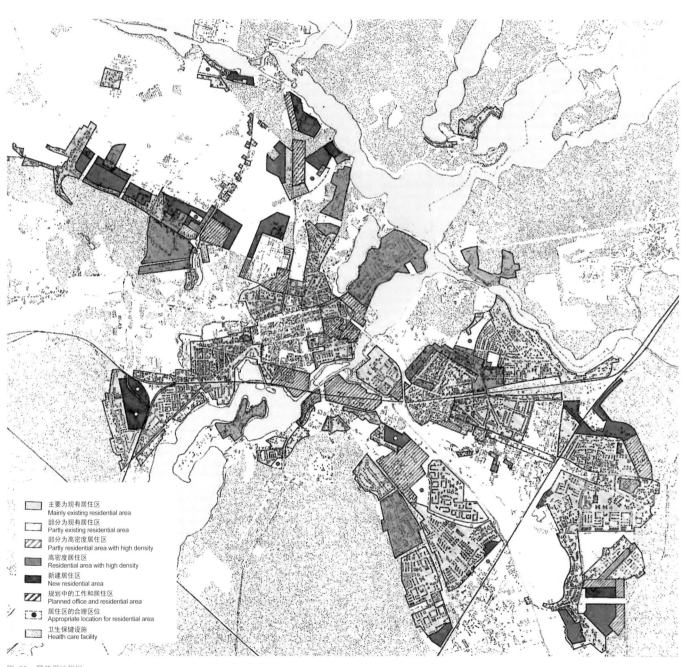

Fig. 09 Partial zoning plan- living

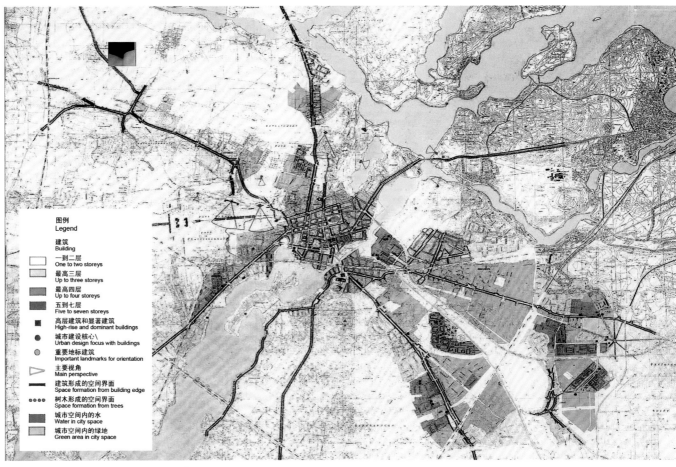

图 10 城市发展规划布局：波茨坦城市形象规划的基础
Fig. 10 Urban development plan, urban shape: the basis for the urban design plan of the city of Potsdam

Urban development planning: land use, traffic system and urban design

The results of the previous planning steps were then summarized in three basic "theme plans": the land use concept, the traffic concept and the design and building mass concept.

The land use concept defined the distribution of land use, based on the existing areas and according to what possibilities were available. New jobs and homes should be the focus of a future urban development; old quarters of town should be rehabilitated and new quarters should be built. There would be a rise in different job opportunities in the service sector, public administration, the federal government and the provincial government, media and design industries, education and science as well as in trade and industry. Special emphasis was placed not only on the preservation of the historic city, palaces and gardens but also on an expansion of areas for recreation and leisure, which was also related to the urban climate. This land use concept was first developed as a guiding principle for the development of Potsdam as a whole and in its individual parts. It was first developed verbally, as a conceptual master plan, and then used as an instructional plan. This was detailed enough that it later served as the basis for the initial zoning plan of the city of Potsdam.

用地、交通和形象

前述规划过程的结果最终将总结成三个基本的专项规划，即用地规划方案、交通规划方案、城市形象规划方案。

用地规划方案追求的是，以现状为基础进行功能划分，且来源于现有的基础，具有良好的可实施性。新的工作岗位和住宅在面向未来城市发展的过程中应引起普遍重视。传统中心城市应尽快进行更新同时建设新城区，缓解城市发展压力。服务业、管理部门、国家和联邦部门、国家行政部门、媒体和设计产业、教育和科研以及商业和工会范围内应形成多样的、不同的工作岗位。不仅重视保留历史城市、宫殿和公园设施，还应扩展休养和娱乐用地，同时也应考虑城市生态与气候。这种用地规划方案以文本和规划图纸的形式首先作为波茨坦城市发展从整体到局部的指导思想，之后不断深化到土地利用规划的深度，成为"转折"以后波茨坦市第一个土地利用规划的基础。

城市发展 URBAN DEVELOPMENT

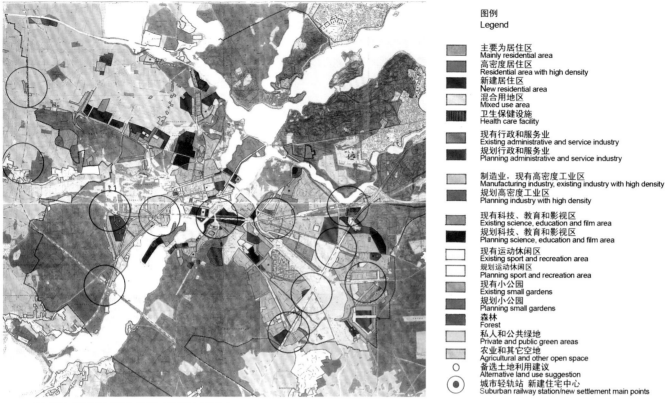

图 11　城市发展规划的用地规划方案：波茨坦土地利用规划的基础
Fig. 11　Land use concept of the urban development plan: the basis for the zoning plan of the city of Potsdam

在交通规划方案中，首先通过SWOT分析，研究了波茨坦内城和城市整体交通系统，以及与大都市区柏林的区域交通联系的优势与劣势。接下来形成了三种未来交通模式的比较方案，并由此制定出了最终的交通规划方案。这一方案不仅包括重要的市内道路和公共交通，同时也包括市域和区域交通系统，并进行了深入的工作，以至于成为波茨坦市第一个综合交通规划的基础。

城市形象规划方案是城市发展规划的另一组成部分。因此形成了对于整个城市的高度和建筑总体发展、城市平面布置、重要城市空间以及城市入口的规划方案。除此以外，还对内城河流两岸的城市中心进行了详细设计，并以文本和图纸形式加以体现。在此基础上也制定了城市形象和建筑控制的整体方案，分别指出了法律许可范围内的建筑高度、高层建筑的可能位置、城市形象的主要特征、重要的视觉联系以及最重要的城市空间序列。这些被看做是城市景观规划的基础。²⁾

最后的工作是研究城市发展规划指引形成的波茨坦市功能、面积、工作岗位以及住宅的潜力。

按照对每个区域可能的建筑密度、现状等级和保护等级的评估，以及每块用地的用地类型，可以得出潜在的最

The traffic concept was initially studied in a SWOT analysis of both the inner city as the entire traffic system of Potsdam, as well as its integration into the metropolitan region of Berlin. Subsequently, three different alternatives of a possible future general traffic concept have been developed, and from these three the final traffic concept was implemented. This concept indicated the most important streets and public transport lines as well as the main transportation arteries and regional transportation systems. They were presented in such detail that it served as the basis for the first general transport plan for the city of Potsdam.

The urban design concept was the third part of the urban development plan. It testified to the building height and mass development of the city as a whole, the city's layout, the main public urban spaces and the city entrances. In addition, alternatives for the heart of the city along the river were also detailed in the form of text and plan drawings. The urban design and building mass concept showed the allowable height of buildings, the possible location of skyscrapers and landmarks, important visual connections and the most important spatial sequences. This served as the basis for the urban design plan. ²⁾

The determination of the urban development plan for Potsdam, which arose from potential land use, land capacity, as well as job and housing units, concludes this part of the work. Depending on the evaluation of potential development density, maintenance and preservation levels of individual areas, and type of land use, potential

前述规划过程的结果最终将被总结成三个基本的专项规划，即用地规划方案、交通规划方案、城市形象规划方案。

The results of the previous planning steps were then summarized in three basic "theme plans": the land use concept, the traffic concept and the design and building mass concept.

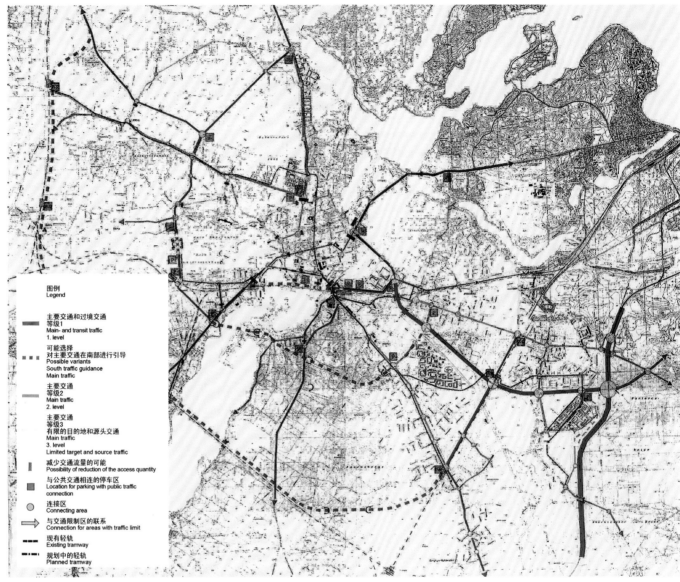

图 12 城市发展规划的交通规划方案：波茨坦市综合交通规划的基础
Fig. 12 Traffic concept plan: the basis for the general transport plan of the city of Potsdam

minimum and maximum values were determined for each plot. The minimum values are to be understood as estimations from the existing conditions, whereas the maximum values illustrate the possible densification, land uses and building structure developments. They were considered legal by the urban development plan in accordance with the building regulations in terms of land use.

The results of this work were stated as an urban development objective and were presented to the Potsdam City Council, the legislative body as "principles of urban development." These urban development objectives were created to provide guidance in local politics, which covered topics that range from the upscale residential town of Potsdam, culture, history and science city, to Potsdam as a media city and capital of the state of Brandenburg. They were finally adopted as the principles of the urban development plan.

大价值和最小价值。最小价值是基于对现状的评估，而最大价值则基于城市发展规划提出的增加密度、功能及建设的可能性，同时应该符合建筑用地法规的规定。

这项工作的结果形成了一系列发展目标，并通过立法机构——波茨坦市议员会议，以"城市发展基本原则"的形式被确立。这一系列城市发展目标被作为地方政策导向的辅助，涵盖了将波茨坦作为高品质住宅城市，军事、文化、历史及科学城市，以及传媒城市和勃兰登堡州的首府等主题，并最终成为城市发展规划的准则。

经验 Experience

波茨坦城市总体规划的进程，是自两德统一开始的十多年间，为波茨坦市所做的一系列城市规划咨询工作的一部分。这一系列规划咨询工作涵盖从欧洲著名新城Kirchsteigfeld的规划准备和规划制定，到波茨坦城市景观规划，再到波茨坦的城市发展规划等层面。此外还包括对如无忧宫3)一类的世界文化遗产周边地区的指导性规划等一系列历史城市相关保护详细规划与设计导则制定、城市分区规划、重要地区巴贝斯堡4)的修建性规划、城市空余用地目录，等等。5)6)

波茨坦城市总体规所划的内容和方法都进行了新的尝试，因为它面临在德国重新统一后统计数据大量缺失的局面，因此几乎是完全建立在一个全新的质量价值评估体系之上。这样就必须为之寻找到合适的基本参数。这些参数可以在人类对其生存环境的基本需求的说明和展开叙述中找到。这个基本问题——人类对其所处的城市环境到底有何要求——在实践中被证明是理解和满足人类对一个城市的多元化需求的核心和取之不尽的源泉。

在规划过程中，基于"场景"所引导出的最重要的需求类型进行了一系列深入研究，转化成详细的专项规划，并对现有的用地潜力进行了对比检测。这些成果最终形成了一个标准的城市总体规划，并将作为土地利用规划、综合交通规划和城市总体规划的基础。即从一个基于质量的城市发展规划，发展成为一个具有量化成果的城市发展规划，最终形成一个综合体现质量和数量参数的、统一的、全面的城市发展规划，这个过程使得波茨坦城市发展规划成为一个具有示范意义的项目，为日后在南美或亚洲等不同文化背景区域的规划工作实践奠定基础。

The creation of an urban development plan was part of ten-plus years of consulting for the city of Potsdam, which began with the unification of Germany. The work ranged from involvement in the preparation and planning of "New Town Kirchsteigfeld," to the urban design and development plans of Potsdam. Part of this consulting activity also consisted of the master plan for the area of World Heritage sites such as "Sansouci" 3), the master plan (Rahmenplan) for subareas of the city, the building code plan for the celebrity neighborhood "Babelsberg" 4), and a catalogue specifying unbuilt lots and many other works that came into being. 5) 6)

The urban development plan is something new in terms of content and method because after the Berlin Wall came down resulting in Germany's reunification, it was only possible to start with a very small quantitative basis of information. Analysis had to be almost entirely based on qualitative values. But this in turn made it necessary to find a fundamental parameter for the collection of such values. This was found in the specification and diversification of basic human needs from his or her environment. This essential question, "what rights does man have towards his urban environment?" turned out to be a central and inexhaustible key in the practical implementation for the understanding and satisfaction of various types, deviations, and links of human needs within a city.

Scenarios were developed for the most important needs, transformed into a precise "theme plan" and examined in terms of existing land use possibilities. The results were finally transferred to an urban development plan, which was concretized by the zoning plan, the general transport plan and the urban design plan. In the beginning, there was consequently a qualitative urban development plan, from which the real necessary quantitative urban development planning was developed. The result was an integrated, holistic urban development plan as a synthesis of qualitative and quantitative parameters, which now serves as a model for urban development plans for different cultures such as South America and Asia.

> 这个基本问题——人类对其所处的城市环境到底有何要求——在实践中被证明是理解和满足人类对一个城市的多元化需求的核心和取之不尽的源泉。
>
> *This basic question, "what rights does man have towards his urban environment?" turned out to be a central and inexhaustible key in the practical implementation for the understanding and satisfaction of various types, deviations, and links of human needs in a city.*

注释

1) 波茨坦城市发展规划：Stadtbauatelier，波茨坦市政府；斯图加特，波茨坦，1992/1993年
2) 波茨坦景观规划：Stadtbauatelier，波茨坦市政府与Wendland, Enzmann, Ettel, Kirschnig的建筑师合作；斯图加特，柏林，波茨坦，1992年
3) 波茨坦城市整体文物保护规划：Stadtbauatelier，波茨坦市政府；斯图加特，波茨坦，1997年
4) 波茨坦世界文化遗址影响地区设计导则：Stadtbauatelier，波茨坦市政府；斯图加特，波茨坦，1997年
5) 波茨坦Babelsberg别墅区建设规划：Stadtbauatelier，波茨坦市政府；斯图加特，波茨坦，1997年
6) 波茨坦瑞恩郊区及耶格尔郊区设计导则：Stadtbauatelier，波茨坦市政府；斯图加特，波茨坦，1998年

References

1) Urban development plan for the city of Potsdam: Stadtbauatelier, municipal authority of the city of Potsdam; Stuttgart and Potsdam in 1992/93
2) Urban design plan Potsdam: Stadtbauatelier, municipal authority of the city of Potsdam in cooperation with architects Wendland, Enzmann and Ettel, Kirschnig; Stuttgart, Berlin and Potsdam in 1992
3) Cultural heritage plan for the whole town of Potsdam: Stadtbauatelier, municipal authority of the city of Potsdam; Stuttgart and Potsdam in 1997
4) Visions for the sphere of influence of the world heritage cities of Potsdam: Stadtbauatelier, municipal authority of the city of Potsdam; Stuttgart, Potsdam in 1997
5) Building code plan for the mansion area of Potsdam "Babelsberg": Stadtbauatelier, municipal authority of the city of Potsdam Stuttgart; Potsdam in 1997
6) Building design guideline for the areas "Nauener Vorstadt" and "Jaegervorstadt": Stadtbauatelier, municipal authority of the city of Potsdam; Stuttgart, Potsdam in 1998

城市发展
URBAN DEVELOPMENT

01 世界文化遗产城市波茨坦城市总体规划
The Urban Development Planning of
World Cultural Heritage City of Potsdam

02 丽江玉龙新城规划
Masterplan of Yulong New Town, Yunnan

03 泉州市城市新区建筑天际轮廓控制规划方法研究
Planning of the Skyline of the City of Quanzhou, Fujian

04 泉州市区和环湾地区空间发展战略研究
Strategic Study on Development of the Urban Districts
and General Area around Quanzhou Bay, Fujian

05 广州市南沙光谷地区城市发展规划
Urban Development of Nansha Guanggu in Guangzhou

06 阳宗海旅游度假区西北部概念规划设计
Conceptual Planning of the Northwestern Part of
Yangzonghai Resort, Yunnan

07 东山蝶岛发展战略规划
Dongshan Butterfly Island in Fujian:
the Strategic Planning of Urban Development

08 十堰市东部新城概念规划及城市设计
Conceptual Planning of the East New Town
in Shiyan, Hubei

09 福州市东部新城中心城市设计
City Planning for the Center of
the Eastern New Town of Fuzhou, Fujian

10 杭州天堂鱼生活:杭州市运河新城概念规划
A Fishing Life Paradise in Hangzhou:
Concept Planning for the Canal New Town
in Hangzhou

11 唐山市南湖生态城概念性总体规划设计
及起步区城市设计
Conceptual Planning and Urban Design
for an Ecological Zone in Tangshan, Hebei

12 大同市御东新区概念性总体规划
及核心区概念性城市设计
Conceptual Planning of Yudong District and
Urban Design of the Core Area, Datong

13 文昌市抱虎角概念规划
Conceptual Planning for Baohujiao Area
in Wenchang, Hainan

14 昆明市绿地系统概念规划
Concept Planning for Kunming Green Space System

02 丽江玉龙新城规划
Masterplan of Yulong New Town, Yunnan

规划面积	AREA	490 hm²
人口规模	POPULATION	40000
完成时间	PROJECT DATE	2005

丽江古城 The old town of Lijiang

Lijiang, the historic trading town and former residence of the kings of the Naxi folk, is located at an altitude of 2400m, in the midst of dramatic mountain scenery in the southeastern foothills of the Himalayas.

A range of rivers, streams and water canals run through the city and create an equally unique and picturesque townscape. The different streams are fed from a mountain river, forming the complex irrigation system of the city, which is still used today and to which the city owes its name. Lijiang means "city on the beautiful river."

The city was built in the 12th Century by the tribe of Naxi, descendants from Tibetan nomads and who possess a distinct and rich culture, even its very own writing, which is the last living symbol-based language in the world.

Lijiang is one of the best-preserved old towns in China and is considered a prime example of traditional urban construction. In 1997, the city of Lijiang, the surrounding landscape and the pictograms were included in the list of World Heritage Sites.

丽江，历史上的贸易城市，亦为纳西族文化的王国，坐落在喜马拉雅山麓，海拔约2400米。

河流、小溪和水渠织成的网穿城而过，形成画卷般的城市意象。山涧汇入不同的河道，形成完整的传承至今并赋予这座城市名字的完整水系。"丽江"的意思就是"美丽河畔的城"。

这座城市在12世纪由藏族牧民的后裔——纳西族人建成，形成了独立而丰富的文化，还有自己的文字。纳西文字亦是世界上仅存的尚具有生命力的象形文字。

丽江是中国保存最好的古城之一，被视为中国传统建筑的典范。1997年丽江及其周边景观同象形文字一起被列入联合国教科文组织世界文化遗产名录。

山涧汇入不同的河道，形成传承至今并赋予这座城市名字的完整水系。"丽江"的意思就是"美丽河畔的城"。

The different streams are fed from a mountain river, forming the complex irrigation system of the city, which is still used today and to which the city owes its name – Lijiang means "city on the beautiful river".

图 01　位于丽江古城南部的玉龙新城
Fig. 01　Location of the new city planning area in the south of the old town

图 02　历史上丽江的市中心
Fig. 02　The historic city center of Lijiang

城市发展 URBAN DEVELOPMENT

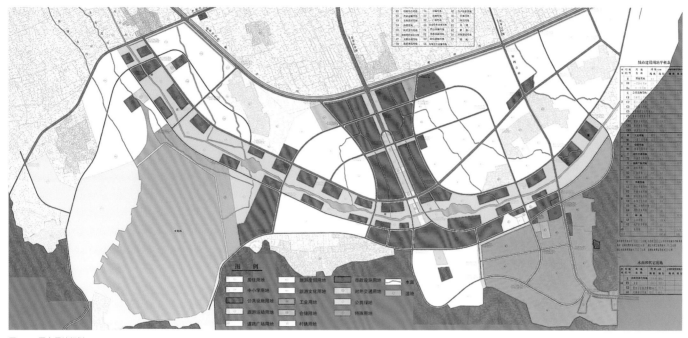

图 03 原有用地规划
Fig. 03 Predetermined structural concept

上位规划 Subordinate planning

为适应人口增长、旅游业的兴旺以及城市化进程等方面的要求，在云南省对丽江市总体规划编制结束后，丽江的扩建规划也提上了日程。扩建的目的是为缓解大量游客的涌入对老城核心区基础设施的损害。为保护城市本来的意象，人们决定以玉龙为名建造一座距丽江古城5.5公里的新城。新城与古城将以四条道路来连接。

新城位于南部高原的末端，它的南部和西部被山川环绕。和古城一样，新城也在5000米高的雪山影响范围内。另一个主导元素是水。当地溪涧众多，一条大河从西向东流过，西边是优美的文笔湖。待建区占地约490公顷，规划常住人口40000人。除了居住区外还需要规划政府所在地和一些度假村。

云南省规划院同时制定了规划结构方案，其中涵盖功能用地和道路体系。该设计的主要象征是两条相交的轴线，南北方向轴线是以政府建筑为端点的大道，东西方向则围绕河流形成被称为"玉带"的轴线。地方规划部门的上级规划结构方案完成后，组织了新城规划城市设计的国际竞赛，德国意厦与韩国BAUM事务所和北京市建筑设计研究院脱颖而出，并成为该项目后续工作的承担者。

According to the development plan of the Yunnan province, an extension to the city of Lijiang was scheduled in order to absorb the population increase due to the development of tourism and the urbanization process. Furthermore, it will relieve the overburdened infrastructure of the historic core because of tourists. In order to protect the largely preserved cityscape, it was decided to build a new city instead of extending it. An independent new city called Yulong, located 5.5 kilometers from Lijiang, was planned, which would be connected to the existing streets via four new ones.

The site for the new town is located at the southern end of the plateau to the southwest, and is surrounded by mountain ranges. As with the old city, this location is dominated by the more than 5000 meters high "Snow Mountain." Another dominant element is water. In addition to numerous small streams, a slightly larger river crosses the area from west to east, while there is also a small lake to the west of the location. The total building area has a size of about 490 hectares and is expected to take up the seat of government in the city, as well as the capacity for 40,000 inhabitants and some vacation resorts.

The Yunnan province provided a structural concept as a largely binding framework, in which the general land use zoning and overall roads were already established. The main feature of the concept is two intersecting axes, a major highway in the north-south direction with the seat of government as an end point, and the area called the "band" around the river, which crosses from west to east. After local authorities initiated development of the overall structural concept, an international competition was commissioned for further development of the plan. These are the submissions from the ISA urban planning studio together with the Korean office TREE and the Beijing planning institute.

为保护历史城市原真性的面貌，人们决定以玉龙为名建造一座距丽江古城5.5公里的新城。

In order to protect the largely preserved cityscape, it was decided to build a new city instead of extending it. An independent new city called Yulong, located 5.5 kilometers from Lijiang, was planned.

从两种文化中学习 LEARNING FROM TWO CULTURES

图 04 丽江古城现状城市肌理（左）、玉龙新城规划城市肌理（右）
Fig. 04 Existing building structure of the old town of Lijiang and planned building structure of the new town Yulong

心心相印：城市设计与建筑设计的理念
Soul mates: a guiding principle for urban planning and architecture

The stunning beauty of the old town – exciting urban spaces with its narrowness, dramatic views and directions; surprising visual links; perfect union of water, plants, landscape and buildings; its wealth of architecture – was the reason for the ISA urban planning studio to develop the new city concept from a reinterpretation of the tradition of the Naxi people.

"Soul mates" was a guiding principle for the draft, as a copy of the historic city was not supposed to be – and could not be – created. Instead, a modern city was the goal, one that keeps the spirit of the Naxi culture alive in our time. By using in-depth analysis principles, the old construction methods, which are also suitable for building in the present time, could be expanded upon both in the fields of urban planning and architecture. In this sense, the planning of new town Yulong altered the following features of the old town, according to the framework of a modern city:

- Mixed use consisting of residential, work, leisure time
- Compactness of the urban structure, resulting in dense and lower buildings
- Closed, space-creating street walls
- Dynamic building structures with a space-enclosing effect
- Lively space successions and sequences
- Perceptibility of mountain scenery, especially of Wenbi Mountain and the Snow Mountain
- Differentiated building groups
- Integration of the natural elements of water and greenery into the city
- Architectural diversity within a common architectural language

古城的美丽在于它那犹如被大自然精细雕琢而成的富于变化的空间、令人惊喜的视角关系，它们构成紧凑的城市空间，景观、建筑、植物和水体组成完美的一体。这些，都是地方建筑的宝贵财富，也是意厦为玉龙新城设计一个融合纳西传统文化的方案的动因。

"心心相印"是设计的主题，所以新城设计不应该是对历史古城的简单模仿，而应当是建造一座传承了纳西文化精神的现代之城。通过深入的分析，可以得出一系列既能满足现代城市建设规范要求，又能体现古城传统风貌的城市规划和建筑设计原则。从这个角度来说，玉龙新城的规划将尊重以下的丽江古城特征，并将之纳入现代城市的框架之中：

- 居住、工作和休闲功能的混合布置
- 城市结构的紧凑性，较低的建筑高度和较高的密度
- 连续的空间界面，围合的空间
- 曲折变化的空间界面
- 生动的空间序列
- 山地场景的体验，尤其是文笔山和雪山风光的体验
- 生动的建筑群
- 自然元素、绿化和水体的融合
- 统一建筑语言框架内的建筑多样性

"心心相印"是设计的主题，所以新城设计不应该是对历史古城的简单模仿，而应当要建造一座传承了纳西文化精神的现代之城。

"Soul mates" was a guiding principle for the draft, as a copy of the historic city is not supposed to be – or could not be – created. Instead, a modern city was the goal, one that keeps the spirit of the Naxi culture alive in our time.

图 05 用地规划方案（左）、景观规划方案（右）
Fig. 05 Land use concept and landscape concept

总体规划设计：绿树和流水掩映下的生动城市空间
Master plan: vibrant urban spaces with greenery and water

基于以上原则，设计保留并拓展了现状河道，由此水体就能和南部景观空间内的绿化一起成为城市的主导元素。从边缘到中心，建筑平缓地融入景观之中，建筑高度由一两层逐渐递增至最高四五层（也有少量的七层建筑）。避免建设高层建筑可以保证城市的天际线效果不会妨碍周边的自然山地景观。

为塑造和维护生动活泼的城市特征，用地功能规划以古城为蓝本，采用功能混合的用地模式。规划重点功能区域中，南北向轴线主要提供行政用地，但也混合有一定的文化商业设施。为了使已规划轴线的纪念性不过于强烈，建议将开放空间整体塑造为城市公园。

作为中央轴线的补充，规划了一条曲折的东西向步行轴线，其上分布有零售业、餐饮业和酒店。这条轴线在南侧穿越整个城区，从东部现状村庄一直延伸到西部的玉龙湖。中央轴线西面的步行路线沿河流展开，在河岸北部形成城市林荫道，河流东段则继续维持生态绿地原状。

新城街道的首要功能不是服务于汽车交通，而是作为城市公共生活的空间。在原有用地规划中边长约为600~800米的大型地块被分割成小地块，以便更精确细致地构建功能网络和公共空间。大部分街道被设定为交通管制和混合功能道路，所有的道路使用者都有同等路权，在该地段汽车只能以接近于步行的速度行驶。

Based on these principles, the overall plan of the ISA urban planning studio aims to not only maintain most of the existing watercourses but to complement them, so that the water is in conjunction with the landscape of invading greenery from the southern region, a prominent element of the city. The area to be settled is gently inserted into the landscape with one to two-story structures along the edges, which rise slowly to a maximum of four to five, or in special cases to a seven-story building in the center. Because they are low-rise buildings, the silhouette does not stand out against the surrounding mountain landscape.

Based on the relaxed zoning laws model of the old town, the structure concept suggested a mixture of the various functions in order to maintain a vibrant and urban character. Yet the main focus was on its usage in distinctive locations. The main north-south axis of the city-planning concept mainly includes administration use; however, it should be supplemented with cultural institutions. To reduce the monumentality of the predetermined axis, an urban park has been suggested for open space.

As a complement to the powerful government center axis, a smaller, repeatedly angled pedestrian axis with retail shops, restaurants and hotels will be provided. That area diagonally extends all the way from the preexisting villages to be preserved in the east, up to the Yulong Lake in the west. The pedestrian link runs along the river, west of the government center, which gets an urban promenade on the northern bank. The east part of the river, however, is left to nature as part of a large green area.

The streets of the new city are primarily not designed for automobile traffic, but for the main areas of public life. The structural concept defined by super blocks of about 600 to 800 meters are subdivided into smaller blocks in order to improve the urban network and to provide more public spaces. Most streets will be designated as mixed traffic areas, which means that all users are fundamentally equal and allowed to drive cars only at walking pace.

避免建设高层建筑可以保证城市的天际线效果不会妨碍周边的自然山地景观。

Because they are low-rise buildings, the silhouette does not stand out against the surrounding mountain landscape.

从两种文化中学习 LEARNING FROM TWO CULTURES

图 06 中央轴线透视图
Fig. 06 Perspective of central axis

从地块到邻里街区 From blocks to neighborhood quarters

In order to create manageable neighborhoods and to ensure a relatively small-scale and diversified development structure, the territory was divided into sections with a dimension of about two and a half hectares. Neighborhood quarters were designed as closed units following a rule: a quarter consisted of six or seven blocks, totaling approximately 3000 square meters. In the center, each quarter had a local or street square.

The system of blocks was in turn based on a modular system of different buildings types in order to take into account all the diversity of structural forms and economic interests. Each block must contain at least four different types of buildings. Repositioning the buildings led to a lively, salient structure line with niches, expansions and contractions, which grant the street a varied, spatial rhythm.

A neighborhood quarter is accessible only from outside drivable streets, and parking is located in garages under the courtyard. In this way, the neighborhood quarter is car-free in the interior and creates a coherent pedestrian and cycle path with small spaces for neighborhood meetings, play areas for children or simply comfortable seats along the streams and pools.

> 为实现建筑形式的多样化，并加之对生态因素的考虑，邻里街区组团建立在建筑类型"模块化系统"的基础上。
>
> *The system of blocks in turn was based on a modular system of different types of buildings in order to take into account all the diversity of structural forms and economic interests.*

为创造清晰具有可掌控尺度的邻里关系和相对细致、多样的建设结构，地块被划分为约2.5公顷的规模。每个新地块都是独立的单元，通常由面积为3000平方米的6～7个邻里街区组团组成。每个单元中央是住宅区广场或者街边广场。

为实现建筑形式的多样化，并加之对生态因素的考虑，邻里街区组团建立在建筑类型"模块化系统"的基础上。每个模块至少包含四种建筑类型。通过建筑类型变化获得灵动跳跃的建筑界面，营造沿街道层次变化丰富的空间节奏感。

每个邻里街区对外部闭合，停车位通过地下车库解决。通过这种方式，住区内部实现了无车辆通行的目标。所有的邻里街区之间的公共空间一起构成步行及自行车网络，其间包括依水而建的休闲小广场、儿童游乐区或者简单的座椅。

 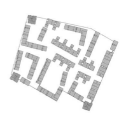    

开放流通的底层公共空间与相对封闭的街坊　　模块化的居住单元　　半封闭的邻里街区　　地标建筑　　6～7个邻里街区　　新地块

图 07 典型街区由6～7种组团和建筑类型组成
Fig. 07 Neighborhood quarter composed of six or seven blocks and building typologies

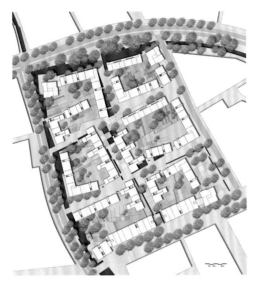

图 08 典型邻里街区建筑布局
Fig. 08 Building structure for a neighborhood quarter

图 09 居住组团内部街道空间透视图
Fig. 09 Typical residential street in a neighborhood quarter

新风格中的建筑传统 Building traditions in a new style

传统的纳西建筑多采用土木结构。厚重的石墙之间穿插轻质、精细的木结构，结合水平与垂直分割的和谐转换，形成有力的、雕塑感极强的立面。与之相呼应的是以白色粉刷墙体为主体材料，结合不同颜色的天然石材，以及漆成红色或棕红色的木材。

新城的建设要用新的形式和新的载体阐释这些传统建筑的基本原则，例如通过类似的材料和方式、同样的立面分割手法，形成有力的、雕塑感极强的立面，同时结合一些细节的借鉴，如檐口边缘以及街区入口空间等。

从这些原则中发展而来的建筑类型具有紧密的亲缘关系，同属一个建筑体系。所以这些变化丰富的建筑并不突兀，而是具有和谐统一的特性，将新城的现代性同古城的传统性巧妙地融合在一起。

The traditional Naxi architecture is a very lively and diverse mixture of massive buildings and wood frame construction. Heavy and plastered stonewalls alternate with light, delicate wooden construction, which forms strong and plastic facades with a balanced exchange of horizontal and vertical layouts. Accordingly, the dominant local materials are walls plastered in white, natural stone in various hues, and wood painted in red and auburn colors.

As for the construction of the new city, the fundamental principles of traditional architecture have been interpreted in a new form and translated into modern buildings. These criteria include similar materials and construction methods, a strong plastic horizontal and vertical language for the facades, and the application of similar detailed solutions like roof edges or an entrance situation.

The principles developed from these types of buildings have a strong affinity with each other and thus merge into a whole subordinate building. In this way, a uniform cityscape would be guaranteed to have diversity within, which is a modern counterpart to the historic old town.

新城的建设要用新的形式和新的载体阐释这些传统建筑的基本原则，将新城的现代性同古城的传统性巧妙地融合在一起。

As for the construction of the new city, the fundamental principles of traditional architecture have been interpreted in a new form and translated into modern buildings.

图 10 典型街区剖面图
Fig. 10 ross section of a typical neighborhood quarter

从两种文化中学习 LEARNING FROM TWO CULTURES

图 11　古城及新城的公共空间
Fig. 11　Public space in the old and the new city

图 12　丽江古城公共空间内的水畔回廊
Fig. 12　Dealing with water in a public space in Lijiang old town

图 13　古城及新城的街区入口空间
Fig. 13　Entrance situation in the old and the new city

公共空间 Public space

In the old town, public space is treated as if it were a building in itself. With stairs, bridges, ramps, built-in plants, and retaining walls, they create a unique architecture of exceptional charm. This pleasantness reconciles the distance between public space and buildings so that houses, streets, bridges and water plants constitute a harmonious unity, as if they are unique to this place.

老城中，公共空间被当做一个独立的建设对象来处理。公共空间与建筑之间的楼梯、小径、斜坡、盆栽和挡土墙等各个要素散发出独特的魅力和房屋、街道、桥梁、水体与植物一起组成和谐的整体，形成独一无二的风格。

老城的公共空间被当做一个独立的建设对象来处理，与建筑组成和谐的整体，形成独一无二的风格。

In the old town, public space is treated as if it were a building in itself. With stairs, bridges, ramps, built-in plants and retaining walls, they create a unique architecture of exceptional charm, which reconciles the distance between public space and buildings.

Again, the historical model served as the inspiration for the new city: the public spaces in the new city are not only used to actuate the remaining areas between the buildings, but they form a distinct character and stand in an equitable relationship with the blocks. The blocks of buildings determine the shape of the public space, but the public space also determines the shape of the blocks and the layout of the architecture. Natural stone covers the waterways. It also defines the streets on one side and the base of the buildings on the other. Distortions in the building line or watercourse are used to create built-in plants at the foundation, which form the base for trees, shrubs or flowers.

在这里历史典范仍然作为新城设计的灵感：新城的公共空间不仅是建筑之间的空隙，也有它自己独特的形象，与建筑同等重要。建筑与公共空间互为图底，建筑确定公共空间的形状，反过来公共空间也决定了街区形态和建筑平面。这一规划中，公共空间整体围绕河流展开。天然石材环绕河道，在河道一侧水平延伸形成街道，另一侧则向上延展成为建筑的基座。建筑界面或河道前后错动形成的小空间也被充分利用起来，用来种植花草树木等。

城市发展 URBAN DEVELOPMENT

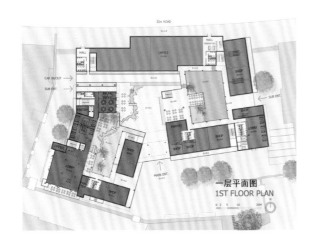

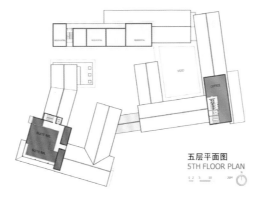

图 14 一期典型建筑竖向功能混合布局
Fig. 14 Vertical mixed-use development

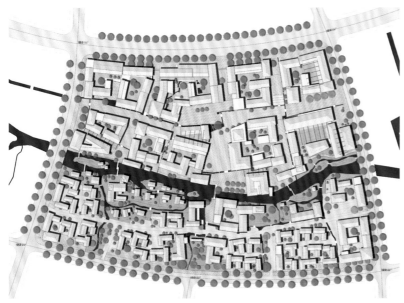

图 15 一期总平面图
Fig. 15 Map of the area of the first construction phase

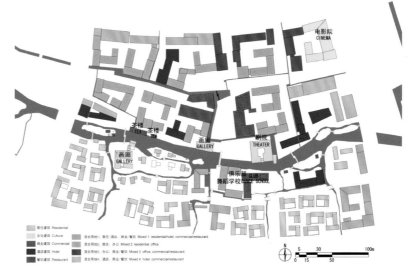

图 16 一期建筑平面功能混合布局
Fig. 16 Horizontal mixed use

图 17 北侧立面序列
Fig. 17 Facade sequences

深化阶段：玉带 In-depth phase: The belt

As a result of the competition, the Beijing Institute and the ISA urban planning studio were commissioned to further develop portions of the city. The Beijing Institute's area includes the region surrounding the government center, while the ISA urban planning studio was put in charge of developing the "belt" along the river. The remaining districts will be realized only in a later phase.

According to the land use concept of the master plan suggested by the ISA urban planning studio, the "belt" area was provided a mixed use in order to develop vivid urban spaces. The western part is designed in an urban way since it connects the central government district with the axis of the Yulong Lake. It has a higher density as well as an increased current trend for city functions such as commercial stores, restaurants, hotels and public buildings. As for the eastern part, however, there is more greenery and it is orientated towards lower density landscaping, as it primarily contains flats for residential use. To the east, at the intersection of the axis and the belt, a five-star hotel is being planned and the ISA urban planning studio has developed its urban form. The final execution planning will come from the Korean office, BAUM Architects.

The main streets of the new town and the government premises were largely finished before the start of construction of buildings. As the first urban district within the belt, the 12-hectare western area is now implemented next to the government center. Upon completion of detailed planning and the accompanying urban permission scheme by BAUM, with assistance from the artistic management of the ISA studio, construction started in the spring of 2006.

The multi-functional uses for this area were then further refined. Structures were kept flexible so that the diverse uses can be mixed within buildings and blocks. This means that different uses can exist side by side and one upon the other.

The focus of the first area was a succession of spaces, which connected the main shopping street with the river and its opposite side. In the square, shops and cafes were to be built, as were cultural facilities such as galleries or theaters. Pictures depicting the construction site as of fall 2007 reflect a first impression of the new town district.

该项目竞赛的结果是，北京市建筑设计研究院以及意厦与BAUM事务所成为了深化设计工作的承接单位。北京市建筑设计研究院负责设计的是环绕中央轴线的地区，意厦与BAUM事务所负责的是沿河流的"玉带"部分。该地区余下的部分将在后续的分期建设中陆续完成。

根据意厦在竞赛阶段提出的用地方案，为了形成生动的富于变化的城市空间，玉带区域采用的是混合用地的模式。玉带的西部地区联系了玉龙湖和中央轴线；中部景观塑造更为城市化，是一个高密度区域，布置了诸如商业、餐饮、酒店和公共设施等城市功能用地。与之相对应的玉带东部地区则以自然景观为导向，广布绿地和低密度住宅。中央轴线与玉带交叉点的东部，计划建设意厦设计的五星级酒店，并与BAUM事务所合作完成了具体的实施设计。

新城干道和政府大楼的设计在规划初期已经完成。中央轴线西侧的一块面积约12公顷的用地作为玉带区域的一期建设用地也已经得以实施。深化设计完成后，意厦作为总体艺术监督，与BAUM共同完成了建筑设计工作，由云南省建筑设计院进行施工图配合并于2006年春季开始进行建设。

在一期用地中，功能混合的概念得到了进一步的完善。建筑的设计更为灵活，便于平面和竖向的功能混合使用。这意味着多种功能可以相邻或相叠。

在一期用地的中央形成了一个广场序列，将主要购物街与河流及河对面的区域联系起来，环绕广场的是小商店、咖啡屋以及画廊、剧院等文化设施。一期已于2008年建设完成。

城市发展 URBAN DEVELOPMENT

图 18　建设中的玉龙新城
Fig. 18　The new town under construction, Yulong

城市发展
URBAN DEVELOPMENT

01 世界文化遗产城市波茨坦城市总体规划
The Urban Development Planning of World Cultural Heritage City of Potsdam

02 丽江玉龙新城规划
Masterplan of Yulong New Town, Yunnan

03 **泉州市城市新区建筑天际轮廓控制规划方法研究**
Planning of the Skyline of the City of Quanzhou, Fujian

04 泉州市区和环湾地区空间发展战略研究
Strategic Study on Development of the Urban Districts and General Area around Quanzhou Bay, Fujian

05 广州市南沙光谷地区城市发展规划
Urban Development of Nansha Guanggu in Guangzhou

06 阳宗海旅游度假区西北部概念规划设计
Conceptual Planning of the Northwestern Part of Yangzonghai Resort, Yunnan

07 东山蝶岛发展战略规划
Dongshan Butterfly Island in Fujian: the Strategic Planning of Urban Development

08 十堰市东部新城概念规划及城市设计
Conceptual Planning of the East New Town in Shiyan, Hubei

09 福州市东部新城中心城市设计
City Planning for the Center of the Eastern New Town of Fuzhou, Fujian

10 杭州天堂鱼生活：杭州市运河新城概念规划
A Fishing Life Paradise in Hangzhou: Concept Planning for the Canal New Town in Hangzhou

11 唐山市南湖生态城概念性总体规划设计及起步区城市设计
Conceptual Planning and Urban Design for an Ecological Zone in Tangshan, Hebei

12 大同市御东新区概念性总体规划及核心区概念性城市设计
Conceptual Planning of Yudong District and Urban Design of the Core Area, Datong

13 文昌市抱虎角概念规划
Conceptual Planning for Baohujiao Area in Wenchang, Hainan

14 昆明市绿地系统概念规划
Concept Planning for Kunming Green Space System

03 泉州市城市新区建筑天际轮廓控制规划方法研究

Planning of the Skyline of the City of Quanzhou, Fujian

规划面积 AREA	11015 km²
人口规模 POPULATION	7280700
完成时间 PROJECT DATE	2008.02

项目介绍 Introduction: The Project

The city skyline reflects a city's physical character, the changes of culture and life style, and the quality of life. In China today, with its rapid pace of urbanization, the characteristics of city skylines are becoming increasingly blurred. The relational dialogue between urban architecture and the natural landscape has been all but ignored, and at this point, there are just too few in depth studies on urban silhouette planning in China and abroad to rely on. This project, done for the city of Quanzhou was a pilot project taking the effect of landscape on urban space as its starting point and reinterpreting the landscape with regard to the city skyline.

城市天际轮廓线——一座城市呈现在人们面前的城市界面形态，在反映城市的风貌与变迁的同时，折射出城市的文化底蕴、风格与品位。在当前中国新城建设飞速发展的阶段，城市天际线作为城市名片和品牌的独特性日益模糊，城市建筑轮廓线与自然山水轮廓线的对话关系也被忽视，甚至相互对抗。而以城市天际线控制规划方法为课题的研究，目前国内外均相对不足并且缺乏针对性。在此形势之下，本课题研究尝试以城市空间景观效果为出发点，以泉州作为研究与实践的对象，给予城市天际景观另一个视角的解读。

图 01　泉州中观天际轮廓形态
Fig. 01 Cityscape of Quanzhou on the medium level

拓展：城市天际轮廓线的概念 Development: Concept of city skyline

According to Professor Michael Trieb, "the skyline is the horizontal line, where the sky and the earth seem to meet. In cities, the skyline is the overall structure composed of high buildings, or local landscape, often consisting of many skyscrapers and creating an artificial line" In his book "The Theory of Cityscape Form", Trieb posits that a city's silhouette forms the point of the harmonious relationship between nature and city.

"天际线所指即为一般的天地相连的交界线。"

"天际线是由城市中的高楼大厦构成的整体结构，或由许多摩天大厦构成的局部景观，天际线亦被作为城市整体结构的人为天际。"

> 城市天际线应该是由自然轮廓与城市建筑群轮廓叠加构成的整体与天空的交界形态面。
>
> **A city's silhouette forms the point of the harmonious relationship between nature and city.**

The approach to the city skyline of Quanzhou was developed from this concept and summarized on two levels.

In a broad macro-level sense, a city skyline should be entirety composed of the natural meeting point between the sky and buildings. For the simple linear elements of a skyline, a city is reduced to the plane element of the

当前城市天际轮廓线的概念多元且相对含糊，定义的角度和层次各有所侧重。本次研究以 Michael Trieb 教授的城市天际景观形态理论为基础，从自然和城市的和谐关系着眼，将天际轮廓线的概念进行了层次和内容上的概括和拓展。

广义上(宏观层面)：城市天际线应该是由自然轮廓与城市建筑群轮廓叠加构成的整体与天空的交界形态面。

城市发展　URBAN DEVELOPMENT

图 02　泉州宏观天际轮廓形态
Fig. 02　Skyline of the Quanzhou on the medium level

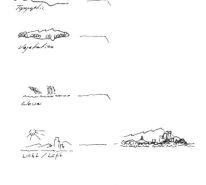

图 03　天际线构成要素
Fig. 03　Elements of the skyline

广义层面将城市天际线由单纯的线性要素推演到了城市的景观面状要素，使其概念得以充实和丰富，同时强调自然与城市建设的联并考虑。

狭义上(中观和微观层面)：根据人所处的城市位置，城市天际线可理解为城市中的建筑和各种构筑物以及自然山水树木等与天空交界的轮廓线。

此外，本次研究过程着重强调的是，不管是广义、狭义，宏观、中观还是微观，都应考虑自然因素在城市中不容忽视的地位和作用，好的天际线更应该尊重自然并与自然山水配合得当，这也与ISA集团长期贯彻的生态规划理念相得益彰。

landscape. The concept is enriched in a broad sense, while emphasizing the relationship between nature and urban development.

In a narrower medium or micro-level sense, a city's skyline should represent the harmonious coexistence of sky, buildings and natural landscape unique to that city's location. Therefore, whether viewed from the macro or micro-level, the position and role of natural factors should be respected throughout the planning process. A strong skyline should coordinate natural elements with man-made elements of the city in a way that is consistent with ecological planning.

113

工作层面	控制要素	景观要素				观景要素			
		自然影响要素	现状人工环境影响要素	重要人文历史要素	未来规划发展前景	自然	现状人工环境	重要人文历史观景点	未来规划前景
总体层面	控制要素选择	大型水体、山体、湿地、绿地	城市入口、重要功能区域、城市重要界面、现状大型基础设施、现状大型公共建筑	重要古代建筑及人文历史、名胜保护区域	重要城市和自然的交界面、规划重要功能区、门户区、城际间市政廊道规划、土地利用规划、道路系统规划、桥梁等城市基础设施规划、城市大型公建的设计和公共开敞空间布局	大型山体制高点、谷地、重要大型水上观景点等可以提供多重观景效果（鸟瞰，远景，仰视等）的观景点	城市级别重要道路、桥梁、大型公建等观景路径和观景点	1.为重要历史观景区域提供整体观赏的可能场所 2.历史上重要的特殊文化活动场所（观海峰，祈风处）	规划重要道路、桥梁、海运线路等观景路径
	工作方法	——确定山体、水体等自然要素与城市的关系；与观景点联合考虑规划景观廊道、景视线面等 ——确定优势与问题后提出发展原则与建议	——分析现状大型人工景界面、节点的整体结构 ——确定优势与问题后提出发展原则与建议	——确定重要历史文化保护区、名胜风景保护区等人文价值突出区域 ——确定优势与问题后提出发展原则与建议	——结合其他影响背景综合确定城市天际线的典型气质、整体塑造原则，空间层次关系、大型景观界面和用地、大型景观廊道、景观视线面等，各个分区建筑高度基本配比关系，重要地标区域位置，以及区域建筑群地夜景照明系统	与景观点联合考虑，确定重要的自然观景廊道	确定城市现状的大型观景廊道与界面：滨海散步道、重要城市林荫道	确定具有特殊历史意义和价值的城市重要观景区域	规划重要的未来城市观景路径和观景点，特别是观景角度（鸟瞰，远景，仰视等）的确定
	评价标准	优秀——整体设计与自然配合得当、城市轮廓层次鲜明、连续，自然特色突出、历史人文特征显著，具有统领性制高点且形式丰富，富有变化。 基本满意——整体层次较为清晰，连续性具有统领性制高点需要改善——自然、城市轮廓远近层次不明，缺乏连续性和统领性制高点 需要改善——自然、城市轮廓远近层次不明，缺乏连续性和统领性制高点				优秀——具有大型连续多样的自然型和城市型观景路径，多角度的观景场所、多视角的历史人文型和设计型观景点 基本满意——具有连续的大型自然和城市观景路径，具有多角度的观景场所需改善——缺乏连续的观景路径和多角度的观景 需要改善——缺乏连续的观景路径和多角度的观景点			
控规层面	控制要素选择	1.上一层面天际线规划提出的区域性自然空间塑造原则与建议 2.本规划区域内重要绿地、水体等相应尺度的廊道等要素	1.上一层面天际线规划提出的区域性自然空间塑造原则与建议 地块具有天际线影响要素：建筑高度、建筑体量、色彩、屋顶形式及风格、高层建（构）筑物、大型公共建筑、公共开敞空间分布。街道空间尺度	历史城市空间或重要景观空间所拥有的气质。由此所形成的地块内建筑群体量、高度、色彩、屋顶形式、风格、文化特性、坐标点和地坪标点、历史公共建筑、历史公共开敞空间等元素特征	1.上一层面天际线规划提出的空间塑造要求 2.本层次其他规划（土地利用规划、道路系统规划、桥梁等城市基础设施规划）产生的影响性要素：入口关系、道路尺度关系、建设强度差异、高层集聚点、大型公共建筑分区等	山体、水体、湿地等可以提供多重观景效果（鸟瞰，远景，仰视等）的观景点	现状可提供观景可能的高层建筑、地标建筑、大型公共建筑、公共空间、主要道路、构筑物等	具备观景条件的重要历史建筑、街道、文化标志点、大型公共建筑、公共开敞空间节点、历史事件发生场所等观景点和观景路径	具有观景可能性的高层建筑、地标建筑、大型公共建筑、开敞空间、建筑物地坪标高（±00的确定）
	工作方法	——控制把握景观廊道宽度、走向、空间节点等，控制基本元素符号等，重要路径两侧山水通透关系	——确定现状景观高度轮廓，以及在整体上的空间关系：现有大型公共建筑、公共空间的天际线的关系；重要径空间塑造 ——确定优势与问题后提出发展原则与建议	——确定人文历史街区的整体风貌和天际形态及文化特征等；结合考虑在其基础上，周边环境建设的要求 ——确定优势与问题后提出发展原则与建议	1.结合由其他影响要素所确定的优势、问题及相关的发展原则与建议，综合确定本层面轮廓线控制设计原则 2.规划控制设计地块建筑高度分布、建筑体量、色彩、屋顶形式、风格、高层建（构）筑物、大型公共建筑、广场等要素，重要建（构）筑物地坪标高、空间节点等。建筑物坐标点和地坪标高（包括±00的确定） 3.明确外部公共空间的尺度、气质、重要临街面天际线的韵律与层次塑造方式	确定多路径、多形式的观景类型以及以此为基础寻求景观塑造	确定现状地块内具有观景条件的重要建筑、观景路径和观景点，以及以此为基础寻求景观塑造	确定具有历史意义的重要观景点和观景廊道	控制具有观景可能的地块建筑高度、观景路径等。控制道路两侧建筑退让红线距离
	评价标准	优秀——设计城市轮廓线与整体自然轮廓相协调，与周边区域城市轮廓线相衔接，具有连续性。在规划范围内，城市轮廓在中观和微观层面上具有连续性和多样性，具有区域内的相对统领性制高点，能够突出规划区内的地域和历史人文特色 基本满意——基本能与周边区域轮廓和自然轮廓相协调，在规划区内具有相对制高点。需要改善——规划区内天际轮廓与周边区域和自然层面轮廓配合不当，缺乏连续性和相对制高点 需要改善——规划区内天际轮廓与周边区域和自然层面轮廓配合不当，缺乏连续性和相对制高点				优秀——区域内具有连续的观景路径，具备有地方和历史人文特点的观景区域，具有对周边区域通透的观景视廊和多角度观景区域 基本满意——区域内具有相对连续的观景路径，具备多角度的观景区域需要改善——缺乏连续观景廊道和通透的景观视廊，区域内不具备多角度观景条件 需要改善——缺乏连续观景廊道和通透的观景视廊，区域内不具备多角度的观景条件			

图 04　总体层面和控规层面的城市天际轮廓线控制设计导则
Fig. 04 Skyline planning principles on the level of overall urban and control planning

总结：城市天际轮廓线的评价标准与方法
Summary: The evaluation criteria and methods of the city skyline

The aim of the evaluation criteria and methods of a city skyline was to create a clear and concise expression based on the in-depth theory of urban design as presented by Professor Michael Trieb, as well as our own years of urban design experience. Corresponding analysis and evaluation of multiple famous cities such as Shanghai, Shenzhen, Nanjing, Chicago and Manhattan was conducted. The applied theories and results were then evaluated as in the following example of Chicago's city skyline.

城市天际轮廓线的评价标准和方法是在Michael Trieb教授的城市设计理论基础上的深化与总结，也是集团多年城市设计经验的结晶。我们力求以实践来检验，于是在该标准的基础上，对国内外诸如上海、深圳、南京、芝加哥、曼哈顿等城市分别进行了对应的分析与评价，同时这也是一个理论与实践的结合点。以对芝加哥的城市天际轮廓线评价为例，见118页表。

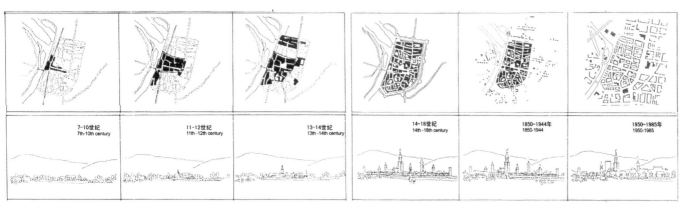

图 05 海尔布隆城市天际轮廓演变历程图：不变的因素是城市景观的重要特征
Fig. 05 Development of the city skyline Heilbronn: tall buildings as a constant factor of cityscape

创新：城市天际轮廓线景观效果的动态视角
Innovation: The dynamic view of landscape's effect on a city skyline

城市天际线作为城市景观极具特殊性，最突出的一点就是它的动态景观效果。当然，这也必然成为本次研究的重点和工作上的一个挑战，同时也是一种全新视角的尝试和摸索。

对于天际轮廓线景观效果的动态性，着重从以下几个方面进行详细的研究与空间模拟：

• 天际轮廓线自身发展的动态演变
• 城市天际线塑造中速度动态变化造成的影响因素
• 城市天际线的观察方位与视角研究
• 城市天际线的夜景效果研究

在此研究过程当中，借助了三维软件的空间效果模拟，并取得了较为满意的研究成果，同时得到了专家和规划界同仁的认可。

通过动态城市景观效果模拟，发现城市天际线的形态并非固定模式，或者说城市设计者和管理者的设计控制意图不过是现实当中天际轮廓线的一种空间形态表象而已，由此认为城市天际轮廓线的控制与设计，必然要以景观点和观景点的相互动态变换关系为基本出发点，这样设计与管理控制工作才是高效而价值最大化的。

A city skyline makes up the specific cityscape; the most prominent element of which is the dynamic overall effect of creating a landscape. This was the focal point and challenge of this project and also a new attempt.

The city skyline has the dynamic effect of landscape.

To create the dynamic effect of a silhouette, detailed research was conducted and space simulations performed analyzing aspects of:
• The dynamic development process due to changes in the skyline itself
• Factors affecting the speed of dynamic changes in the skyline
• Observer position and view corridors
• The effect of the nightscape on the silhouette of the city

With the help of three-dimensional space simulation software throughout the research process, promising results were achieved that were also recognized by experts and planning colleagues alike.

For an example, it was assessed that the biggest dynamic changes to the skyline would occur mostly as a result of observer's changing position. Through the simulation of dynamic cityscape effects, it was discovered that the form of the city skyline is not fixed. In reality, the design intentions of urban designers and planners can only change static spatial patterns. We proposed that city skyline planning should be based on the dynamic relationship of mutual exchange between the observer and landscape, in order to maximize the effectiveness of the planning and final result.

> 城市天际轮廓线的控制与设计，必然要以景观点和观景点的相互动态变换关系为基本出发点，这样设计与管理控制工作才是高效而价值最大化的。
>
> *City skyline planning should be based on the dynamic relationship of mutual exchange between the observer and landscape, in order to maximize the effectiveness of the planning and final result.*

图 06 洛秀组团滨海天际轮廓线改造示意图
Fig. 06 Remodeling of the waterfront skyline of Luoxiu

- 自然山体轮廓线 Outline of the natural mountain
- 自然水体轮廓线 Outline of the natural water
- 城市建设轮廓线 Outline of the urban construction

白天城市的天际轮廓线型　City skyline in the daytime
夜晚强化的城市天际线型　Reinforced city skyline at night

尝试：建议性城市天际线改造设计　Attempt: Designing a new skyline

Throughout this project, theories regarding the skyline and general urban planning practice were fully utilized. The overall urban design of Quanzhou became an interesting link in our application of theory and display of research results. The remodeling of the skyline from different vantage points throughout the city provided new ideas for the overall urban design for all of Quanzhou.

One example was the remodeling of the skyline of the waterfront area Luoxiu

The existing skyline is close to the East Sea groups, with a relatively successful overall visual effect. Yet, as a medium to long-term goal, changing some aspects, e.g. colors used around the bay could improve the overall perception of the area.

(1) Features: The texture of composition is rich and fine. On the left, architectural forms are varied, but the overall picture is quite poor, lacking in unified planning and design. There are elements of the landscape at work however, including the mountains in the middle that naturally form two planes in the distant foreground and background. The right side of Xiutu Harbor's hard coastline also lends characteristic coastal features to the city's backdrop. The coloration of buildings, scale and urban layout are in harmony with the mountains, but the quality of most buildings is poor, and the bare section of mountains causes a diminished overall visual effect.

(2) Planning program:
Left: The unique roof forms would be strengthened to create a rich and gentle skyline.
Middle: The number of tall buildings would be increased at the site of overlapping hills to form three levels based on the characteristic skyline of the original villages. The view of the mountain must be emphasized through protection of vegetation and reforestation. Cutting into mountains for the construction of roads should be carefully controlled and allowed only to a certain degree and under the assumption that the exposed mountain be replanted over time. The large mountain areas would also be linked with the natural green belt areas to the right.

Right: The existing natural hills and green spaces would be used as urban parks including many resorts and cultural facilities. A promenade is planned along the coast to enhance access to the area's natural ecology for tourists and to strengthen its function as a buffer zone for urban expansion on both sides. In the central planning area of the city, the height of some buildings would be increased in order to selectively raise the skyline in some areas.

The final segment: Tall buildings serving as landmarks would be carefully designed to enrich the skyline and to stress the city's maritime function.

理论研究与城市设计穿插互动的工作方法在本次课题探讨过程中得以充分运用，针对泉州的实践性设计工作成为我们运用研究理论与展示研究成果的一个有趣环节，同时对于泉州不同界面的城市天际轮廓线的改造设计也为泉州城市景观控制提出了一个新思路。

以对泉州洛秀界面城市天际轮廓的改造设计工作为例：
洛秀界面与东海组团之间距离较近，视线通视效果较好，其环湾界面肌理、质感、色彩均能较好识别（属于中—远景）。

(1) 特征：界面肌理丰富、细腻，组团建筑形式丰富多样，但其整体性较差，缺乏统一规划和设计；中段天然山体条件较好，自然形成远近两个层次，建筑色彩、尺度和布局与山体较融合，但山体断面裸露，视觉效果不良，大部分建筑品质欠佳；组团以秀涂港硬质岸线为背景，海岸风格特征突显。

(2) 控制方法：强化独特的屋顶建筑形式，形成平缓中丰富的天际轮廓线。

·中段：远近山峰重叠处，增加高层建筑，保证视廊通畅，形成第三个天际线层次。延续原有村落肌理，进行新的城市建设；山体绿化作为组织视线的要素，其重要性必须加以强调，山体植被的保护应引起足够的重视，开山采石行为应控制，已经裸露的山体要及时进行绿化栽植，对大面积的山体断面作景观处理，与右段自然绿带衔接；保持原有自然坡地和绿地作为城市公园，零星布置部分度假村落或文化建筑等，沿岸设景观步行道，强化其自然、生态的旅游品质，并作为两侧城市扩张的缓冲带。在城市中心规划区，提高高层建筑密度，拉高天际线走势。

·末段：布置海洋性特征显著的标志性建(构)筑物，在丰富天际线轮廓的同时强调滨海城市的标志塑造。

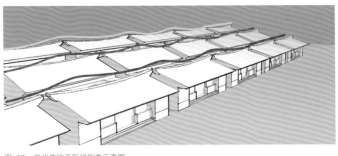
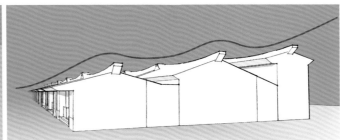

图 07 泉州传统天际线形态示意图
Fig. 07 Diagram of the typical skyline of the Quanzhou

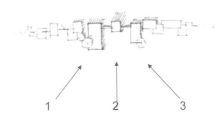

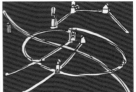

图 08 重点建筑照明对天际线效果的影响效果
Fig. 08 Effects of the nightscape on the city skyline

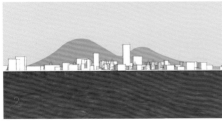
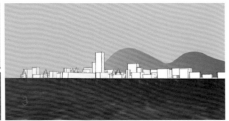

图 09 观察者不同方位的天际轮廓动态变化图
Fig. 09 Dynamic changes to the skyline resulting from the observer's point of view

突破：以观景点和景观点为出发的城市天际线控制方法
Breakthrough: Skyline planning through observation of the natural and cultural landscape

城市天际轮廓线的设计和形成是一个极其复杂的过程，为使城市天际轮廓线的研究成果更为切实地落实在具体的城市规划设计和管理工作当中，以天际线的动态效果研究为基础，景观控制作为出发点，分别从景观点和观景点两大要素对城市天际线的控制要素选择、控制方法指导、设计成果评定三个方面进行具体的列表要求。在各个层面的城市规划和城市设计中给予设计者一个设计方法的选择，同时以表格的形式查询依据。

此外，为将研究成果进一步向更为具体的城市设计当中渗透，在总体规划层面、控规层面、修规层面之外的详细规划设计层面也同样以相应的原则对设计师的工作加以指导和评定，从而在城市天际轮廓线的控制设计中，使设计师和管理者在各个不同的设计层面均能在设计表格中得到相应的设计指导原则、控制方法和评定标准。这也是本次研究工作在控制规划方法上的一次重要尝试与突破。

The design of the city skyline is an extremely complex process. In order to create effective planning tools, that would produce a positive result and enable successful implementation, we prepared a list of specific requirements divided into in three categories that could inform the choice of planning factors for the skyline. The guideline of planning methods and assessment of skyline design would be a starting point for final control of the landscape. This could be a basis for research and useful in choosing the design methods at all levels of urban planning and design.

In addition, the research results permeate the more specific aspects of urban design, guiding designers in the overall planning, control planning, renovation planning and even detailed planning and design. So, through the process of carefully planning the skyline, designers and planners may discover appropriate guiding principles, control methods and criteria at different levels of design and planning. This represents a bold attempt at reaching breakthroughs via the planning process itself.

在此城市设计过程当中，景观点和观景点的相互影响关系成为关注的重点要素。

Through the process of urban design, the interactive relationship between the observer's position and landscape was respected as a basis for inquiry.

The same principles and methods are applied in the urban design of the Quanzhou East Sea Groups. Through the process of urban design, the interactive relationship between the observer's position and landscape was respected as a basis for inquiry. These two ingredients of skyline were integrated respectively into the overall characteristics and level of the city. In this way, the effects of the landscape on the city skyline come into focus, working in harmony with scenic points.

This represents a very new area of research in regards to city skyline planning that is ripe with challenges and opportunities. Theoretical arguments, practical analysis and urban design guidelines served as a basis for planning the Quanzhou skyline. This also enhanced the practical meaning and value of theoretical study that drove the urban design.

Finally, the achievements of this research were recognized and approved by the experts of the evaluation committee. The results from the study were published as a specialized book under the subtopics "study on the continuation and development of South Fujian's (Quanzhou) traditional architectural culture in respect to new construction" and "the study of city skyline planning in the New Urban District".

同时，这样的原则和方法亦在泉州市东海组团城市设计中得以运用。在此城市设计过程当中，景观点和观景点的相互影响关系成为关注的重点要素，在对两景要素提取的过程当中，综合确定城市整体景观特质与层次，最终形成景观点和观景点相互配合的城市天际景观效果。

以上是本次课题研究当中在城市天际线景观规划控制上颇有新意的研究点。在此研究过程中，挑战与乐趣并存，最终研究理论成果和实践分析以及城市设计导则成为泉州天际轮廓线控制管理的理论基础依据，也使得理论研究工作的实践意义得到提升，更见证了理论研究工作在城市设计当中的意义与价值所在。

最终，本研究成果在评审会上得到专家的一致认可与好评，并作为"闽南（泉州）传统建筑文化在新区建设中的延续和发展研究"中的子课题——《城市新区天际轮廓线控制规划方法研究》，以专业书籍出版的形式作为本课题研究的终结。

对于芝加哥的城市天际轮廓线评价　　　　　　　　　　　　　　　　　　　　　　　　　表1

评价要素	宏观层面			宏观、中观和微观层面		
	地方性	文化标志性	统领性制高点	连续性、层次感和节奏变化	多样性	整体性
芝加哥	新城各种形状新奇、色彩各异的高层建筑构成曼妙的城市天际线，独具艺术特色，人文气质突出	现代感十足的高层建筑群代表着整个城市的先进与文明	西尔斯大厦与怡安中心是城市天际线的两个制高点，形成了两个驼峰，未来将建成的福登螺旋之尖将成为城市天际线的第三制高点	天际线以波峰、波谷的形式连接，富有韵律与层次感，节奏抑扬顿挫，却不失连续性	芝加哥的建筑形体各异，有"建筑博物馆"之称，多变建筑屋顶和建筑群天际线融合于整体的波浪形天际线中，和谐而统一	大量的超高层建筑群落降低了远山背景层次的作用，但与海面取得了良好的配合，尤其是夜景，更加富有现代城市的魅力
评价结果	城市天际线自然与人文因素均得到完美的展现，天际线形式现代而独具一格，可以说是较为完美的天际线					

图 10　建筑群与山体的近景关系示意、建筑群与山体的远景关系示意
Fig. 10　Close view of the relations between the buildings and mountains, Distant view of the relations between the buildings and mountains

城市发展 URBAN DEVELOPMENT

 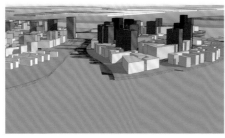 

图 11 东海组团不同角度的城市天际轮廓
Fig. 11 City skylines of Quanzhou East Sea Groups from different points of view

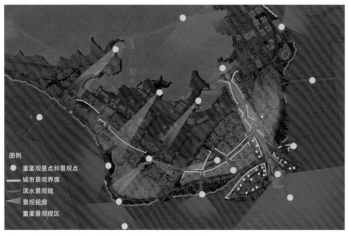

图 12 东海组团观景点、景观点、景观面、景观视廊分析
Fig. 12 Analysis of observation points, landscape points, landscape surfaces, landscape sight corridors of Quanzhou East Sea Groups

图 13 东海组团不同区位层次天际线层面示意
Fig. 13 City skylines of Quanzhou East Sea Groups on different levels

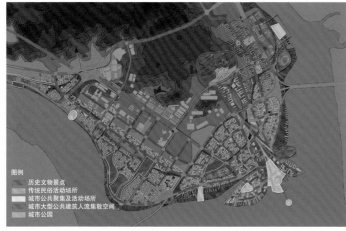 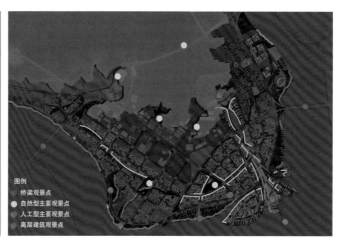

图 14 东海组团景观点及观景场所提取
Fig. 14 Design of the observation points and landscape points of Quanzhou East Sea Groups

119

城市发展
URBAN DEVELOPMENT

01 世界文化遗产城市波茨坦城市总体规划
The Urban Development Planning of World Cultural Heritage City of Potsdam

02 丽江玉龙新城规划
Masterplan of Yulong New Town, Yunnan

03 泉州市城市新区建筑天际轮廓控制规划方法研究
Planning of the Skyline of the City of Quanzhou, Fujian

04 泉州市区和环湾地区空间发展战略研究
Strategic Study on Development of the Urban Districts and General Area around Quanzhou Bay, Fujian

05 广州市南沙光谷地区城市发展规划
Urban Development of Nansha Guanggu in Guangzhou

06 阳宗海旅游度假区西北部概念规划设计
Conceptual Planning of the Northwestern Part of Yangzonghai Resort, Yunnan

07 东山蝶岛发展战略规划
Dongshan Butterfly Island in Fujian: the Strategic Planning of Urban Development

08 十堰市东部新城概念规划及城市设计
Conceptual Planning of the East New Town in Shiyan, Hubei

09 福州市东部新城中心城市设计
City Planning for the Center of the Eastern New Town of Fuzhou, Fujian

10 杭州天堂鱼生活：杭州市运河新城概念规划
A Fishing Life Paradise in Hangzhou: Concept Planning for the Canal New Town in Hangzhou

11 唐山市南湖生态城概念性总体规划设计及起步区城市设计
Conceptual Planning and Urban Design for an Ecological Zone in Tangshan, Hebei

12 大同市御东新区概念性总体规划及核心区概念性城市设计
Conceptual Planning of Yudong District and Urban Design of the Core Area, Datong

13 文昌市抱虎角概念规划
Conceptual Planning for Baohujiao Area in Wenchang, Hainan

14 昆明市绿地系统概念规划
Concept Planning for Kunming Green Space System

04 泉州市区和
环湾地区空间发展战略研究
Strategic Study on Development of the Urban Districts and General Area Around Quanzhou Bay, Fujian

规划面积 AREA		980 km²
人口规模 POPULATION		4100000
完成时间 PROJECT DATE		2009.07

引言 Introduction

With deepening reform and increasing openness to economic development, many new smaller county-level cities are forming around the central cities in the coastal provinces of China. These cities are not only maintaining their original resources, but also injecting a new hegemony into the region as a whole. The development of these cities is having a great impact on the spatial structure and status of the existing central cities, with competitive ability being shifted by the construction and development of new towns.

Such urban phenomena have also arisen throughout the diverse urban districts of Quanzhou and its surroundings. The independent spatial structure of each district, standing in great numbers around Quanzhou Bay, has caused multiple contradictions in the area's urban construction as a whole and decreased the quality for residents. If an overall cohesive functional structure is not provided, the healthy development of Quanzhou's future may inevitably be constrained.

This study was divided into three stages: the analysis of basic index, the determination of development goals and methods, and the elaboration of development policies. First, an analysis of basic index was performed using the SWOT method to ascertain the challenges and opportunities of urban development in Quanzhou. Next, the development goals of Quanzhou were defined, and development methods determined. In the final stage, implementation strategies were presented.

伴随改革开放的不断深化，在我国沿海经济发达省份，许多中心城市周边涌现出许多既保留了原有资源，同时还注入了新的权力的县级城市。这些市县的发育对城市区域空间结构以及中心城区的地位产生了巨大冲击，并使得中心城市和周边市县以及市县之间逐步发展成地区的竞争对手。这也是目前泉州市区与周边地区呈现的关系。环泉州湾地区"诸侯林立"的空间结构与自上而下的经济发展模式导致了城市建设中多重矛盾的产生与人居质量的整体下降。如果不能提供一个具有凝聚力的整体城市功能格局，那么泉州未来的健康发展必然受到制约。

本次研究工作将分三个阶段加以论证，即基础数量与质量指标分析、确定发展目标和模式以及研究发展政策。泉州各项基础指标的分析结果将通过多样化的评估模型加以量化分析，从而寻找泉州城市发展的问题和优势。然后，借助相似地理和产业背景下空间发展模式的比较，制定泉州的发展目标，同时建立若干相适应的空间发展模式供委托方进行选择。在研究工作的最后阶段，还对于前面制定的空间发展模式的实施策略展开了论述。

新的县级城市的发育对城市区域空间结构以及中心城区的地位产生巨大冲击，并使得中心城市和周边市县以及市县之间逐步发展成地区的竞争对手。

The development of these cities is having a great impact on the spatial structure and status of the existing central cities, with competitive ability being shifted by the construction and development of new towns.

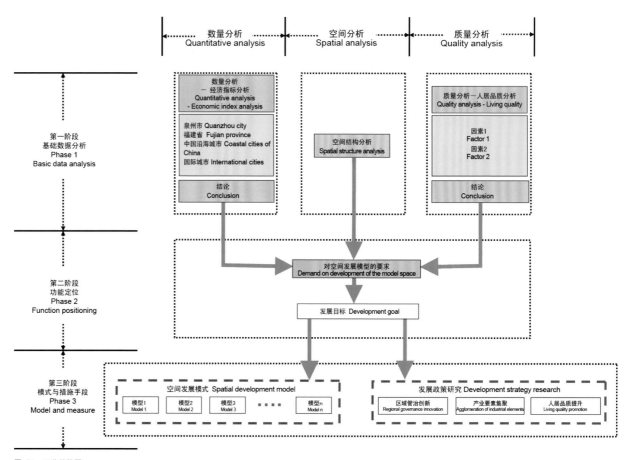

图 01　工作结构图
Fig. 01　Work structure chart

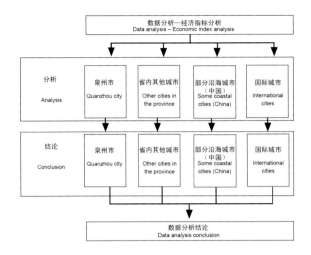

Fig. 02 Quantitative analysis of structure

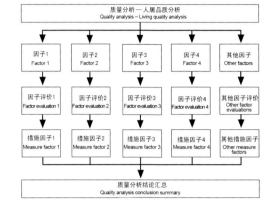

Fig. 03 Qualitative analysis structure chart

泉州基础指标分析 Analysis of Quanzhou basic index

在第一阶段的论述中，将分别从三个方面对泉州城市发展现状加以分析，即数量化分析、质量化分析和空间结构分析。数量化分析的对象是可用数据表达的城市综合发展指标，其分析目的是对泉州综合发展现状进行量化表达和考核。该分析从市域、省域、沿海和国际等四个层面将泉州城市建设现状、城市综合竞争力、其在省内的城市定位和地区经济发展状况等问题加以剖析，多角度地对泉州经济发展现状和潜力、基础设施建设状况、教育与环境等问题加以分析和考核，并利用其分析结果来指导未来泉州城市发展目标的制定。

相对数量化分析，质量化分析则是适用于那些难以借助数据形式表达的考核指标的分析方法，如对城市人居品质指标的分析。针对这一指标，数量化分析仅勾勒出了相对模糊的轮廓，难以再作进一步的描述。因此，借助质量化分析方法中的现场调研、走访等手段，借助完善的评估体系可对城市发展建设的现状作出更为具体和清晰的展示。数量化和质量化两种方法的配合使用，最终充分地展示了泉州城市发展建设的综合情况。

与量化分析结果相结合的空间发展分析，则重点反映了不同历史时期泉州经济、政治发展背景下，结合其特定地理区位条件所揭示的泉州城市空间拓展脉络。空间分析不但给量化分析结果提供了背景依据，而且还概括性地勾勒出泉州空间拓展的潜在方向。

At this stage our group analyzed the current conditions of Quanzhou's development from the perspectives of quantity, quality and spatial structure.

The object of this quantitative analysis was to evaluate the current development situation in Quanzhou, and identify potential development needs that could be indicated through data. The city's existing construction conditions were used to analyze Quanzhou's overall competitiveness in terms of location within the province and regional economic development at the city, provincial, coastal and international levels.

Through this analysis and assessment, potential was identified for Quanzhou's economic development, construction of basic infrastructure, education and environmental problems. The analysis results proved useful in determining the future development goals of Quanzhou.

Qualitative analysis was also used to identify indicators, such as analysis of residential quality, which might not otherwise be expressed through quantitative data. The index from quantitative analysis can only provide a relatively general representation and often does not describe a situation fully. With the assistance of qualitative analysis methods such as on-the-spot research, spot coverage etc., and an excellent assessment system, our team was able to make clearer and more specific diagrams outlining the current state of the urban construction. The integrated use of both quantitative and qualitative analysis helped to fully articulate a comprehensive view of the state of construction and development in Quanzhou.

The analysis of urban development, integrated into the quantitative analysis, focused on Quanzhou's development principles with respect to their economic and political background vis-à-vis geographic location and throughout different historical periods. This not only provided a more robust background in which to view the results of the quantitative analysis, but also helped to outline a potential direction for urban development in Quanzhou.

数量化和质量化两种方法的配合使用，最终充分地展示了泉州城市发展建设的综合情况。

The integrated use of both quantitative and qualitative analysis helped to fully articulate a comprehensive view of the state of construction and development in Quanzhou.

从两种文化中学习　LEARNING FROM TWO CULTURES

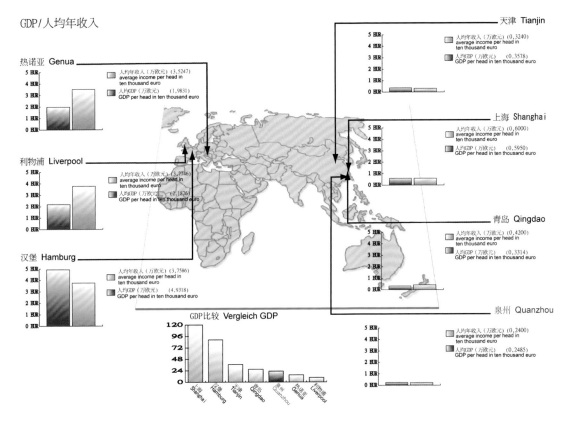

图 04　国际城市数量化分析之经济水平
Fig. 04　Quantitative analysis of internaitonal economic level

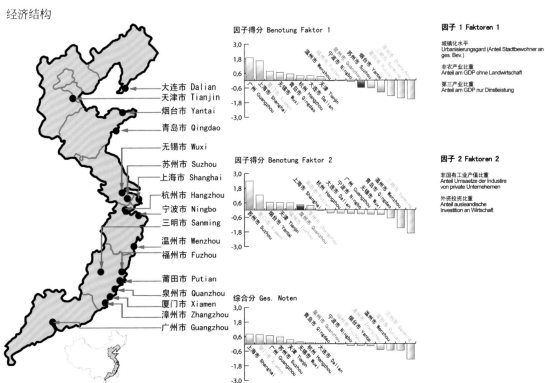

图 05　沿海城市数量化分析之经济结构
Fig. 05　Quantitative analysis of coastal economic structure

经济实力

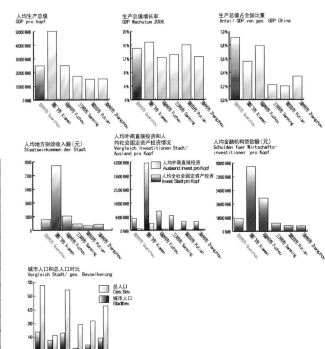

2006年度		泉州市 Quanzhou	厦门市 Xiamen	福州市 Fuzhou	三明市 Sanming	莆田市 Putian	漳州市 Zhangzhou
城市人口(万人)	Stadtsbevölkerung in 10 tausend	193	109	182	28	59	130
总人口(万人)	Gesamt Bevölkerung in 10 tausend	670	160	623	269	307	459
生产总值占全国比重%	Anteil GDP von ges. GDP China	0,91	0,56	0,79	0,22	0,20	0,34
生产总值增长率%	GDP Wachstum 2006	15,00	16,00	12,00	13,00	16,00	12,50
人均生产总值(元)	GDP pro kopf	24847,00	50130,00	24874,00	17181,00	15051,00	15221,00
人均全社会固定资产投资(元)	Investitionen der Stadt pro Kopf	7380,00	41283,00	11755,08	8981,88	5381,13	5214,86
全社会固定资产投资占全国比重%	Anteil Investitionen der Stadt von ganz China	0,45	0,60	0,67	0,22	0,15	0,22
人均外商直接投资(元)	Ausländische Investition pro Kopf	961,36	4166,52	742,35	132,89	215,61	610,81
当年合同外资金额占GDP比重%	Anteil ausländischer Vertragssummen an GDP	7,54	18,60	6,01	2,64	5,69	8,56
人均地方财政收入额(元)	Staatseinkommen der Stadt (Steueretc.)	1388,55	8490,17	185583	819,54	602,21	763,49
地方财政收入占全国比重%	Anteil Staatseinkommen der Stadt (Steuer etc.) an ges. Staatseinkommen China	0,24	0,35	0,29	0,06	0,05	0,09
人均金融机构贷款额(元)	Schulden fuer Wirtschaftsinvestitionen pro Kopf	16833,37	85395,97	43653,19	10673,52	7446,97	7093,96
金融机构贷款额占全国比重%	Anteil Schulden Wirtschaftsinvestitionen von Schulden ges. C.	0,47	2,35	1,14	0,12	0,10	0,14

图 06　省域数量化分析
Fig. 06　Quantitative analysis at the province level

历年地区GDP

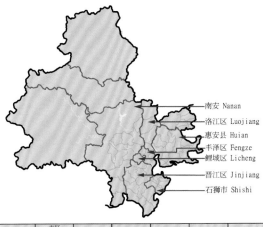

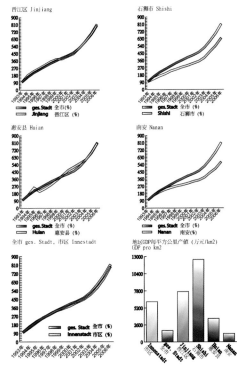

	市区 Innenstadt ges.	晋江区 Jinjiang	石狮市 Shishi	惠安县 Huian	南安 Nanan	全市 Stadt ges.
1993年	410663	618228	315147	275061	439196	2339246
1994年	621750	1119383	417554	447298	695378	3759745
1995年	835043	1422719	525048	692289	874642	4963895
1996年	1042058	1662330	612206	572575	1006967	5882771
1997年	1186722	1916521	725011	676321	1111676	6822648
1998年	1341370	2172424	797512	801997	1199568	7735916
1999年	1451243	2342733	876851	953874	1265839	8498241
2000年	1562206	2378290	941929	1116953	1346515	9310705
2001年	1714996	2580197	1022321	1297242	1445310	9977311
2002年	1847207	2776674	1102740	1409987	1520011	10807497
2003年	2055179	3216048	1242060	1529852	1664083	12141077
2004年	2345940	3698413	1479027	1655009	1927613	14051247
2005年	2732059	4227518	1703261	1890854	2155466	16263034
2006年	3252226	4924963	2005221	2210150	2516742	19007599

图 07　市域数量化分析
Fig. 07　Quantitative analysis at the city level

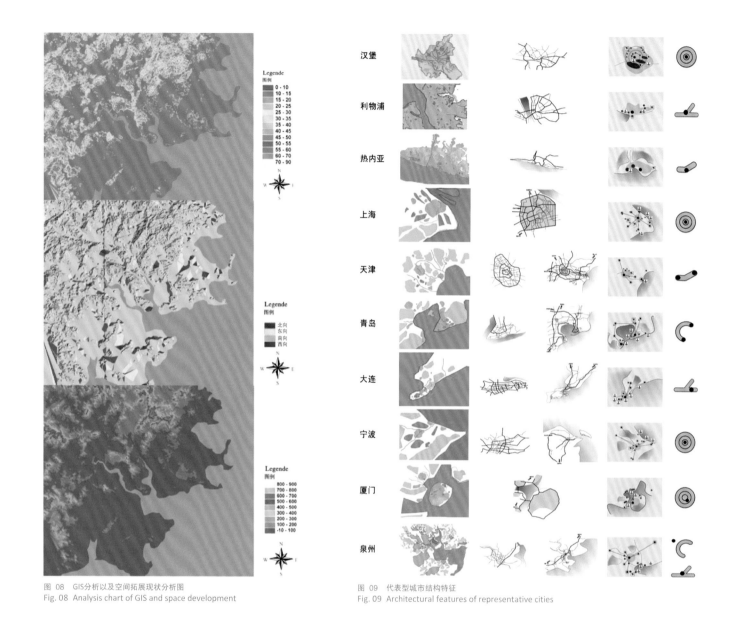

图 08　GIS分析以及空间拓展现状分析图
Fig. 08 Analysis chart of GIS and space development

图 09　代表型城市结构特征
Fig. 09 Architectural features of representative cities

发展目标和模式的确定
Determination of development goals and methods

Based on the results of the analysis, the future development goals of Quanzhou could be determined through several scenarios. A number of development scenarios derived from the development goals have been illustrated and presented, regarding different development conditions and government intervention.

借助基础指标的分析结果，通过论证，泉州未来城市发展目标得以确定。此外，我们针对相似地理、产业和历史背景下系列滨海城市与海滨城市的空间发展轨迹进行了分析，最终确定了若干典型空间模式，在与系列的显示情况比较下，形成具有显示意义的参照性建议，以总体发展目标为导向，考虑对于不同情况下政府可能提供的政策性干预力度的差异，分别制定多个在不同实施条件下的城市空间发展模式，并有侧重地加以说明。

城市发展 URBAN DEVELOPMENT

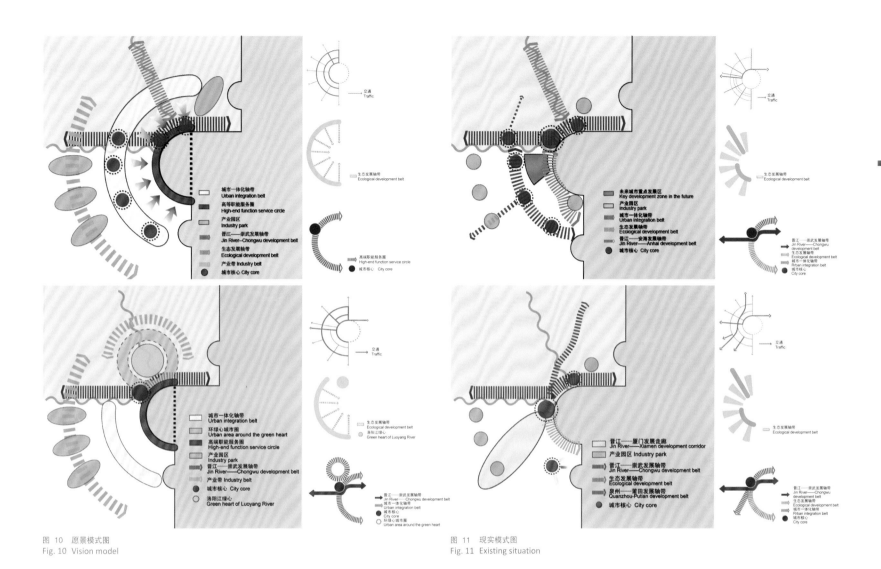

图 10　愿景模式图
Fig. 10　Vision model

图 11　现实模式图
Fig. 11　Existing situation

发展政策研究 The study of development policy

针对发展模式实施中所必需的特定政策扶植条件的区别，分别就不同模式发展所需政策措施手段的具体落实途径加以论述。并着重在区域管治创新、产业要素集聚和人居品质提升等操作方法上提出具体要求。

We searched for specific strategies for each development scenario taking into account the distinction of supporting terms of necessary policies during the implementation of development phase. These strategies stress specific requirements for the innovative operation in regional administration, concentrating on the industrialized building elements and the improvement of residential quality.

通过富有针对性的途径就不同的发展方式实施政策措施。

There are specific ways to implement political measures for the different development methods.

城市发展
URBAN
DEVELOPMENT

01 世界文化遗产城市波茨坦城市总体规划
The Urban Development Planning of World Cultural Heritage City of Potsdam

02 丽江玉龙新城规划
Masterplan of Yulong New Town, Yunnan

03 泉州市城市新区建筑天际轮廓控制规划方法研究
Planning of the Skyline of the City of Quanzhou, Fujian

04 泉州市区和环湾地区空间发展战略研究
Strategic Study on Development of the Urban Districts and General Area around Quanzhou Bay, Fujian

05 广州市南沙光谷地区城市发展规划
Urban Development of Nansha Guanggu in Guangzhou

06 阳宗海旅游度假区西北部概念规划设计
Conceptual Planning of the Northwestern Part of Yangzonghai Resort, Yunnan

07 东山蝶岛发展战略规划
Dongshan Butterfly Island in Fujian: the Strategic Planning of Urban Development

08 十堰市东部新城概念规划及城市设计
Conceptual Planning of the East New Town in Shiyan, Hubei

09 福州市东部新城中心城市设计
City Planning for the Center of the Eastern New Town of Fuzhou, Fujian

10 杭州天堂鱼生活·杭州市运河新城概念规划
A Fishing Life Paradise in Hangzhou Concept Planning for the Canal New Town in Hangzhou

11 唐山市南湖生态城概念性总体规划设计及起步区城市设计
Conceptual Planning and Urban Design for an Ecological Zone in Tangshan, Hebei

12 大同市御东新区概念性总体规划及核心区概念性城市设计
Conceptual Planning of Yudong District and Urban Design of the Core Area, Datong

13 文昌市抱虎角概念规划
Conceptual Planning for Baohujiao Area in Wenchang, Hainan

14 昆明市绿地系统概念规划
Concept Planning for Kunming Green Space System

05 广州市南沙光谷地区城市发展规划
Urban Development of Nansha Guanggu in Guangzhou

规划面积 AREA		783.08 hm²
人口规模 POPULATION		50000
完成时间 PROJECT DATE		2008.06

研究背景 Background

The Nansha District is located in the south of Guangzhou, an area being continuously expanded with mainly heavy industry, some high-tech, and modern service industries. However, cooperation with Guangzhou and Hong Kong should be leveraged to take full advantage of Nansha's existing transport hub in order to bolster the region's existing industries and attract new ones.

The promotion of industrial upgrading, improvement of the overall industrial structure and increasingly efficient use of land and resources were important objectives for the development of Nansha. The planning and urban design was based on a study of technically innovative industries and modern service industries in the Western Industrial Parks and the Information Technology Park south of the Humen highway.

Nansha Guanggu is a hotbed of scientific and technologically innovative industries as well as modern service industry. The size of the area is about seven square kilometers and contains two distinct zones: a fairly large scale high-tech development zone, and the majority of the land dedicated to agricultural use, containing little or no industry.

南沙区是广州市"南拓"的端点，其现状产业结构以重型工业为主，产业结构清晰，发展势头良好。但是，目前南沙区的高新技术产业和现代服务业发展处于起步阶段，需要在现有产业的基础上，进一步发挥南沙交通枢纽和穗港合作的优势，推动产业升级。

西部工业园区和资讯科技园进行科技创新产业与服务业基地的规划与城市设计研究工作，试图回答以下几个问题：1)南沙现有产业基础是否适合于高新科技的发展？2)高新科技产业在南沙的现有背景下应当具体以何种功能形式进行推动？3)这一产业的初步空间单元模式是什么？

规划区域南沙光谷（科技创新产业与现代服务业基地）包含两片基地，总面积约7平方公里。对于一个高科技开发区来说，它的用地规模较大，现状已有一定的工业用地，绝大部仍是空置或可置换用地。

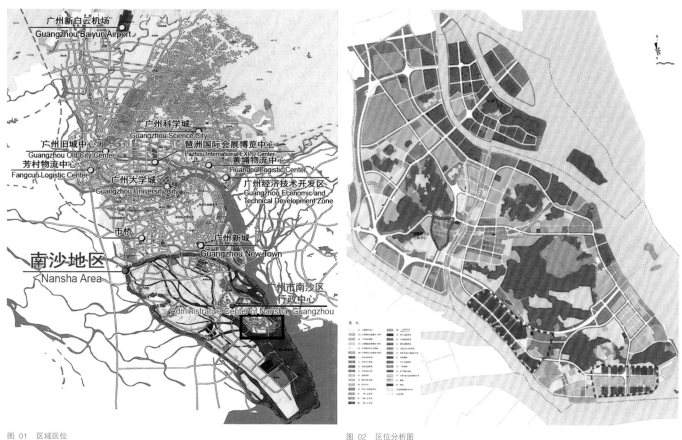

图 01 区域区位
Fig. 01 Regional location

图 02 区位分析图
Fig. 02 Land use

城市发展　URBAN DEVELOPMENT

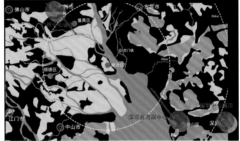

图 03　周边设施——珠三角高科技园区分布
Fig. 03　Surrounding equipment, distribution of the Pearl River Delta's Hi-Tech Park

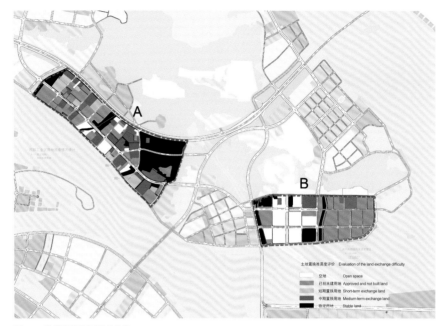

图 04　现状用地及可建设性分析
Fig. 04　Current land use and development potential

在空间规划前的产业定向　Development goal of industries before space planning

在空间规划之前，其难点在于对南沙未来的产业发展方向作出判断。

南沙现状产业结构明晰，有良好的重工业基础。但光谷区块并没有形成明确的产业链和规模企业，高科技产业的总量很小。空间规划需要具体的产业类型来进行支撑。

规划从南沙的周边区位条件及优劣势分析、高科技产业和服务业本质与分布情况、园区发展策略等几个方面入手，因地制宜，强调了园区开发和规划的可实施性。高新技术产业和现代服务业从产业组织上不应该孤立平行发展，而应采取彼此支撑和融合的一体化发展模式，和南沙地区现有重型产业形成更加紧密的有机联系。

此次产业研究从南沙、珠三角和国际三个视野对高新科技产业和现代服务业进行发展评估和预测，并引入效益和约束双重评价系统，在大量的数据收集基础上，使用DEA模型、Weaver-Thomas模型和钻石模型等多种成熟的产业选择模型来确定和校核区域内的主导产业，同时引入线性规划模型和蒙特卡洛随机模型，进行产业用地构成预测。作为理想化状态的数学模型，我们给予一定的弹性，可以在允许的范围内，根据产业发展状况和投资状况进行适量的调整。

Without space planning, it is difficult to determine the future development direction of Nansha's various industries. The current industrial structure of Nansha's heavy industries is unclear and difficult to anticipate. It is therefore essential to define the direction of the area's industrial development, utilizing planning that can promote the development of certain special industries.

There were a number of important attributes to keep in mind such as urban conditions, the advantages and disadvantages of the surrounding area, the quality and distribution of high-tech industries and services, the development strategy of the industrial park, etc. Our plan for Nansha stressed the development of its industrial park in relation to local conditions. The development of the high-tech and modern service industries should not be considered separately but should compliment and integrate with each other, while also organically combining with existing heavy industries. This would determine the timing of the industrial park development.

The future development of high-tech and service industries was assessed via three perspectives: Nansha, the Pearl River Delta and the international perspective, while a double evaluation system of effectiveness and constraints was introduced.

Using a large volume of data, we adopted the DEA (Data Envelopment Analysis) model, the Weaver-Thomas model and the Diamond model along with many other mature industrial models to determine and calibrate the leading industries. The linear programming model and the Monte Carlo random model was also introduced into the forecast of industrial land use constitute. The forecast of industrial land use constitute is an idealized mathematical model, which can be adjusted to an extent according to industrial development and investment with a certain amount of flexibility.

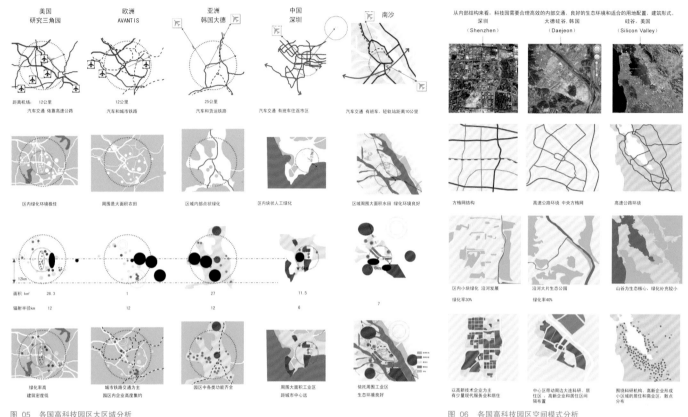

Fig. 05 Analysis of large areas in all development stages of the high-tech park

Fig. 06 Analysis of all development stages of the high-tech park

从国内外成熟的科技园区吸取经验
Lessons from mature domestic and foreign science and technology parks

There are different types of high-tech Parks in the United States, Europe and Asia. China has 53 states run high-tech parks; which for the most part are still in the developmental stages. This study investigated mature high-tech parks from a number of countries with similar conditions to Nansha, as well as several well-developed high-tech parks in China. The contents of the study were as follows:

- Urban and regional development: location, traffic, ecological situation and land use in the areas surrounding the high-tech park which provide educational, scientific research, industrial production, related professional, business and residential, functions, etc.

- Inner structure of high-tech park: communication, ecological situation, land use and building structure

The various elements required for the industrial park, drawn from this analysis, are combined with space patterns to form an abstract space model. A suitable space model for further research was chosen according to the conclusions of the industry-oriented needs and the actual needs of Nansha.

The various elements required for the park, drawn from the analysis, are combined with space patterns to form the abstract space model. According to the conclusions of industry-oriented needs and the actual needs of Nansha, we should choose a suitable space model for further research.

高科技园区在美洲、欧洲、亚洲有不同的构成模式。中国已有的53个国家级高科技园区，大多尚处在发展阶段。本次研究对一批和南沙有一定相似性的、成熟的各国高科技园区以及中国部分发展较好的高科技园区进行较为深入的分析，内容包括：

• 区域空间发展模式——和高科技园区相关，为其提供教育科研基地、工业生产、相关配套、商业、居住等功能的周边区域的区位、交通、生态绿化和土地使用。

• 内部结构——高科技园区内部的交通、生态绿化、土地使用和建筑形式。

通过分析得出发展高科技园区所需的各个要素，并将其和空间形式相结合，得到抽象的空间模型。再根据得到的产业定向结论和南沙的实际情况需求，选择合适的空间模型，并进行进一步深化。

城市发展 URBAN DEVELOPMENT

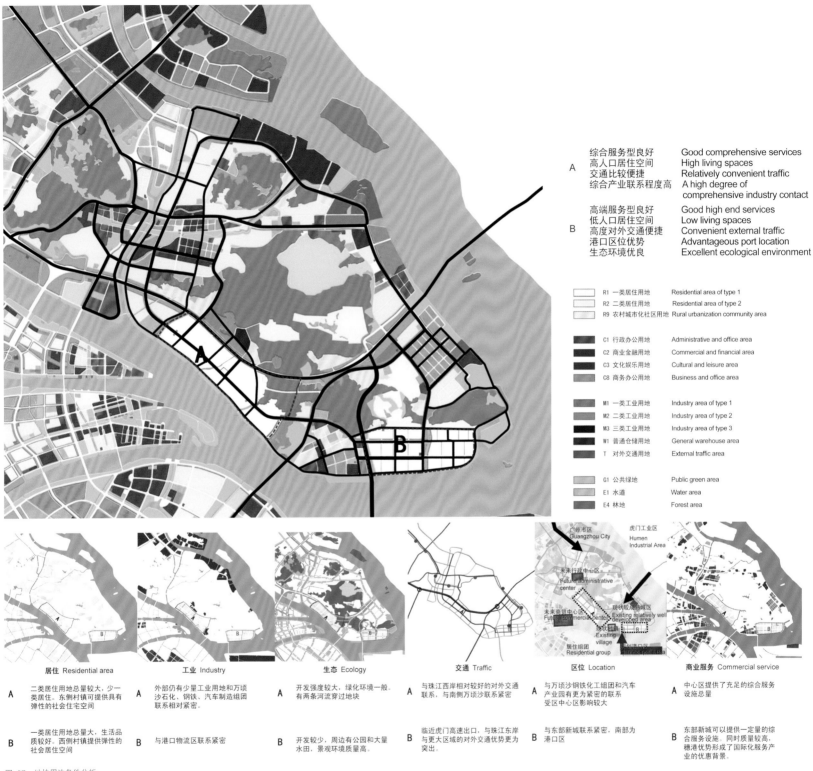

图 07　地块周边条件分析
Fig. 07 Analysis of surrounding area

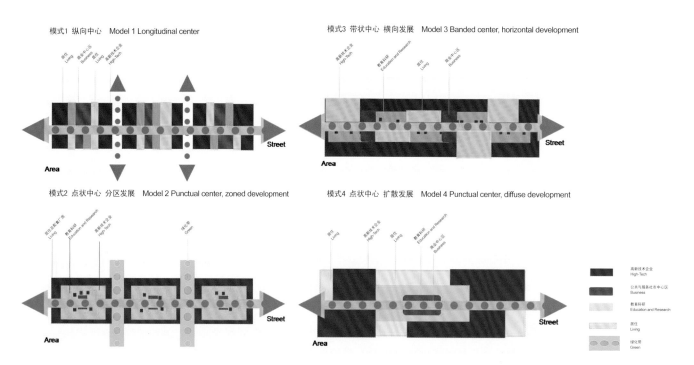

图 08 区内发展模式列举
Fig. 08 Example of development's patterns in the region

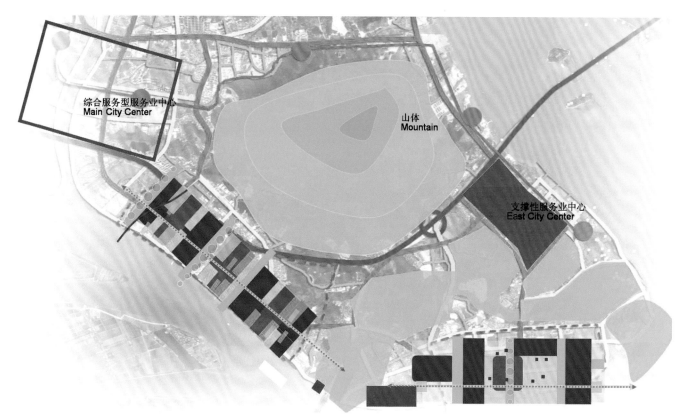

图 09 区内空间模式
Fig. 09 Regional development pattern in the region

城市发展 URBAN DEVELOPMENT

确定弹性发展的空间模式：多样灵活的结构
Determining the development model: diverse and flexible structures

高科技园区用地在国内大多为空白基础，现状产业发展并不明晰，其未来走向有可能因为国际形势和投资情况有所调整。在根据周边情况分析预测未来产业模式的同时，高科技园区亦希望能拥有较为灵活的空间模式，可以根据具体情况进行调整，适应不同产业的发展需求。将科技园区结合相应的商业、生活、服务设施，分成适当的区块，这样既有利于灵活地分段开发，又同时保证了大区域的连续性和完整性。

首先进行大区域分析，以前面的分析结论为基础，推断南沙光谷的大概辐射范围，并对整个区域进行分区分析，确定其可以给基地提供的人口、教育设施、服务设施等，综合分析交通条件和自然条件，以此确定两块规划基地的不同发展目标和发展模式。

在此基础上，提出多种空间发展模式，根据现状进行整合，得出最终结论。A区的现状较为复杂，已建区域多。建议采用纵向条形中心，横向连接居住和高新技术企业组团。组团间以沿河流产生的带状绿地分隔，便于分期开发和改造。B区的现状结构清晰，多科研机构。有大量空地，交通便利，建议以主干道交叉点为中心，向两侧扩散科研机构和服务机构。

在两个月的工作时间内，我们从国际、珠三角地区、南沙三个视野和层面解析和预测了南沙的发展方向，在分析现有成熟科技园区的基础上，结合本地情况，确立了灵活且可行的发展模式，得到当地政府和评委的一致好评，进行下一步城市设计任务。

In general, land use of domestic high-tech parks is mostly open, the development of existing industries unclear, and future development may depend heavily on international relationships and investments. While analyzing and predicting future industrial models according to the surrounding circumstances, the high-tech park should have a more flexible space, which can be adjusted to the specific conditions it is faced with in order to meet the evolving development needs of various industries. Combined with the corresponding facilities for commerce, life and service, the science and technological park should be divided into appropriate blocks, helping it to develop in stages and safeguard the integrity of the continuity and the region as a whole.

The entire region was analyzed to infer the emissions radius of Nansha and to determine its population, education facilities, service facilities, traffic and natural conditions. As a result the planning objectives were divided into two groups. Diverse possibilities for space development were illustrated based on the above research, and final conclusions were drawn in accordance with the current situation.

During a period of two months, our group analyzed and predicted the development direction of Nansha at the international level, the Pearl River Delta level and the Nansha local level. Based on existing mature science and technical parks, and in line with local conditions, we established a flexible and possible model for development. Our plan was approved by the local government and judges, and we will continue to conduct the urban design.

将科技园区结合相应的商业、生活、服务设施，分成适当的区块，这样既有利于灵活的分段开发，又同时保证了大区域的连续性和完整性。

The science and technological park should be divided into appropriate blocks, helping it to develop in stages and safeguard the integrity of the continuity and the region as a whole.

模式1
地块以生产型高科技企业为主，两侧用的提供居住条件和商业设施
B地块以研发类高科技企业为主，西测结合南沙旧镇的中心区设立服务中心和现代服务业中心

优势：研发企业和东部新城关系较好，东部新城能提供高质量住宅
与周边地块结合紧密，带动周边发展
劣势：生产型高科技企业区块较大，缺乏生活气息。和南沙旧镇结合的中心区不易提高档次

模式2
A地块以研发型高科技产业为主，结合南沙中心区设置商业中心和服务中心
B地块以生产型高科技企业为主

优势：生产型高科技企业可以为东部新城和A地块研发高科技企业提供从设计到生产的能力
劣势：生产性高科技企业的周围多为东部新城附近的高档住宅，其企业中大量员工较难利用。研发型科技企业和东部新城的研究机构联系不够密切

模式3
A地块为生产型高科技企业和研发型高科技企业，
B地块为现代服务业，中心区和部分配套住宅

优势：分工明确，各区块设置集中
劣势：高科技企业与现代服务业和中心区关系不佳

模式4
A地块为现代服务业，中心区和部分配套住宅。
B地块为生产型高科技企业和研发型高科技企业

优势：生产型高科技企业与南沙旧镇关系西良好，可以获得充足的廉价住房
劣势：B地块现状较为复杂，全部改为中心区和现代服务业有一定的困难

图 10　区域发展模式列举
Fig. 10　Example of regional development pattern

城市发展
URBAN DEVELOPMENT

01 世界文化遗产城市波茨坦城市总体规划
The Urban Development Planning of World Cultural Heritage City of Potsdam

02 丽江玉龙新城规划
Masterplan of Yulong New Town, Yunnan

03 泉州市城市新区建筑天际轮廓控制规划方法研究
Planning of the Skyline of the City of Quanzhou, Fujian

04 泉州市区和环湾地区空间发展战略研究
Strategic Study on Development of the Urban Districts and General Area around Quanzhou Bay, Fujian

05 广州市南沙光谷地区城市发展规划
Urban Development of Nansha Guanggu in Guangzhou

06 阳宗海旅游度假区西北部概念规划设计
Conceptual Planning of the Northwestern Part of Yangzonghai Resort, Yunnan

07 东山蝶岛发展战略规划
Dongshan Butterfly Island in Fujian: the Strategic Planning of Urban Development

08 十堰市东部新城概念规划及城市设计
Conceptual Planning of the East New Town in Shiyan, Hubei

09 福州市东部新城中心城市设计
City Planning for the Center of the Eastern New Town of Fuzhou, Fujian

10 杭州天堂鱼生活：杭州市运河新城概念规划
A Fishing Life Paradise in Hangzhou: Concept Planning for the Canal New Town in Hangzhou

11 唐山市南湖生态城概念性总体规划设计及起步区城市设计
Conceptual Planning and Urban Design for an Ecological Zone in Tangshan, Hebei

12 大同市御东新区概念性总体规划及核心区概念性城市设计
Conceptual Planning of Yudong District and Urban Design of the Core Area, Datong

13 文昌市抱虎角概念规划
Conceptual Planning for Baohujiao Area in Wenchang, Hainan

14 昆明市绿地系统概念规划
Concept Planning for Kunming Green Space System

06 阳宗海旅游度假区西北部概念规划设计

Conceptual Planning of the Northwestern Part of Yangzonghai Resort, Yunnan

规划面积	AREA	13.05 km²
人口规模	POPULATION	2695
完成时间	PROJECT DATE	2006.07

规划方案 Planning Concept

The Yangzonghai Tourist Resort is located on the development axis of Kunming and Yiliang. Together with the Stone Forest and Jiuxiang scenic areas, it constitutes a tourist traffic network, which represents a part of the heart or "golden route" of regional tourism in Yunnan. In recent years the Yangzonghai Tourist Resort has seen rapid and advantageous regional development thanks to its abundant tourist resources and its own advantages in policy.

Based on the four main keywords "lake, mountain, sunshine and city", a planning system dubbed "R.M.P.S." was devised, with "R" indicating resources, "M" indicating market, and "P" indicating production; all of which would be integrated into "S" indicating space.

旅游规划的重要核心工作是资源的评估与空间的落实。

The main work of the planning was the assessment of resources and the development of implementation strategies.

The development goal for the northwest area of Yangzonghai was find a way to combine ecology with tourism. Adequate touristic functions and cultures had to be organized and developed through the ecological planning without burdening the landscape. The resort had to be operating on par with its international counterparts, offering entertainment and sports such as sailing and golf as well as other facilities like spas. Based on the current ecological conditions of Yangzonghai, a protective development area was defined in the center of the area, with tourist facilities arranged in a circular

阳宗海旅游度假区位于昆明市与次级市宜良的发展轴线上，与国家级风景区石林、九乡构成紧密的旅游交通网络，处于滇中旅游圈黄金线和中心地带，有较好的城市区域发展优势。近年来，阳宗海旅游度假区凭借其丰富的旅游资源及诸多自身和政策优势迅速发展。

规划通过提取“湖”、“山”、“阳光”、“城”等关键词，将RMP框架在旅游规划中进行提升型全面应用。在传统旅游策划的R-资源、M-市场、P-产品格式中，紧密结合S-空间规划，形成连续整体的工作逻辑。整体确定了西北片区“生态+旅游”——以生态为前提塑造旅游景观、组织功能、发展文化，和以旅游为媒介鼓励促进生态景观规划的发展模式，将其定位为包括水上帆船运动、SPA休闲、高尔夫球运动等，以国际康体休闲娱乐为主的高标准国际化度假区。针对阳宗海地区生态现状，确定了以阳宗海为中心，“圈层式”保护性发展旅游度假设施的原则，对现有自然岸线、山地特征进行了保护、更新和选择性开发。

阳宗海地区旅游产品定位定量分析中，在旅游承载力

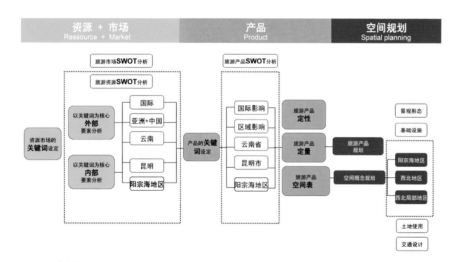

图 01 工作框架
Fig. 01 Framework

	镇域人口	集镇人口
汤池镇	34186	8545
草甸镇	18000	4499
阳宗镇	22410	5602
七甸乡	18907	
云南铝厂		3000
阳宗海地区总计		35147
(汤池集镇+草甸整镇+阳宗集镇+铝厂)		

阳宗海地区现状人口
Current population of Yangzonghai area

	卢加诺	林岛
当地面积	48	33
居民人数	53000	24560
床位数	17460	16389
床居比	0.33	0.67
游客人数（人次）	395831	740000
游居比	7.47	30.13
日游客人数	1084.47	2027.40
人均游客旅游面积	22.59	61.10
日床位使用率	0.06	0.12

旅游设施比较
Comparison of tourism facilities

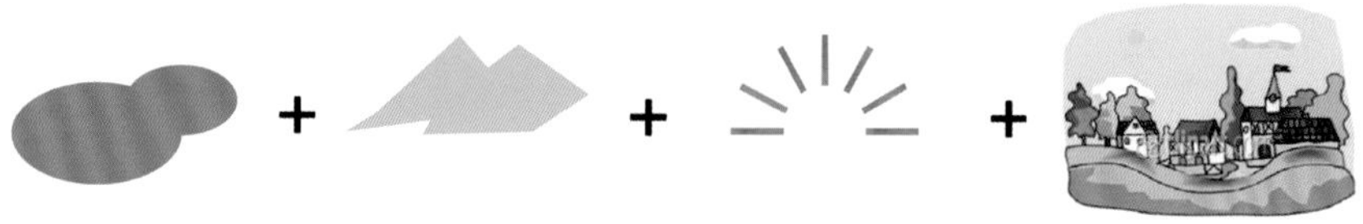

ake
湖

36平方公里大型湖面
二级清洁湖水
多处其他小型湖面
丰富的高温地热温泉

Mountain
山

高海拔
山体形态优美陡峭
覆盖植被
土壤色泽特殊

Sun
阳光

日照时间长而均匀
温度舒适，峰值较少
干燥清爽
空气清洁

City
城

云南大旅游环境
昆明旅游环境
区域内旅游环境
旅游设施

图 02 评价资源的核心要素
Fig. 02 Main factors of resource assessment

"Carrying Capacity"的确定上，开创了建立在比较学基础上，结合加权系数Weighting的计算方法。选取与阳宗海周边具有相似特征的两个欧洲城市——意大利卢加诺、德国波登湖畔林岛。这几个城市规模相当，自然资源相似，产业类型相似，旅游产品类型相似，和阳宗海地区有着相似的环境空间容量、生态容量和旅游资源容量。综合横向比较确定旅游承载力，并通过社会环境及心理容量的差异，进行比较定量计算。针对欧洲城市与中国城市的差异性，结合加权系数比较计算，将误差降至最小，从而计算出年游客总量及床位数，进而确定水环境承载能力，并从市场角度验证计算结果。

工作内容贯穿了旅游策划、空间整体规划、西北片区概念规划、西北片区控制性详细规划和帆船小镇修建性详细规划等系列工作，为上级城市、地方城市与开发集团共同提供了满意的工作成果。

pattern around it, with selective development of the existing natural coastline and mountain features to be carried out.

To analyze of the quantity and position of Yangzonghai tourist facilities and predict tourism capacity, we created a method of calculating combined weight on the basis of comparison. We selected two European cities, Lugano in Italy and Lindau in Germany at Lake Constance, that had similar characteristics to the surrounding areas of Yangzonghai including city size, natural resources, types of industry and tourist facilities. The areas also had similar capacity of the environmental space, ecology and capacity of tourism resources. Yet the differences between the social and psychological environments in each county were estimated and considered in terms of tourist capacity. The total number of tourists and beds needed each year was calculated in order to estimate the capacity of water needed, then verified by the value point of market.

The planning covered the travel scheme, the overall planning of space, the concept planning for the northwest area, the planning of the Aozi yacht harbor and villages, which would enable satisfactory performance for the city and the developer group.

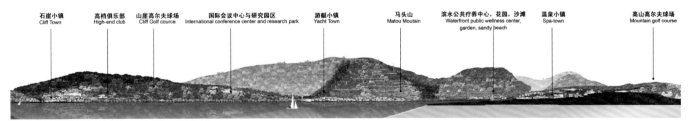

图 03　西北方向滨湖立面
Fig. 03　Lakeside view in the northwest part of Yangzonghai

图 04 温泉小镇
Fig. 04 Simulation of spa village: Bird's eye view and street perspective

城市发展 URBAN DEVELOPMENT

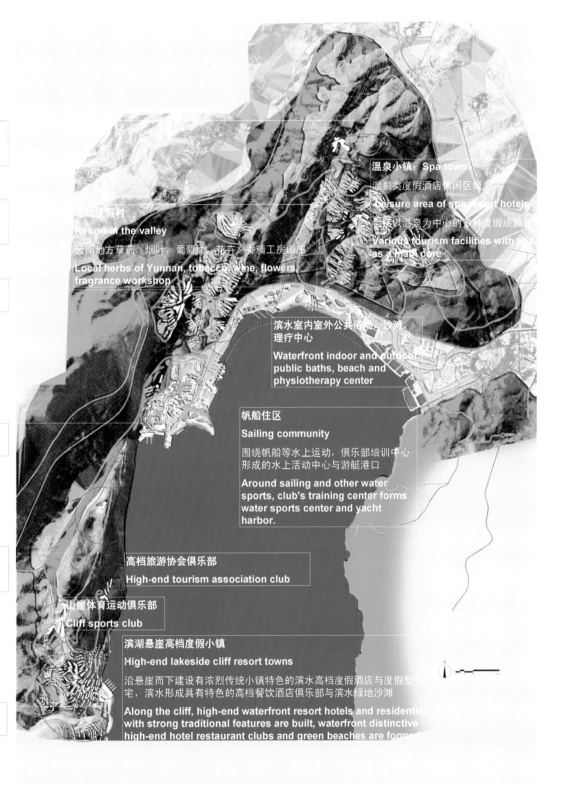

图 05　云南省宜良县阳宗海旅游度假区西北部概念规划设计
Fig. 05　Concept Planning of northwest part of Yangzonghai Resort, Yunnan

城市发展
URBAN DEVELOPMENT

01　世界文化遗产城市波茨坦城市总体规划
　　The Urban Development Planning of
　　World Cultural Heritage City of Potsdam

02　丽江玉龙新城规划
　　Masterplan of Yulong New Town, Yunnan

03　泉州市城市新区建筑天际轮廓控制规划方法研究
　　Planning of the Skyline of the City of Quanzhou, Fujian

04　泉州市区和环湾地区空间发展战略研究
　　Strategic Study on Development of the Urban Districts
　　and General Area around Quanzhou Bay, Fujian

05　广州市南沙光谷地区城市发展规划
　　Urban Development of Nansha Guanggu in Guangzhou

06　阳宗海旅游度假区西北部概念规划设计
　　Conceptual Planning of the Northwestern Part of
　　Yangzonghai Resort, Yunnan

07　东山蝶岛发展战略规划
　　Dongshan Butterfly Island in Fujian:
　　the Strategic Planning of Urban Development

08　十堰市东部新城概念规划及城市设计
　　Conceptual Planning of the East New Town
　　in Shiyan, Hubei

09　福州市东部新城中心城市设计
　　City Planning for the Center of
　　the Eastern New Town of Fuzhou, Fujian

10　杭州天堂鱼生活：杭州市运河新城概念规划
　　A Fishing Life Paradise in Hangzhou:
　　Concept Planning for the Canal New Town
　　in Hangzhou

11　唐山市南湖生态城概念性总体规划设计
　　及起步区城市设计
　　Conceptual Planning and Urban Design
　　for an Ecological Zone in Tangshan, Hebei

12　大同市御东新区概念性总体规划
　　及核心区概念性城市设计
　　Conceptual Planning of Yudong District and
　　Urban Design of the Core Area, Datong

13　文昌市抱虎角概念规划
　　Conceptual Planning for Baohujiao Area
　　in Wenchang, Hainan

14　昆明市绿地系统概念规划
　　Concept Planning for Kunming Green Space System

07 东山蝶岛发展战略规划
Dongshan Butterfly Island in Fujian: the Strategic Planning of Urban Development

规划面积 AREA		194 km²
人口规模 POPULATION		211000
完成时间 PROJECT DATE		2008.08

从两种文化中学习 LEARNING FROM TWO CULTURES

研究背景 Background

The strategic planning of urban development does not belong to statutory planning, which is the macro guiding principle sought by local governments to cope with the rapid development of construction. Strategic planning deals with present and future urban development from a higher level, conducts scientific analysis and research on the major problems of urban development, and identifies the advantages and disadvantages from a unified view of the regional space and economic environment. It creates the direction and policy of the development and provides the guiding strategic principles for the establishment of good operability and feasibility.

As a relatively independent geographical unit, the ecological environment of Dongshan Island is fragile. Moreover, it is also particularly meaningful because of its political and military value. When creating the strategic plan for an island development, one should not only focus on its own resources but also its relationship with the surrounding continents and oceans.

岛屿作为一个相对独立的地理单元，因其生态环境的脆弱性，以及一些特殊区位岛屿的政治、军事价值，有其特殊性。

As a relatively independent geographical unit, the ecological environment of the island is fragile. Moreover, it is also particularly meaningful because of its political and military value.

Dongshan Island is located at the south-eastern end of Fujian Province, the border of the East China Sea and South China Sea, and is part of the southern tip of South Fujian "Golden Triangle" economic open zone. It is within a county of 200 islands between the Xiamen and Shantou Special Economic Zones, facing Taiwan across the strait. Dongshan Island has a good natural environment and excellent beaches, as well as unique human resources.

In recent years, the economic zone on the west side of the strait located mainly in Fujian – based on the China ASEAN Free Trade Area – creates a strategic economic and market environment for Dongshan. With the development of cross-strait relations, Dongshan has even greater space for development. In addition, external factors such as the overflow of industries in the economic development zone and the closeness to Xiamen, Quanzhou and Taiwan are decisive for the development of Dongshan. So the island faces another new large-scale development, and therefore needs a strategic growth plan to ensure its healthy and orderly evolution.

Through regional development research, the strategic development plan affirmed the feasibility and practicality of being an international tourist port city, which then became the primary development goal for Dongshan. We conducted a further study of urban industrial and spatial development, as well as an in-depth study of tourism planning as an important central industry. Furthermore, we designed new districts, tourist areas and the major road axis, covering all the key points of the Dongshan Island development.

发展战略规划不属于法定规划，是地方政府为了应对快速的地区发展建设而寻求的宏观指导性原则。战略规划可以站在更高的层次上处理城市现在和未来的发展问题，对城市发展重大问题进行科学分析和研究，从区域的空间、经济环境的综合视角，找出地区的优劣所在，制定出发展的方向和政策，给城市一个具有良好操作性和可行性的发展战略指导原则。

岛屿作为一个相对独立的地理单元，因其生态环境的脆弱性，以及一些特殊区位岛屿的政治、军事价值，有其特殊性。为一个岛屿作发展战略规划，不仅要着眼于岛屿本身的资源，还要关注其和周边大陆、海洋的关系。

规划区东山岛位于福建省东南端、东海与南海交汇处，处于闽南"金三角"经济开放区的最南端，是厦门、汕头两个经济特区之间的一个200平方公里左右的海岛县，与台湾岛隔海相望。东山拥有良好的自然环境和优良的海滩，以及富有特色的人文资源。

本次发展战略规划通过对东山进行区域发展研究，综合讨论了现在东山提出的发展目标——作为国际旅游港口城市——的可行性和实际可操作性，并进一步研究城市产业发展与城市空间发展两部分内容，且将旅游规划作为重要的城市中心产业进行专题研究。此外，还示范性地对新城区域、旅游区域以及重要城市道路轴线进行了城市设计工作，涵盖了东山岛近期发展的各个关键点。

图 01 区位图
Fig. 01 Location map

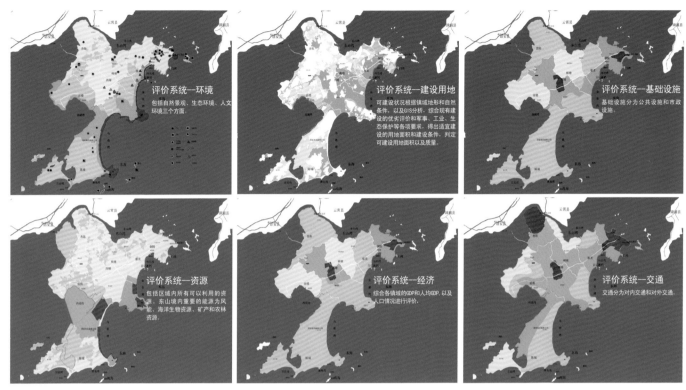

图 02 分区资源评价图
Fig. 02 Resource assessment map

整体区域发展规划 Overall development planning

福建省岛屿众多，东山岛在面临一定竞争的同时，也和各岛屿存在协调发展的可能。分析大区域背景，将东山岛和周边类似岛屿进行比较，如厦门市、湄州岛和广东南澳等拥有良好的生态资源和丰富的旅游资源的岛屿，得出东山岛相对的优势和劣势，从而确定下一步发展方向。东山岛整体面临的是一个短期发展与长期发展的平衡问题。

随后确定了东山岛的近期和远期发展阶段。整体而言，东山岛未来的发展定位为：充分利用东山县现有的区位和资源优势，调整现有产业布局，协调城镇体系的发展，保护和利用自然生态环境，把东山县建设成为生态品质优良、近期面向国内市场、远期面向国际的旅游港口城市。

There are numerous islands in Fujian Province, including Dongshan, which compete against each other; but they can all profit from a coordinated development with the other islands. Analyzing the large regional background, we compared Dongshan to the surrounding islands – such as Xiamen, Meizhou and Guangdong Nanao, which have rich ecological and tourism resources – in order to draw out its advantages and disadvantages and to determine the direction of the next development stage.

Subsequently the short-term and long-term development phases of Dongshan Island were created. As a whole, the objectives of future development on Dongshan Island were: to make the best use of the existing advantages of location and resources; to adjust the layout of the existing industries; to coordinate urban development; to protect and utilize the natural ecological environment; and to make Dongshan County into a tourist port city of good ecological quality, of which the immediate goal is the domestic market while the long-term goal is internationally based.

产业发展分析 Analysis of industrial development

东山岛产业发展的最大制约条件在于交通受限，城市腹地不足，不宜发展规模较大与生态环境负担较重的工业项目。针对其约束条件和现状发展状况，对产业进行进一步调整：

The greatest challenge to the Dongshan industrial development project is a limitation of approval caused by the lack of urban hinterland, so it would not be allowed to develop large-scale industry. According to the restrictions and current development conditions, we should make the following modifications to local industry:

- To expand fishery, economical agriculture and long-term ecological industries, thus forming an industrial chain of agricultural production.

- To develop port industry and clean energy industry, a new and expanding field, which can drive the economy, yet also pay attention to its relationship with the environment.

- To introduce yacht industries – which greatly promote the development of secondary and tertiary industries, cause a minimal burden to the environment, and develop more tourism and high-end sports leisure industry – in order to effectively create the dynamic international image of Butterfly Island.

- To restrict the location of non-metallic mining industry in order to preserve resources.

- To focus on the development of tourism, combined with the glass arts, along the outer regions of the island, in order to enhance the overall image of Dongshan and its cultural qualities.

- To develop the two Dongshan industry wings of ocean-based business and tourism, thus promoting prosperity and growth for the economy.

- 长期生态化产业为渔业和经济农业，应进行进一步扩展，形成农副产品加工产业链。

- 拉动经济的新兴工业为港口工业和清洁能源产业，应在发展产业使其快速拉动经济的同时注重其与生态的关系。

- 重要引入产业为游艇产业，可带动的第二产业与第三产业总量巨大，生态负担较小，与旅游以及高端体育休闲产业关联性较好，同时有效塑造了活力蝶岛的国际形象。

- 限制发展产业为非金属矿采选业，作为资源储备。

- 主力发展产业为旅游服务业，与此结合发展玻璃高端艺术加工，充分利用现有资源，结合岛域的外部空间，提升东山的整体形象，塑造文化气质。

- 以海洋综合型产业和旅游业等优势产业为东山的产业双翼，推进经济的繁荣与腾飞。

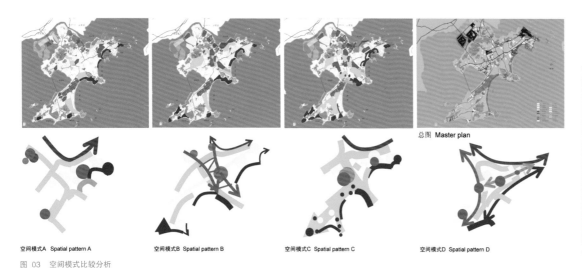

图 03　空间模式比较分析
Fig. 03　Comparison and analysis of spatial pattern

图 04　岸线评价
Fig. 04　Coastline evaluation

空间发展研究 Study on space development

Restricted by its historical background, the existing downtown Dongshan does not have the appearance of a gulf city because it is smaller than the other big gulfs. Four different land use patterns came into being after the assessment and analysis of 12 districts, according to five criteria: comparison of the ecological protected areas, tourist areas, industrial areas, downtown areas and residential areas.

The spatial layout consisted of one core and three axes: a central green ring and three functional axes along the coast, designed for Dongshan Butterfly Island to eventually create a beautiful central region facing the sea with the hills behind.

受到历史背景的制约，现有东山中心城区不具备海湾城市的面貌，与各大海湾可达性较差，经过对全岛12个分区在6个方面进行的评估分析，对生态保护区、旅游区、工业区、城市中心区与居住区进行了权衡对比，最终形成4种不同的土地使用模式。

城市发展 URBAN DEVELOPMENT

选取"三轴一心"的布局,即一个中央绿环和沿海湾发展的三条功能轴线,为未来的东山蝶岛形成一个背山面海的优美临海中心区域以及滨海岸线利用的明确指导。以分区方式进行独立发展,形成充分利用现有资源优势,彼此功能互不干扰,充分利用海湾资源,各自交通和基础设施配套完善的城市功能系统。

Every district should be developed independently, resulting in its own urban system, which would have good transportation and well-equipped facilities that do not interfere with each other. This would also optimize the use of the existing resources and gulf resources.

旅游规划 Tourism planning

分析东山本地的旅游优势,借鉴国际知名的海岛、山林旅游区域的经验,重点对突破其季节性旅游的方法进行研究。将可行的旅游产品分为投资小、有一定经济效益,中等投资、经济效益较大,投资大、经济效益大三类,对东山的不同地域提出不同的旅游产品和模式。将东山打造成一个四季不同的活力蝶岛。

其遵守的规则如下:
- 充分利用现有资源,以海湾为单元进行开发,各海湾突出不同特色。
- 由大众旅游开始带动城市建设,但严格控制开发强度以及对生态环境的破坏。
- 逐渐发展高端旅游,以优美滨水生态岛屿的建设为背景,水上康体休闲运动为带动点,在岛体的南侧与东南侧发展高端旅游。

We placed an emphasis on the study of seasonal tourism by analyzing the local Dongshan attractions and using the experience of internationally renowned island and forest tourism regions for reference. The tourism products will be divided into three types: small investment for moderate economic benefits, medium investment for considerable economic benefits and large investment for great economic benefits. There will also be different tourism products for the different regions in Dongshan. To make Dongshan a dynamic island for all four seasons, these rules should be observed:

- To make good use of existing resources and develop the gulf as a unit, each region respectively highlighting its own characteristics.
- To control the development of mass tourism as well as the damage to the ecological environment caused by it.
- To develop high-grade tourism on the south and southeast side of the island, constructing a beautiful eco-island and leading to sustainable recreational water sports.

将可行的旅游产品分为三类,对东山的不同地域提出不同的旅游产品和模式。

The tourism products will be divided into three types. There will also be different tourism products for the different regions in Dongshan.

马銮湾片区旅游产品选择目录 Catalogue of tourism products at the Maluan Bay

图 05 旅游产品目录
Fig. 05 Catalogue of tourism products

图 06 马銮湾片区旅游概念性规划
Fig. 06 Conceptual planning of Maluan Bay Resort

城市设计 Urban design

Under the guidance of the general planning principles, we developed exemplary designs for the Maluan Bay Resort, Xipu Central District and the area along Xitong Avenue.

Maluan Bay Resort

In addition to the existing tourist facilities, a large-scale water park, a commercial promenade, one large high class resort hotel, and an area of high class villas will be designed – with a marine area, spa and fitness studio facing the sea on the east side – which then becomes the dynamic coastal resort tourist area for the public.

Xipu Central District

Based on the existing site plan, the district was divided into three urban development regions, of which the core group of urban functions connected conveniently with the existing Xipu residential region and further extended towards the gulf. The administrative center was located at the end of the development axis, facing the gulf, opposite a large central park. Every region shared the central park and service facilities, which was split by the existing waterways, forming a new, beautiful coastal district with a healthy ecology and convenient transportation.

The area along Xitong Avenue

A system of paths was designed to connect each area that ran parallel to Xitong Avenue. This was done to preserve a large quantity of landscape with unique spatial characteristics between the two areas, as well as outside the coastal region. Some paths acted as walking corridors with a leisurely atmosphere for people on vacation. We designed the landscape along the Avenue with local characteristics, making use of the existing mountain, farmland and sandy topography, combined with the planning of large public green spaces, such as Forest Park, Wetland Park, City Park and so on. The various sections have different characteristics according to their different functions.

在整体规划原则的指引下，对马銮湾度假区、西埔湾城区和西铜大道沿线进行示范性设计。

马銮湾度假区

结合现有旅游设施，补充性提供大型水上乐园、步行商业街、高档大型度假酒店各一处。与高档别墅区结合，在东侧临海区域，建设大型海洋SPA健身区、功能区各一处，整体形成生机勃勃的大众滨海度假旅游区域。

西埔中心区

以现有盐田分布形态为基础，分为三个综合城市发展组团。核心城市功能组团与现有西埔居住组团联系便利，并进一步向海湾发展。行政中心位于发展轴线端点，面对内向海湾，与大型中央公园岛屿相对，各组团之间以现有水面分割，分别具有中心公园与服务设施，整体形成生态环境优美、交通便捷高效的优美滨海新城。

西铜大道沿线

在各个景观点之间建立步行栈道系统，与西铜大道平行发展，彼此间形成大量具有空间特色的大地景观区域，并成为滨海区域之外，第二条为度假休闲旅游人群服务的散步廊道。利用现有山体、农田及沙地景观，结合规划的森林公园、湿地公园、城市公园等大型公共绿地，打造具有当地特色的道路沿线景观界面。各路段根据使用功能的不同呈现不同的气质。

城市发展　URBAN DEVELOPMENT

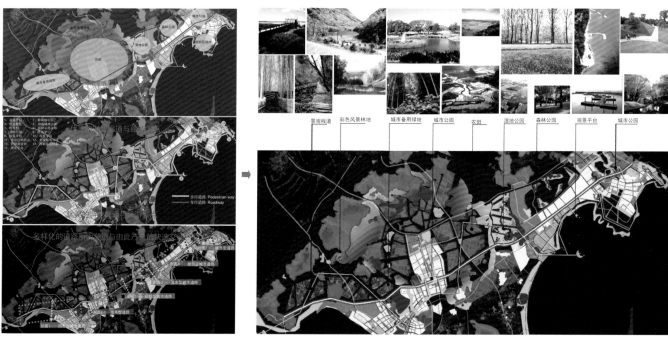

图 07　西铜大道总体设计
Fig. 07　Master plan of Xitong Avenue

图 08　西埔湾效果图
Fig. 08　Rendering of Xipu Bay

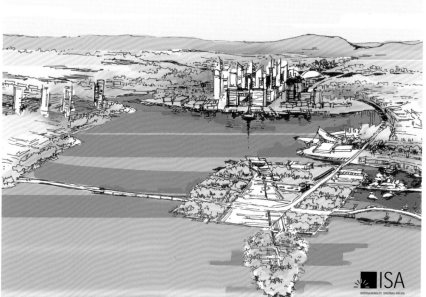

图 09　西埔湾城市设计
Fig. 09　Urban layout of Xipu Bay

149

城市发展

URBAN DEVELOPMENT

01 世界文化遗产城市波茨坦城市总体规划
The Urban Development Planning of World Cultural Heritage City of Potsdam

02 丽江玉龙新城规划
Masterplan of Yulong New Town, Yunnan

03 泉州市城市新区建筑天际轮廓控制规划方法研究
Planning of the Skyline of the City of Quanzhou, Fujian

04 泉州市区和环湾地区空间发展战略研究
Strategic Study on Development of the Urban Districts and General Area around Quanzhou Bay, Fujian

05 广州市南沙光谷地区城市发展规划
Urban Development of Nansha Guanggu in Guangzhou

06 阳宗海旅游度假区西北部概念规划设计
Conceptual Planning of the Northwestern Part of Yangzonghai Resort, Yunnan

07 东山蝶岛发展战略规划
Dongshan Butterfly Island in Fujian: the Strategic Planning of Urban Development

08 十堰市东部新城概念规划及城市设计
Conceptual Planning of the East New Town in Shiyan, Hubei

09 福州市东部新城中心城市设计
City Planning for the Center of the Eastern New Town of Fuzhou, Fujian

10 杭州天堂鱼生活：杭州市运河新城概念规划
A Fishing Life Paradise in Hangzhou: Concept Planning for the Canal New Town in Hangzhou

11 唐山市南湖生态城概念性总体规划设计及起步区城市设计
Conceptual Planning and Urban Design for an Ecological Zone in Tangshan, Hebei

12 大同市御东新区概念性总体规划及核心区概念性城市设计
Conceptual Planning of Yudong District and Urban Design of the Core Area, Datong

13 文昌市抱虎角概念规划
Conceptual Planning for Baohujiao Area in Wenchang, Hainan

14 昆明市绿地系统概念规划
Concept Planning for Kunming Green Space System

08 十堰市东部新城概念规划及城市设计
Conceptual Planning of the East New Town in Shiyan, Hubei

规划面积 AREA	37.65 km²
人口规模 POPULATION	200000
完成时间 PROJECT DATE	2007.03

背景与十堰发展需求
Background and development needs of Shiyan

The city of Shiyan is located in the northwest of the Hubei province, adjacent to the Shaanxi province, Henan province and Chongqing city. The New East Town is located to the east of the existing city center. The focus of area development was on residential construction, along with comprehensive functions such as industrial services, trade, office space and high end shopping. Additionally the planning area was intended to offer a touristic function with hopes of animating the economic power of Shiyan. In this way new business branches could be created in the whole city, especially in its downtown.

Based on the study and analysis of local conditions, history and landscape, comprehensive development strategies were defined to function as a basis for the long-term development process.

十堰市位于湖北省西北部,与陕西、河南、重庆三省市相邻。十堰市东部新城将在中心城区东侧开辟城市新的发展空间,以居住为主体,综合性建设高品质工业服务业、商业贸易办公、购物休闲等综合功能,同时作为十堰市的旅游基础设施基地,为十堰市及其中心城区提供新的产业与经济动力。

十堰市东部新城占地约39平方公里,预计人口规模为20万人。

目标:一个可持续性的生态山地新城
Planning goal: A sustainable ecological mountainous city

The New East Town, which covers an area of about 39 square kilometers, was designed for a population of about 200,000 people. The overall planning of the development strategy was founded on the urban planning principles of humane cities and sustainable development.

The conceptual planning and urban design of the New East Town was geared to meet existing urban development demands. In this context, previous experience developing other mountainous cities internationally provided a framework for developing the conceptual strategies for Shiyan. A major part of the job was putting in place a strategic plan for geographical development, which would define the long-term development process and its overall direction. Medium and long-term measures for the New East Town were devised taking into consideration the city's economic strengths as well as the all-important objective of implementation. The planning goal was to develop an independent micro ecosystem, which could contribute to realizing a beautiful, high quality mountainous city in terms of both residential areas and ecology.

新城总体发展战略的制定基于下述城市规划原则:人性化城市规划+可持续发展战略规划。

以现有城市发展需求为依据,借鉴国际山地城市发展经验,意图为城市东部地区提供一个全新的城市与自然相结合的空间模式,一个经济性与景观性兼备的功能模式。以城市经济实力为背景,以最大限度的可实施性为目的,建立拥有独立的微观高品质生态系统的城市空间规划模型,由此塑造一个具有高度人居与生态品质的优美的山地城市发展框架。

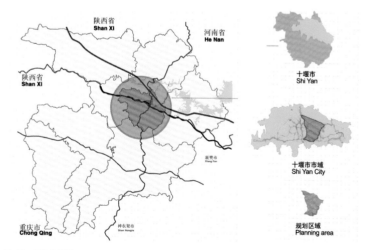

图 01 区位分析图
Fig. 01 Location of the planning area

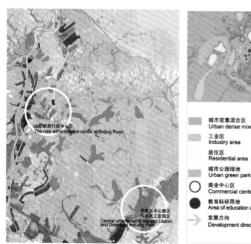

图 02 用地功能现状
Fig. 02 Current land use

城市发展 URBAN DEVELOPMENT

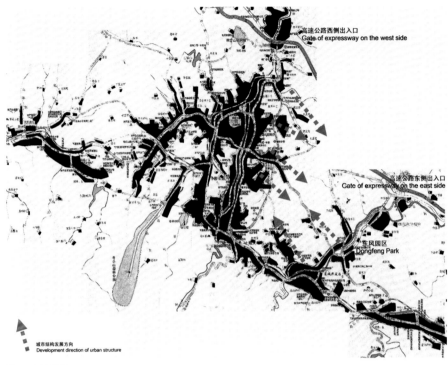

图 03 城市肌理
Fig. 03 Urban texture

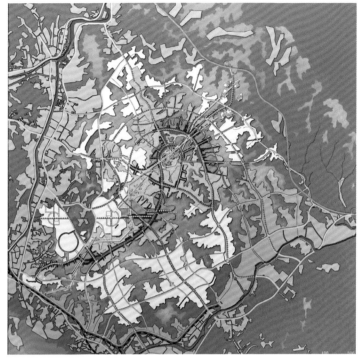

图 04 总平面图
Fig. 04 Master plan

城市规划理念
Urban design concept

在本次规划中，城市设计部分以地形特征分析为切入点，通过GIS高程分析确立新城区域内10座高于400米的山峰作为生态休闲公园及文化用地予以保留，建立步行体系联系10座山峰形成独特的山地城市游步公园，借此作为新城的内在文化溯源，使其成为新城的对外宣传名片——凸显其山地城市特色，彰显城市魅力，同时"十峰"的概念与原有老城区"十堰"以"堰"为名的历史渊源相得益彰，在历史文脉上延续了旧城的内涵，在文化脉络上实现了与原有旧城的联系。

十堰原有城市建设结构为沿铁路与公路建设的带状结构。在十堰中心城区长期发展的影响下，城市结构逐渐发生改变，在中心形成了类指状结构，这一结构在高速公路西侧出入口建成后将更加突出，这一城市建设动力在规划区域中直接形成了西北东南向的联系动力。

规划区域西侧为十堰城市中心功能区，围绕北京路将形成一个新的行政中心区。东部为茅箭区中心城区与东风工业园区，发展已经相当成熟。两者将形成东部新城发展的重要启动点，分别具有的中心商业地位以及工作岗位密度，将形成东部新城赖以发展的两个功能极核与交通目标。

The urban design took into consideration the area's topographical features as a starting point. It was imperative that the ten, 400meter mountain peaks be preserved, serving as ecological parks with cultural facilities. The peaks were linked by a path system to create a distinctive mountainous hiking trail, intended to showcase the characteristic charm of mountainous Shiyan, becoming a source of internal culture for the new district and a calling card for the city. The "ten hills" concept would also serve to strengthen the historical and cultural context of this enduring city.

The initial spatial structure of Shiyan would be a series of bands along the railway and highway. Under the long-term influence of development in the city center of Shiyan, the urban structure would gradually change, forming a finger-like structure in the center; particularly after the introduction of a highway entrance in the west connecting the southeast and northwest sections of the planning area.

In the western part of the planning area and around the Beijing Road, a new administrative center would be built. In the east, the already highly developed centers of the Maojian District and Dongfeng Industrial Park would be strengthened. Both hold a central position in the area's commercial sector, offering diverse opportunities for employment and making them an important starting point for urban and traffic development in the New East Town.

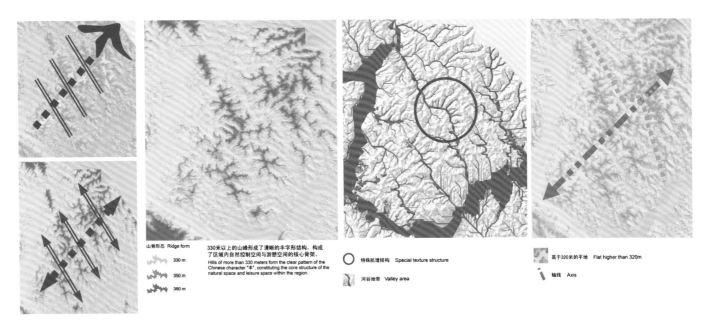

图 05 景观分析: 山脊结构
Fig. 05 Landscape analysis: ridge structure

手段：一个落实在地理与现状功能结构上的空间方案
Measures: spatial concept focusing on the current conditions

Based on analysis of area canyons, hills and mountains, two physical axes were defined which integrated the existing topography and cultural facilities into a comprehensive planning concept. One of these axes was a functional boulevard which spans both sides of the new eastern district; and the other, an ecological pedestrian landscape axis that runs through the hills, linking important area cultural facilities and universities and connecting to the tourism information center to the north. These two axes were interwoven to form the core of city life.

The functional axis developed from the northeast and to the southwest linked the various land uses of the existing two mountains, beginning at the railway station in the south and ending in the north. It was designed as an urban boulevard and development axis offering shopping, office space and other diverse service facilities. In contrast to the functional axis, the landscape axis was designed as a view corridor. It includes a public pedestrian path, which runs around the hills and links to cultural facilities. There are many cultural and public buildings along this axis, including a railway station, the Yunyang College, a Medical College, the Shiyan University and the Hubei College of Automotive Industries.

This axis was designed to be quiet and beautiful, partially divided by mountains but also connected by bridges crossing the ridges and the valleys. From this vantage point, one could overlook the East New Town from a majestic coulisse in the mountains. In addition to the view from this pedestrian axis, a cultural center consisting of a theater, library and museum was located in radial form to the north. This location would act as a "heart" and link the surrounding ten peaks through visual corridors. The edge of the city in the east was developed as a green buffer zone separating the industrial and residential areas.

The above planning design was based on the concept of urban ecology using specific research of the ecological requirements of the area. A detailed ecological concept was de-

通过对山峰的分析及意向性概念结合自然地形的沟壑与高地平台，以Brainstorming"丰"字形方案为基础，结合现有的主城区建设情况，确定了以两条轴线为骨架的规划结构方案，一条是功能轴线，通过林荫大道的车行路径，贯穿东部新城的南北两侧，另一条是虚轴线——文化生态轴线，主要峰体之间的步行轴线，主要联系重要的公共设施，并最终与北侧的旅游基础设施联系在一起，形成市民的散步道。这两条轴线的相互交织形成了十堰东部新城的城市生活核心。

功能轴线向东西两侧，一方面功能性地联系各个组团，另一方面通过两组西北东南向的山体生态性地联系了各个组团，形成了一个更加良好的生态与城市整体环境。以火车站为中心，这一新城轴线与现有的十堰中心商业城区发展轴线与工业城区发展轴线形成了完整优美的指状山地城市结构。在东边的城市边缘区域是一条绿色轴线，适当地分离了工业与城市居住区域。

应用城市生态学针对本次规划的生态要求所作的专项性研究，是新城概念规划及城市设计的重要依据之一。基于对当地自然环境要素的充分调查研究分析，例如水体、土地、植被、野生动物以及相关的城市生态理论，制定详细的生态规划，并在概念规划以及城市重点地区的设计中具体体现。

在景观系统规划中，进行了对城市景观和城市空间结构的深入设计，充分体现了十堰作为山地城市的特征和魅

城市发展 URBAN DEVELOPMENT

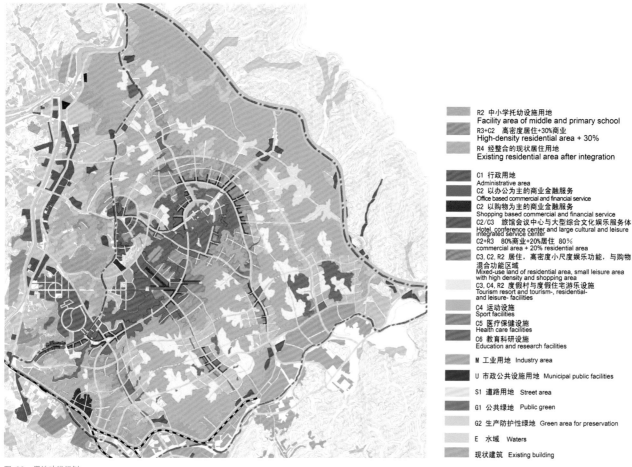

图 06 用地功能规划
Fig. 06 Land use planning

力。设计侧重于创造高品质的城市空间和现代的山地城市形态，并且以具有很高的可实施性为基本前提。本次规划中50%的山地将被保留下来，作为生态保护林地或者是城市游憩与公园绿地。

同时本次规划还对十堰市域及东部新城范围内的城镇体系、交通系统、基础设施、生态系统、风景旅游资源开发进行了合理布置和综合安排。确定了城市用地发展方向和布局结构、对外交通系统的结构和布局，规划了城市内部交通体系，进行了生态系统的比较分析并制定了可实施性建议。

依据绿色交通和公共交通优先的原则，建立了行成网络化综合交通系统，重点解决新旧区域之间的交通协调问题、山地交通的网络化问题、轨道交通系统的建设、公共交通系统的建设和道路景观建设问题。

veloped based on research and analysis of the local natural elements such as water, land, vegetation, wildlife and related urban ecological theories, which would be evidenced in the concept planning and the urban design in key areas.

The urban landscape and the urban spatial structure were also carefully crafted, with a green system plan that embodied the characteristics and charm of the mountainous Shiyan region. The aim was to create a mountainous city with high quality urban spaces implemented as widely as possible. Nearly 50% of the mountainous regions would be preserved as either protected ecological forest areas or urban green spaces.

At the same time a comprehensive urban system for Shiyan and the East New Town was developed that included a transport system, technical infrastructure, ecological system, and the development of tourism resources. This determined the development axis and guided urban land use, the internal and external transport system, the development of ecosystems and recommendations for implementation.

A comprehensive transport network system was developed according to the priority of ecological and public transport systems. The coordination of regional transport elements between new and old districts, especially the mountain traffic network, lead to an efficient rail transit system for public transport.

城市结构
City structure

原有城市肌理
Original city texture

星状结构
Star like structure

有机结合
Organic combination

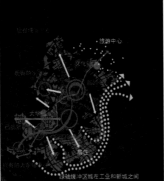

图 07　功能轴线
Fig. 07　Functional axes

图 08　文化生态轴线
Fig. 08　Cultural and ecological axes

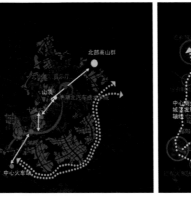

图 09　城市整体结构
Fig. 09　City integral structure

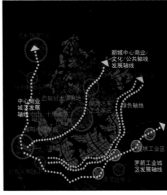

图 10　景观视线轴
Fig. 10　View axes

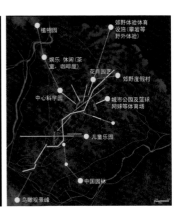

图 11　绿色景观规划
Fig. 11　Green concept

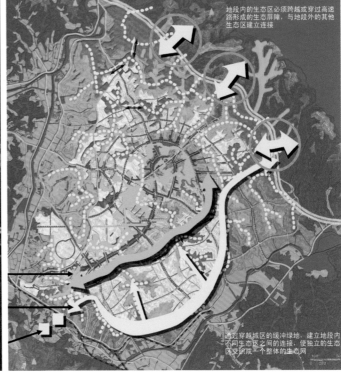

图 12　多个自然生态区之间的衔接规划
Fig. 12　Connecting planning of the nature areas

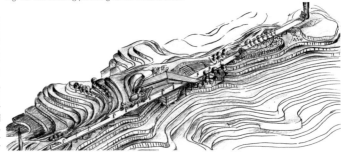

图 13　山脊上的景观轴线：领略新城生态、文化与生活的代表性结构
Fig. 13　Landscape axes on the ridge: overview of the new town ecology, representative structure of the culture and life

城市发展 URBAN DEVELOPMENT

图 14　文化教育中心区平面图
Fig. 14　Master plan of Cultural and educational area

图 15　中心区水资源经济的应用原理
Fig. 15　Application principle of water economy in the centeral area

图 16　模型照片
Fig. 16　Model photos

157

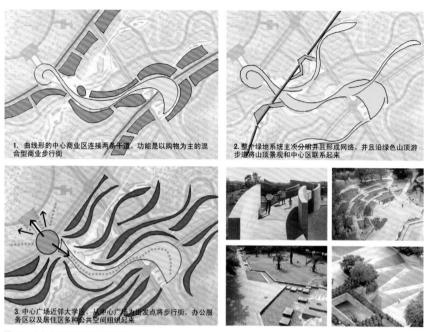

Fig. 17 Basic structure of the shopping mall in the south

Fig. 18 Atmosphere of the shopping mall in the south

成果：一个山地城市的高品质城市生活与生态环境
Result: a mountainous city with high quality city life and ecological environment

The planning selected representative focus areas in which to implement detailed designs. The specific patterns of the mountainous city and the character of the existing landscape were all applied and respected, while the central business zone of the new district, the culture and entertainment center, the urban sub-center, the typical residential area, the pedestrian paths on the mountain peaks, the landscape axis, etc. were upheld as important design principles.

There is no question that China is a mountainous country. According to a summary of materials published by the Ministry of Construction and the National Bureau of Statistics, by the end of 2003 there were 660 cities and 19,811 towns in China, of which more than 300 cities and 10,000 towns are located in mountainous regions. This means that more than half of China's cities are situated in mountainous areas.

山地城市是一种特殊的规划类型，在生态、交通、城市设计等领域都应进行相应的探索。

Cities in mountainous regions are unique. Aspects of ecology, traffic, urban design and so on should be explored accordingly.

A great deal of useful exploration into the functional aspects of the city, the layout of central areas and planning of residential areas was conducted throughout the planning of the new eastern district of Shiyan. The exercise also provided information that could prove invaluable in the planning of mountainous cities in the future.

本次规划还选择有代表性的地区进行了详细设计，包括新城商业中心区、文化娱乐中心区、城市副中心区、典型居住区、山顶游步道、景观轴线等，重点表现山地城市建设原则在城市局部地区建设中的具体使用方式，以及其他一些重要的生态与景观原则。

中国是一个多山的国家，根据建设部和国家统计局公布的统计资料，截至2003年底，中国共有设市城市660个，建制镇19811个，根据有关专家的总结，其中300多个设市城市和10000多个建制镇位于山地区域。也就是说，中国有一半以上的城镇属于山地城市的范畴。本次十堰市东部新城规划工作，结合城市功能组团布局、中心区布置与居住区规划等方面，在山地城市规划领域进行了大量有益的探索，并且为今后山地城市规划工作总结了宝贵的经验。

城市发展 URBAN DEVELOPMENT

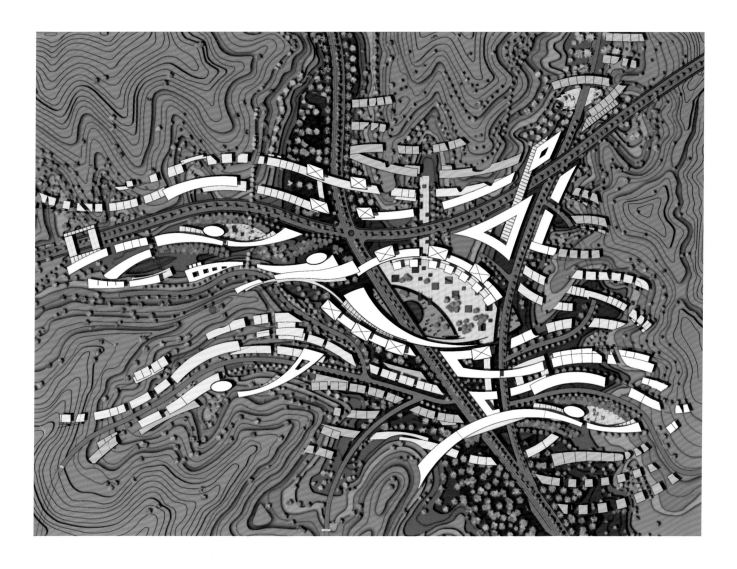

图 19　山地居住区平面图
Fig. 19 Site plan of mountain residential area

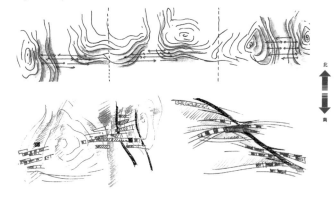

图 20　山地居住区基本构思和效果图
Fig. 20 Basic design and rendering of mountain residential area

城市发展
URBAN DEVELOPMENT

01 世界文化遗产城市波茨坦城市总体规划
The Urban Development Planning of World Cultural Heritage City of Potsdam

02 丽江玉龙新城规划
Masterplan of Yulong New Town, Yunnan

03 泉州市城市新区建筑天际轮廓控制规划方法研究
Planning of the Skyline of the City of Quanzhou, Fujian

04 泉州市区和环湾地区空间发展战略研究
Strategic Study on Development of the Urban Districts and General Area around Quanzhou Bay, Fujian

05 广州市南沙光谷地区城市发展规划
Urban Development of Nansha Guanggu in Guangzhou

06 阳宗海旅游度假区西北部概念规划设计
Conceptual Planning of the Northwestern Part of Yangzonghai Resort, Yunnan

07 东山蝶岛发展战略规划
Dongshan Butterfly Island in Fujian: the Strategic Planning of Urban Development

08 十堰市东部新城概念规划及城市设计
Conceptual Planning of the East New Town in Shiyan, Hubei

09 福州市东部新城中心城市设计
City Planning for the Center of the Eastern New Town of Fuzhou, Fujian

10 杭州天堂鱼生活：杭州市运河新城概念规划
A Fishing Life Paradise in Hangzhou: Concept Planning for the Canal New Town in Hangzhou

11 唐山市南湖生态城概念性总体规划设计及起步区城市设计
Conceptual Planning and Urban Design for an Ecological Zone in Tangshan, Hebei

12 大同市御东新区概念性总体规划及核心区概念性城市设计
Conceptual Planning of Yudong District and Urban Design of the Core Area, Datong

13 文昌市抱虎角概念规划
Conceptual Planning for Baohujiao Area in Wenchang, Hainan

14 昆明市绿地系统概念规划
Concept Planning for Kunming Green Space System

09 福州市东部新城中心城市设计
City Planning for the Center of the Eastern New Town of Fuzhou, Fujian

规划面积 AREA	Min River Area 410 hm²
	Sanjiangkou Area 540 hm²
人口规模 POPULATION	43000
完成时间 PROJECT DATE	2006.08

前期分析 Preliminary analysis

Fuzhou is a city with a cultural history dating back 2000 years. With the construction of a high-speed railway and central terminal station, urban development has jumped across the Min River, expanding to Naitai Island south of the central urban area, forming a new town with offices, commerce, administration, residences and entertainment.

This new town is located in the center of the economic zone in Fuzhou and at the joint of a strategic development axis. The planning goals were to expand toward the east and south, developing along the river and toward the sea. The area was once a wetland area downstream of the Min River that included farmlands, villages and scattered industrial facilities with special ecological value. In addition, the ecological waterfront landscape areas in the north and south form an important urban landscape framework of Fuzhou and also function as important arteries for automobile traffic.

生态之城、山水之城、团圆之城。
A city of ecology, Landscape and unity as planning goals.

The project area was divided into four areas: an administrative center on the north bank of the Min River, a public service area along the River's south bank, and a CBD Center and residential area in the south. The area crosses the Min River and connects the economic development belt between the Gushan District and the east side in an east-west direction. The business area forms a functional traffic node due to the high density of buildings and jobs.

福州是一座具有2000多年悠久历史的文化名城。伴随高速铁路与中心客运站的建设，城市建设跨越闽江，向中心城区南侧南台岛区域进行扩张，形成一座集行政、办公、商业、娱乐和居住功能为一体的新城。

规划新城处于整个大福州中心经济区的中心地带，是福州总体规划确定"东扩南进，沿江向海"发展战略轴线上的节点地带。这一区域为历史上的闽江下游湿地区域，包括农田、村镇用地以及少量工业用地，形成了特殊的生态价值背景。此外北侧与南侧的滨江景观轴线是福州市重要的城市景观骨架，也是重要的交通道路，同时肩负景观与生态意义。

规划将区域分为四个片区：闽江北岸的行政中心区、南岸沿江的海峡公共服务区、CBD商务中心区和南侧的住宅区。用地跨越闽江两岸，东西方向连接鼓山区与城市东部经济发展带，加之建筑的高密度和就业的高密度，极可能在商务办公区形成交通颈节点。

图 01 生态之城、山水之城、团圆之城
Fig. 01 A city of ecology, landscape and unity as planning goals

规划定位 Plan orientation

On the basis of the analysis of Fuzhou and its surroundings, the eastern new city of Fuzhou should be repositioned in the following ways:

- A city of ecology: Preserving the existing landscape quality, improving the ecological quality and combining with urban function to form a self-cleaning ecological system.

作为一个具有一定交通限定的新城，生态特色将是新城形态的核心，与轨道交通节点支持下的中心商务区形成鲜明的品质补充。
As a new town with certain traffic limit, ecological features will form the core of the new town, with the support of rail transit node, it is in stark contrast to the Central Business District.

- A city of landscape: using the existing cultural environment and resources, respecting the traditional planning culture of "three mountains and one lake", which fully represents the civic style and cultural characteristics of Fuzhou.

- A city of unity: forming a comprehensive urban center with the core of the administrative center, expanding its vitality to the surrounding areas and leading urban development.

综合对福州及其周边的分析，给予福州东部新城新的定位：

- 生态之城：保存现有的景观质量，充分改善现有生态环境，提高生态品质，与城市功能结合，形成一个自洁型的生态系统。
- 山水之城：利用现有城市优良资源文化环境，重建福州传统城市规划文化中的三山一湖格局——一个充分体现福州代表性城市风格与文化特征的意象。
- 团圆之城：跨越海峡两岸形成一个以行政中心为核心的综合性城市功能中心，并以此为基础向周边城市环境辐射其城市活力，带动城市经济发展。

城市发展 URBAN DEVELOPMENT

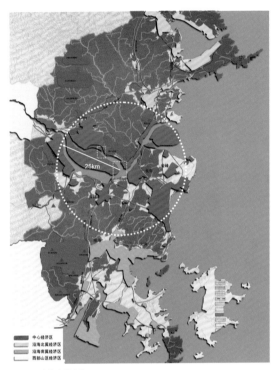

图 02　城市发展走向
Fig. 02　Urban development

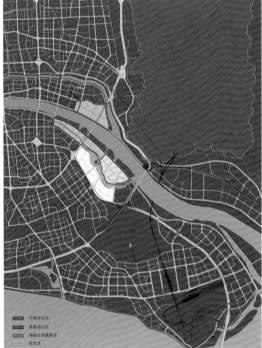

图 03　区域划分
Fig. 03　Regionalism

图 04　大区域交通
Fig. 04　Traffic system on the regional level

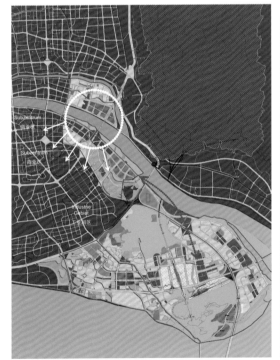

图 05　区域功能联系
Fig. 05　Connection with surroundings

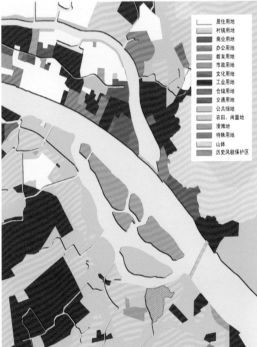

图 06　用地现状
Fig. 06　Current land use

图 07　交通瓶颈及节点
Fig. 07　Traffic nodes

163

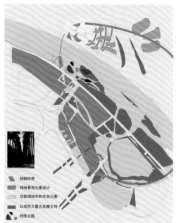

图 08 污水净化分析图
Fig. 08 Water system analysis

图 09 雨水净化分析图
Fig. 09 Rainwater system analysis

图 10 风向规划分析图
Fig. 10 Wind plan

图 11 绿地系统分析图
Fig. 11 Green areas

生态规划理念和总体规划 Ecological concept and master plan

The four ecological elements that constitute the planning framework were derived from an analysis of the ecological conditions present in the planning area: water, greenery, wind and earth. The shape of the water space and shorelines were basically preserved. The water space in the north was in use mainly as an urban water space, while the other regions were mainly wetland terrain, forming an ecological balance within the built-up areas. The water would be controlled by sluices to help prevent external flooding and to ensure a level of internal water and adequate current speeds.

Green areas would penetrate into the urban center to help make sense out of the new ecological city. The urban and natural landscapes would coexist, with agricultural and natural landscape elements integrated into the inner city.

以生态背景为前提分析规划区域水、绿、风、土四项要素，形成规划工作框架。规划基本保留区域内的水体结构与岸线形态。北部以都市型水体为主，其余地区以原有湿地地形为主，形成建设环境的内部新的生态平衡。通过排水闸将水域进行划分，对外可以防治洪水，对内保证内部较高的水质与水平位置。

将绿色元素渗透到城市主导功能中，构建并诠释全新的生态城市功能；强调人工景观和自然景观的共生，力求将农业及自然景观元素以景观塑造的原则融入城市内部。

> 本次规划面临的最大问题是如何联系山水分隔的三个核心功能区。
>
> *The key planning question was how to combine the three core functional areas divided by the mountains and waterways.*

The key planning question was how to combine the three core functional areas divided by the mountains and waterways.

Land use concept: The north bank of the Min River was planned to be the administrative and office area, the south bank the commercial, financial, cultural and entertainment area, and finally the residential area was to be located at the edges.

Landscape concept: Water from the Min River would be brought into the inner city, forming a network of diverse waterways. The ecological water landscape would pass through the planning area with green landscape axes crossing the Min River and connecting both sides. These "axes of reunion" were intended to link the natural landscape nodes including Gushan, forming a footpath into nature.

Urban open space concept: In addition to the axis of reunion as natural landscape, two pedestrian streets within the urban landscape were designed with important landmarks and city squares. They were intended to be other "axes of reunion" in the form of circles, which connect the two sides of the river bank and serve as the main public urban space. Important landmarks and nodes were scattered throughout the two circular spaces creating a dynamic spatial sequence in combination with other smaller green spaces.

本次规划面临的最大问题是如何联系山水分隔的三个核心功能区。

用地功能规划：规划闽江北岸为行政办公用地，闽江南岸滨江为商业金融业用地、文化娱乐用地，居住用地处于地段南侧。文化用地作为一种结合型的绿地，以休闲娱乐用地的形式嵌入中心商务区的结构，以大型文化公共设施的形式与居住和会展用地充分结合起来。

景观系统规划：以现有的水体架构为基础，引入部分闽江水，形成活水体系，生态水景景观廊道穿越规划区域。以一条景观兼绿化轴将闽江南北两岸连接起来，是本次规划的"团圆之轴"之一，串联了包括鼓山在内的各个自然景观节点，形成了一条自然环境中的散步道。

城市公共空间规划：除了自然景观的团圆之轴之外，一条串联城市景观的散步道，连缀了各个重要的标志性建筑与城市广场，是本次规划的另一"团圆之轴"，两条连接南北两岸的圆轴，塑造主要城市公共空间，主要的城市地标点与重要节点以点状分布在这两条圆轴上，结合其他小型树状城市局部空间形态营造出动态的空间序列。

城市发展 URBAN DEVELOPMENT

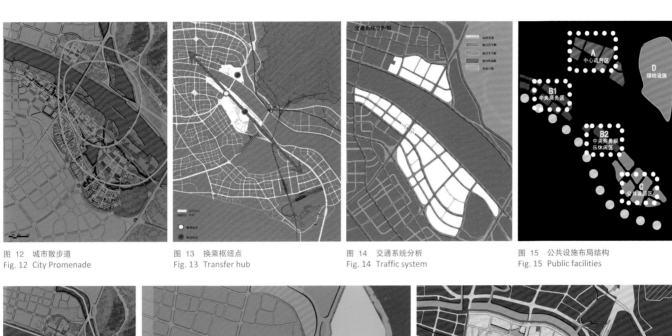

图 12 城市散步道
Fig. 12 City Promenade

图 13 换乘枢纽点
Fig. 13 Transfer hub

图 14 交通系统分析
Fig. 14 Traffic system

图 15 公共设施布局结构
Fig. 15 Public facilities

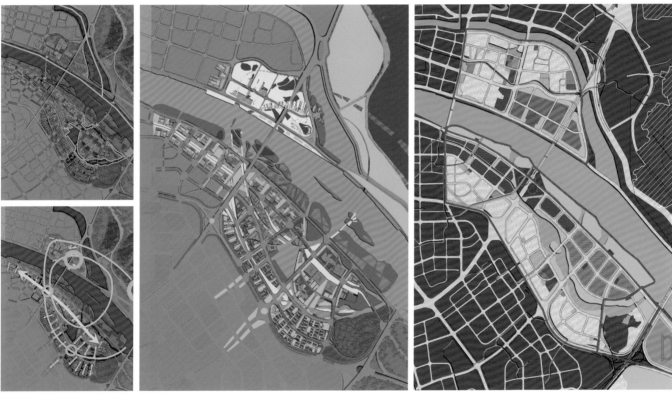

图 16 绿地系统规划图
Fig. 16 Green and water space concept

图 17 土地功能
Fig. 17 Land use concept

图 18 城市设计总平面图（三江口地块）
Fig. 18 Master plan of the Sanjiankou Area

图 19 城市设计总平面图（闽江地块）
Fig. 19 Master plan of the Min River Area

图 20 模型照片
Fig. 20 Model photos

城市发展 URBAN DEVELOPMENT

图 21 广场分析
Fig. 21 Square analysis

图 22 附属建筑分析
Fig. 22 Annex analysis

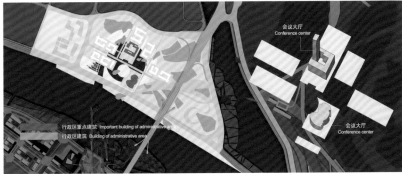

图 23 核心建筑分析
Fig. 23 Important building analysis

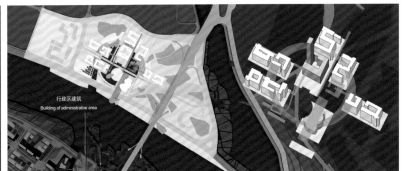

图 24 屋顶形态
Fig. 24 Roof form

节点区域规划 Node concept

The government center was located to the north of the Min River in the form of a ring. It holds the complex of offices and includes other spaces such as conference centers. Landscape design was characterized by the planting of banyan trees, which symbolize that Fuzhou is a 'banyan city' offering beautiful, open public spaces. The landscape structure transits gradually to the east into a large green park and links the ecological and cultural landscape with Gushan Mountain.

The central business area would serve as a rail transfer hub, thereby fulfilling one of the core functions of the city - embodying economic utility thanks to its fully developed transport system. Accessibility to traffic would be enhanced by building small blocks, while giving pedestrians priority in the commercial and financial areas. The entertainment area on the other hand, was planned to be a more compact series of green corridors and rivers that would help to strengthen the overall quality of life. All of these elements would be located within walking distance, creating an active urban life with access to a variety of services.

A medium scale public building area would be located at the center of the city with large public facilities more compactly organized. This area would be combined with the residential areas to enhance the complementary functions between itself and the transport, services, entertainment and other areas.

政府中心区：位于闽江北岸，以环状结构控制了包括会议中心在内的各个机关建筑综合体。正对政府中心建筑的水体中为一球状白色汉白玉岛屿，一株大榕树矗立其侧，暗示榕城，并形成了优美的市民休憩空间。这一景观结构向东侧逐渐过渡为大型的绿地公园结构，在生态上、文化和景观上与鼓山形成有力的联系。

中央商贸区：建筑形态集约，轨道换乘枢纽作为城市功能核心，充分体现了综合交通优势的经济效用。通过小型紧密的街区结构加强车行交通的可达性，商贸金融区域内全线步行优先。娱乐片区空间更为紧凑，系列人性尺度的绿廊与河流进一步加强了城市生活品质。混合式的功能布置形成了理想的功能背景，步行范围内得以形成各种服务联系与活跃的城市生活。

海峡公共建筑区：以中等尺度的中心生态湿地水面为核心，相对紧密地组织大型公共设施。并与居住区域相对设置，以加强公共建筑区域与居住功能在交通、服务、游憩生活方面的系列相互功能补充。

从两种文化中学习　LEARNING FROM TWO CULTURES

图 25　行政中心区鸟瞰图
Fig. 25　Bird's eye view of the administrative center

图 26　行政中心区设计总图
Fig. 26　Master plan of the administrative center

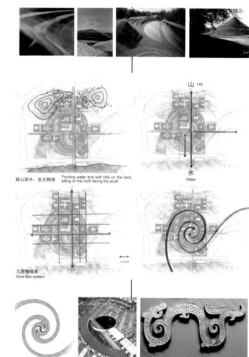

图 27　传统中国城市规划思想
Fig. 27　Traditional Chinese urban planning concepts

168

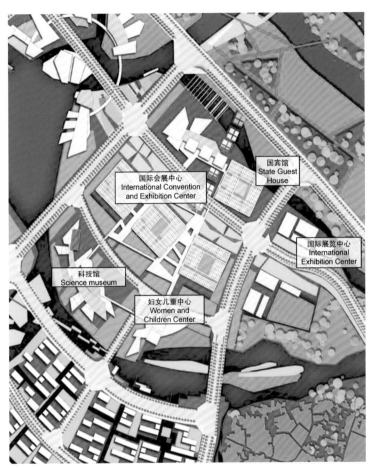

图 28 海峡公共设施区
Fig. 28 Public building area

城市发展
URBAN DEVELOPMENT

01 世界文化遗产城市波茨坦城市总体规划
    The Urban Development Planning of
    World Cultural Heritage City of Potsdam

02 丽江玉龙新城规划
    Masterplan of Yulong New Town, Yunnan

03 泉州市城市新区建筑天际轮廓控制规划方法研究
    Planning of the Skyline of the City of Quanzhou, Fujian

04 泉州市区和环湾地区空间发展战略研究
    Strategic Study on Development of the Urban Districts
    and General Area around Quanzhou Bay, Fujian

05 广州市南沙光谷地区城市发展规划
    Urban Development of Nansha Guanggu in Guangzhou

06 阳宗海旅游度假区西北部概念规划设计
    Conceptual Planning of the Northwestern Part of
    Yangzonghai Resort, Yunnan

07 东山蝶岛发展战略规划
    Dongshan Butterfly Island in Fujian:
    the Strategic Planning of Urban Development

08 十堰市东部新城概念规划及城市设计
    Conceptual Planning of the East New Town
    in Shiyan, Hubei

09 福州市东部新城中心城市设计
    City Planning for the Center of
    the Eastern New Town of Fuzhou, Fujian

10 **杭州天堂鱼生活：杭州市运河新城概念规划**
    A Fishing Life Paradise in Hangzhou:
    Concept Planning for the Canal New Town
    in Hangzhou

11 唐山市南湖生态城概念性总体规划设计
    及起步区城市设计
    Conceptual Planning and Urban Design
    for an Ecological Zone in Tangshan, Hebei

12 大同市御东新区概念性总体规划
    及核心区概念性城市设计
    Conceptual Planning of Yudong District and
    Urban Design of the Core Area, Datong

13 文昌市抱虎角概念规划
    Conceptual Planning for Baohujiao Area
    in Wenchang, Hainan

14 昆明市绿地系统概念规划
    Concept Planning for Kunming Green Space System

10 杭州天堂鱼生活：
杭州市运河新城概念规划

A Fishing Life Paradise in Hangzhou:
Concept Planning for the Canal New Town in Hangzhou

规划面积 AREA		12.38 km²
人口规模 POPULATION		100000
完成时间 PROJECT DATE		2008.11

分析与定位 Analysis and positioning

The canal new town is located on the gateway of the north Hangzhou. Its development had to fulfill complex requirements and achieve a unique charm. The existing Beijing-Hangzhou Grand Canal also has the special background of being a world heritage site applicant, and therefore has great potential for urban development along the canal as well as a tourism driven market.

In this context tourism facilities on the canal area of the Hangzhou could give the new town the power to be a world-class tourist destination as a unique node of traditional canal business and culture. Therefore, tourist service facilities should be added to the canal landscape of the new town and the canal industries modernized.

Hangzhou: The development of the "20+1" new towns was a good chance for general urban development. The renewal of formal industries also brought with it the renewal of urban space and service facilities. In this context the municipal government of Hangzhou planned a canal new town in the north, intended as a model area for urban renewal and development, and an engine of the tourism resources for the region. It was meant to be a new starting point, a driving force, and a core area of integrated services, forging a new city image in northern Hangzhou.

杭州北部门户多层次的需求与独有的潜力，是运河新城规划的深层背景。

京杭大运河：特殊的"申遗"背景，沿线的旅游发展现状与市场潜力，特别是杭州段城市建设、旅游产品的状况使其有条件形成一个凸显运河旅游资源优势的世界级旅游产品，一个继承"传统运河商旅文化"、保存"现代运河工业文化"的特色节点，一个具有运河景观特色、标志性的现代旅游休闲新城。

杭州："20+1"新城发展的优良契机，产业更新带来的城市空间更新，需要完善的旅游产品格局等因素，使得杭州需要一个北部城市门户，一个城市更新发展的示范区域，一个北部地区旅游资源的开发动力点。杭州主城北部地区：北部的现状、更新与规划背景使得其需要一个更新的启动力、策动点，一个未来的综合服务核心区，一个全新的北部城市形象。

总体规划 Master plan

The planning area would feature a characteristic spatial structure, which would bring a dynamic development structure for the whole northern area of the city of Hangzhou. In combination with the canal resources and land use structure, the following structure was recommended:

- The existing center coupled with the Jinhengde Bussiness Center in the west would be developed into a functional center of northern Hangzhou
- 2 sub-centers in the south and north of the planning area were planned to promote regional development
- The Oil Refinery and Forum Island in the west should be reorganized as the functional core of a central park and ecological green area
- The canals and ecological axes should connect the surroundings in the west and the east.

不同空间模式不仅仅对内部功能造成影响，而且为北部空间带来不同的动态发展格局。结合运河资源与北部用地格局最佳化需求形成推荐格局形式：

- 主中心与西侧金恒德商贸中心联动发展形成整体杭州北部功能区中心；
- 2个副中心分别带动南北侧规划区域发展；
- 炼油厂与西侧论坛岛作为中心公园与生态绿地功能核心；
- 多条运河及其生态功能轴线联系东西两侧周边区域。

图 01　大生态环境布局体系
Fig. 01 Macro-ecological environment

图 02　推荐格局形式
Fig. 02 Recommended spatial structure

城市发展 URBAN DEVELOPMENT

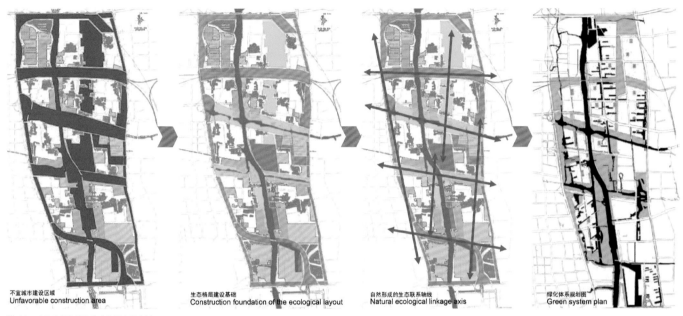

图 03　生态空间格局为基础的可建性分析
Fig. 03 Analysis of the development potentials based on the ecological spatial structure

生态环境布局形成

结合未建设而具有生态品质的区域防护绿地、特殊工业设施造成的地标型地段，形成本区域不宜进行城市建设的区域。作为本区域生态格局建设的基础，这一结构同时与区域生态结构比对，从而在量级上形成合理的区域性生态支持。

生态结构框架约束形成的8个楔形功能区组成的4个主要建设区域如岛屿般被绿地环绕。

4个城市岛——"渔水生活"、"现代水城"、"工业时代"、"休闲港市"，作为独立社区，结合环境拥有自己的特殊品质与独特的生活特色。

3个中心区与东北侧组团中心共同塑造了4个城市岛的中心，并向南北侧展开城市功能轴线。3条运河与其平行发展，形成了现代运河水视界的感受性空间骨架。炼油厂与北岸新运河论坛岛形成了工园港——整体空间结构的绿心。3条主要轴线将这些楔形连接成一个组群，每条主要轴线都与水系紧密相连，并各自打造不同"水景世界"。

建设结构中居住区之间的"生态楔子"密集而紧凑，将高层建筑规划在绿地包围下。社区界限明确，每个城市岛都可以在5分钟到达公共绿地，并向外通向杭州北部生态绿地体系。住户更有可能享受到公园绿地带来的生态利益。

Formation of the ecological environmental layout

The protected green areas must be kept free from urban development, because of their positive ecological quality and position as landmarks with special industrial facilities. This should serve as a starting point of the ecological network in the planning area and region, as well as support the regional ecosystem.

Eight wedge-shaped blocks were defined as the building sites and grouped into four "city islands" surrounded by green spaces. The areas were designated a "fishing city", "modern water city", "city of industry" and "harbor city". Each island with its green environment has to be characterized by its own special qualities and unique lifestyle.

Four urban centers were planned with three functional urban axes from south to the north, and accompanied by three parallel canals which would also serve as a modern tourist attraction. These three main urban landscape axes connected these wedge-shaped areas into a group, and every major axis was closely linked to the water in some way, creating different aquatic landscape features. The Oil Refinery and Forum Island on the north shore formed the industrial park harbor, the core landscape axis, serving as the central green and water space of the larger spatial structure.

The wedge-shaped building sites were designed as dense "ecological wedges" with a requirement that high-rise buildings be surrounded by green spaces. Each city island consisted of several neighborhoods, each with accessible public green spaces no more than five minutes away on foot. These "islands" should contribute to the ecological green system in the north of Hangzhou, making Inhabitants more likely to enjoy the ecological benefits brought by the parks and green areas.

生态框架约束形成的城市岛。

City islands were formed under the constraints of the ecological framework.

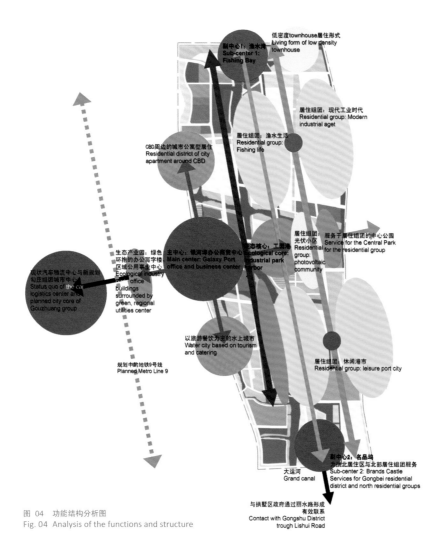

图 04 功能结构分析图
Fig. 04 Analysis of the functions and structure

Taking advantage of the existing resources, the development of sub-centers in the north and south was first started in order to promote their developmental impact at the regional level. The main center of the core area was linked with Jinhengde logistic center, bringing about the overall development of northern Hangzhou.

The functional urban axis of the city, the three canal axes of the water city and the horizontal and vertical green axes contacted the public spatial structure of the areas in the east and west of the new town, shaping a modern water city with garden city quality.

南北侧副中心利用现有资源首先启动，南北并举带动核心区域开发；核心区域主中心与金恒德物流中心联动，进而带动杭州北部整体发展。

城市功能轴线与3条水城运河轴线以及横纵向绿化轴线，联系新城东西两侧周边区域公共空间结构，共同塑造具有花园品质的现代水城。这一结构同时支持了一个良好的生态城市空间体系。

> 城市功能轴线与三条水城运河轴线以及横纵向绿化轴线，共同塑造了具有花园品质的现代水城。
>
> *The functional axis of the city, the three canal axes of the water city and the horizontal and vertical green axes contacted the public spatial structure shaping a modern water city with garden city quality.*

The canals, running from north to south, would offer a series of landmarks and distinguishing scenery. The public canal and green areas would be intertwined, forging a balance between the urban life along the canal and the ecological benefits, a variety of spatial variations and different open space nodes.

Along the central landscape axis, the sequence and location of the landmarks and the space between them were measured and decided according to pedestrian speed and vehicle speed. The dynamic experience process was extended about 6 km through a series of small nodes.

本段落运河将由此形成自南向北一系列地标点与地标景观区域，公共空间岸线与绿化岸线相互交织，形成沿河城市生活与生态利益的整体平衡、丰富多彩的空间变化韵律与多样化的开放空间节点。

沿运河两岸核心轴线，根据行人速度和车速来测量并决定序列和地标的位置、间距；动态感受过程整个延伸约6公里。一系列的小型空间节点，使其具有丰富的空间感受效果。

城市发展 URBAN DEVELOPMENT

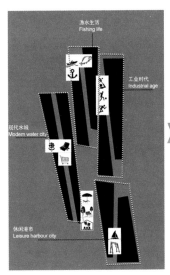
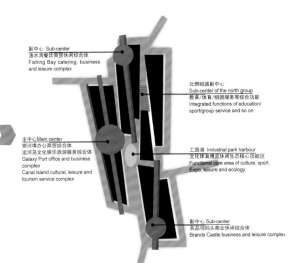

图 05 城市建设结构："绿地与水体环绕的城市岛"
Fig. 05 Urban construction structure: "Green areas and water around the city island"

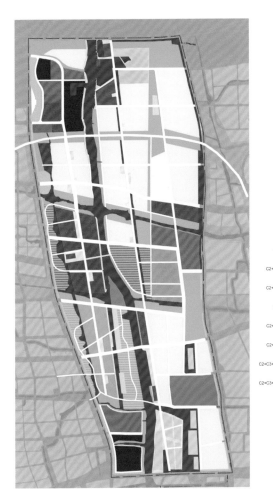

R1	一类住宅用地	Residential area of type 1
R2	二类住宅用地	Residential area of type 2
C1	行政办公用地管理	Administrative and office area
C2	商业金融用地	Commercial and financial area
C3	文化娱乐用地	Cultural and recreation area
C4	体育设施用地	Sport facilities area
C5	医疗卫生用地	Health care area
C2+C3 C2(50%)+C3(50%)	商业文化娱乐混合用地	Mixed-use land of commercial, cultural and recreation area
C2+C3+G1 C2(20%)+C3(30%)+G1(50%)	高绿化率旅游用地	Tourism area with high green rate
C2+C6+G1 C2(20%)+C6(20%)+G1(60%)	教育科研及商业混合用地	Mixed-use land of education, research and commercial area
C2+R2 C2(30%)+R2(70%)	商住混合用地	Mixed-use land of commercial and residential area
C2+C3+R2 C2(30%)+C3(30%)+R2(40%)	文化居住村	Cultural and residential village
C2+C3+R2 C2(60%)+C3(20%)+R2(20%)	文化旅游度假村	Cultural and tourism resort
C2+C3+R1+G1 C2(10%)+C3(10%)+R1(10%)+G1(70%)	高绿化率文化度假居住村	Cultural, tourism and residential resort with high green rate
C2+C3+R2+G1 C2(20%)+C3(10%)+R2(10%)+G1(60%)	旅游用地及其配套度假村	Tourism area and supporting tourism resort
M	工业用地	Industry area
T	交通用地	Traffic area
S1	道路用地	Street area
S2/S3	广场、停车场用地	Square and park area
U	市政设施用地	Municipal utilities area
W	仓储用地	Warehouse area
G1	公共绿地	Public green
G2	生产防护绿地	Green space for protection
E	水域	Waters

图 06 用地功能规划图
Fig. 06 Land use concept

从两种文化中学习 LEARNING FROM TWO CULTURES

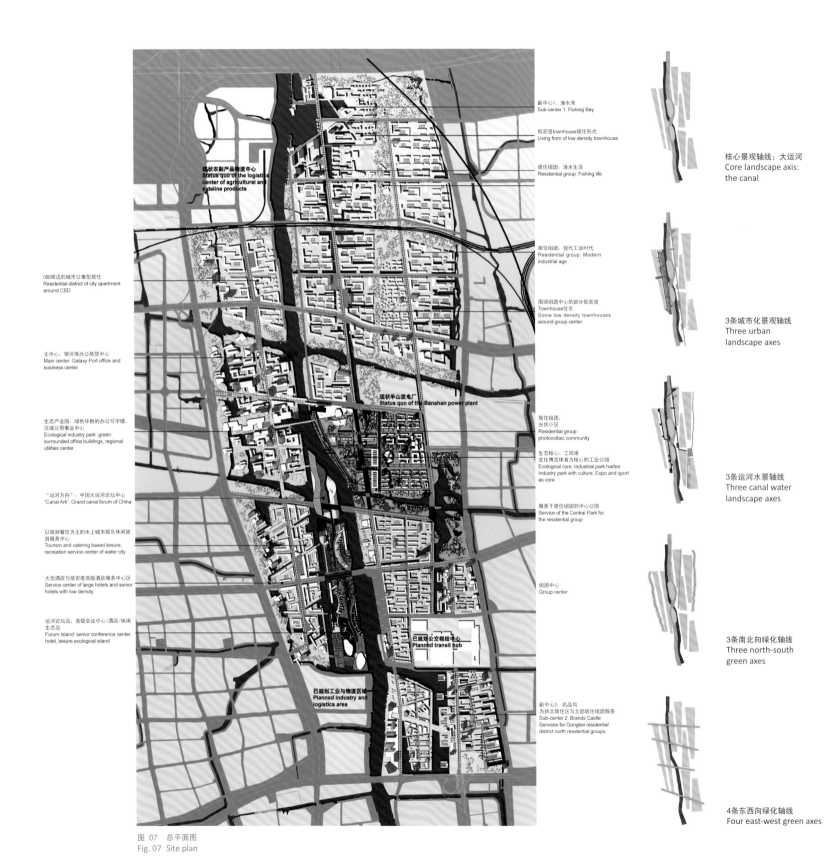

图 07 总平面图
Fig. 07 Site plan

城市发展 URBAN DEVELOPMENT

图 08 鸟瞰图
Fig. 08 Bird's-eye view

图 09 沿运河岸线形态
Fig. 09 Current situation of the canal

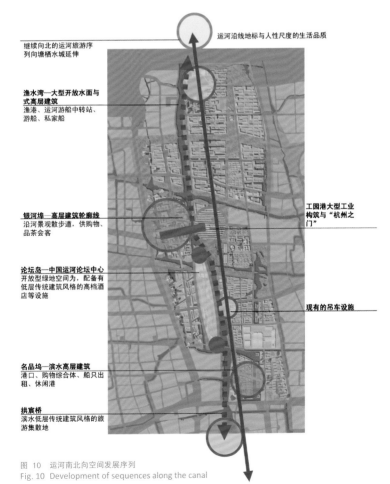

图 10 运河南北向空间发展序列
Fig. 10 Development of sequences along the canal

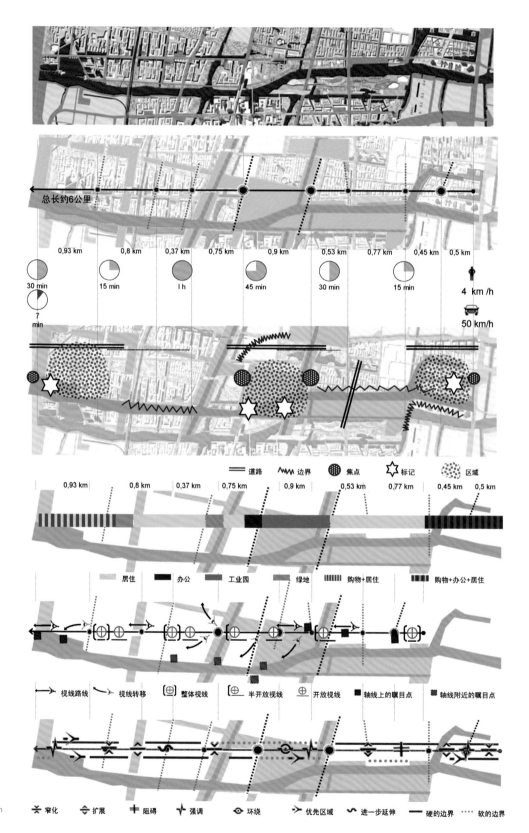

图 11 运河两岸核心轴线时间序列和空间序列推演方案
Fig. 11 Time and space sequence of the central landscape axis with canal

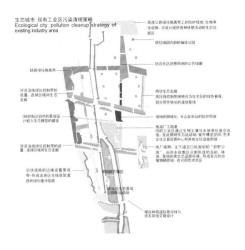
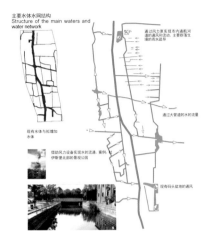
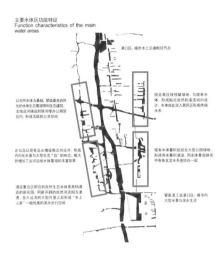

图 12 生态城市空间体系
Fig. 12 Ecological spatial system of the city

具有高度发展弹性与独特空间品质的局部区域
The local regions with high development elasticity and unique spatial quality

- 主中心：银河埠，以办公商贸、城市公寓为主体；位于基地地理核心位置，快速路与主干路交汇处，对外交通便捷。结合运河新城开发，建议将轨道交通9号线推移至巨洲路，形成两个城区的缝合线。
- 运河新城东北副中心：渔水湾，依托农副产品物流中心、绕城高速与现有大型开放水面作为杭州和运河新城的水产餐饮休闲综合体与水陆交通集散中心。
- 运河新城东南副中心：名品坞，结合管家漾内港，为拱北地区提供滨水商业休闲综合体一处。
- 工园港：炼油厂公园与运河论坛岛形成了整体空间结构的绿心。

借助运河支流体系的梳理，形成了兼具生态功能与多样化滨水体验的水视界，以不同的滨水形态为特色形成渔水湾、论坛岛、名品坞、银河埠、工园港等城市综合功能体，结合特色的水上交通形式满足"吃、住、行、游、购、娱"六大旅游需求，塑造以运河现代水城为背景的天堂杭州鱼生活。

- The main center "Galaxy Port" was a mixed-use area of offices, businesses and city apartments. It was located at the junction of the high-speed road and main-street in the geographic center, to make transportation very convenient. With the development of the Canal New Town, it was proposed, that metro line 9 be moved to Juzhou Road connecting to the two city districts.
- The sub-center in the northeast of the Canal New Town in the fishing bay will serve as the leisure center of Hangzhou Canal New Town and the hub of water and land transportation. It will offer a rich variety of seafood and water leisure facilities through the agricultural logistic center, an expressway around the city and the existing large-scale open water space.
- The sub-center in the southeast of the Canal New Town "Brands Castle" provides the Gongbei area; a waterfront business and leisure complex in combination with the Yangnei Harbor.
- The oil refinery and the Forum Island are transformed into an industrial park harbor.

Diverse waterfront experiences were created with the help of a canal tributary system, which enables healthy ecological use. Characteristic waterfront patterns informed the various urban functional complexes such as the Fishing Bay, Forum Island, Brands Castle, Galaxy Port, Industrial Park Harbor and so on. In combination with the characteristic water transport, the six basic tourist demands of "food, shelter, transportation, travel and shopping and entertainment" were satisfied. This is how the Canal New Town in Hangzhou was shaped.

从两种文化中学习 LEARNING FROM TWO CULTURES

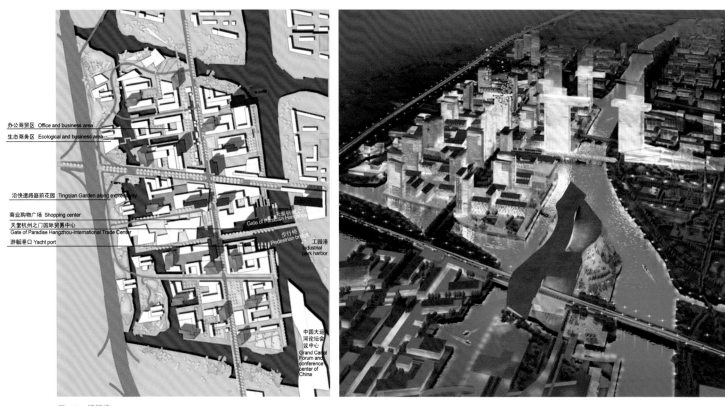

图 13 银河埠
Fig. 13 Site plan and bird's eye view of Galaxy Port

图 14 名品坞
Fig. 14 Bird's eye view of Brands Castle

城市发展 URBAN DEVELOPMENT

图 15 渔水湾
Fig. 15 Site plan and bird's eye view of Fishing Bay

 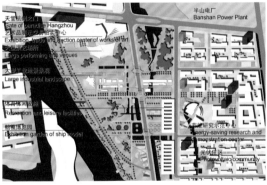

图 16 工园港
Fig. 16 Site plan and bird's eye view of Industrial Park Harbor

图 17 论坛岛
Fig. 17 Forum Island

城市发展
URBAN DEVELOPMENT

01 世界文化遗产城市波茨坦城市总体规划
The Urban Development Planning of World Cultural Heritage City of Potsdam

02 丽江玉龙新城规划
Masterplan of Yulong New Town, Yunnan

03 泉州市城市新区建筑天际轮廓控制规划方法研究
Planning of the Skyline of the City of Quanzhou, Fujian

04 泉州市区和环湾地区空间发展战略研究
Strategic Study on Development of the Urban Districts and General Area around Quanzhou Bay, Fujian

05 广州市南沙光谷地区城市发展规划
Urban Development of Nansha Guanggu in Guangzhou

06 阳宗海旅游度假区西北部概念规划设计
Conceptual Planning of the Northwestern Part of Yangzonghai Resort, Yunnan

07 东山蝶岛发展战略规划
Dongshan Butterfly Island in Fujian: the Strategic Planning of Urban Development

08 十堰市东部新城概念规划及城市设计
Conceptual Planning of the East New Town in Shiyan, Hubei

09 福州市东部新城中心城市设计
City Planning for the Center of the Eastern New Town of Fuzhou, Fujian

10 杭州天堂鱼生活：杭州市运河新城概念规划
A Fishing Life Paradise in Hangzhou: Concept Planning for the Canal New Town in Hangzhou

11 唐山市南湖生态城概念性总体规划设计及起步区城市设计
Conceptual Planning and Urban Design for an Ecological Zone in Tangshan, Hebei

12 大同市御东新区概念性总体规划及核心区概念性城市设计
Conceptual Planning of Yudong District and Urban Design of the Core Area, Datong

13 文昌市抱虎角概念规划
Conceptual Planning for Baohujiao Area in Wenchang, Hainan

14 昆明市绿地系统概念规划
Concept Planning for Kunming Green Space System

11 唐山市南湖生态城概念性总体规划设计及起步区城市设计

Conceptual Planning and Urban Design for an Ecological Zone in Tangshan, Hebei

规划面积 AREA	91 km²
人口规模 POPULATION	1000000
完成时间 PROJECT DATE	2008.11

起点 Starting point

The South Lake Ecological New District in Tangshan is an important area of development with strategic location and ecological advantages for the city. Yet, the urban development structure is too far south in order to function. The ecological potential is located between the downtown region and high quality ecological areas in the south. The main intensive traffic corridors with access from the south of the city are located in the southwest corner. Today, Tangshan is transforming from an industry-based city to a city of comprehensive structures and functions, so the South Lake ecological district is significant for regional research.

Studies of the Ruhr Region in Germany showed that the total quality of a city could be strengthened by the ecological transformation of cultural and educational facilities and businesses. Culture, leisure and entertainment industries are not only supplemental for the economy but also promote the cultural identity and urban image. The South Lake Ecological District possesses the following functions in the transformation planning of Tangshan: an eco-green center, roles for culture, leisure, entertainment, tourism, combined with large-scale educational institutions and research centers for the development of high-tech industries, as well as different types of high-quality residences supplemented by clean industry parks. The ultimate goal was to build a "cultural and ecological Nan Lake District" between the central regions in the north and new industrial district in the south.

唐山市南湖生态新城是唐山市兼具区位优势与生态优势的重要战略发展空间。功能结构上决定了城市的南向发展结构，生态结构上位于唐山中心城区与南侧系列高品质生态功能区域之间，其西南角交通廊道密集，是城市重要的南向出口位置。在唐山从工业主导型城市向综合产业结构与城市功能转换的今天，南湖生态新城具有区域层面的研究意义。

德国鲁尔区的系列研究表明，以生态化改造为前提，多元化产业的引入，尤其是教育文化产业的发展，将极大地提升城市综合品质。文化休闲与娱乐产业不仅仅是经济的补充，更是增强居民文化认同与提升城市对外形象的手段。南湖生态新城在唐山的转型中规划将形成下述功能组成：生态绿心，多样化的文化休闲娱乐功能，旅游休闲功能，结合高科技产业研发的大型教育机构与研究中心，整体区域范围内应当形成高品质、多种类的居住类型，同时辅以部分清洁工业园区。最终目标是在北部综合城市与南侧工业新城之间形成一个"南湖文化生态新城"。

> 文化休闲与娱乐产业不仅仅是经济的补充，而且是增强居民文化认同与城市对外形象的手段。
>
> *Culture, leisure and entertainment industries are not only supplemental for the economy but also promote the cultural identity and urban image.*

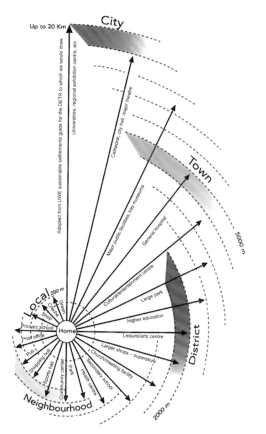

图 01　人力流动辐射关系
Fig. 01　Radiation relationship of human flow

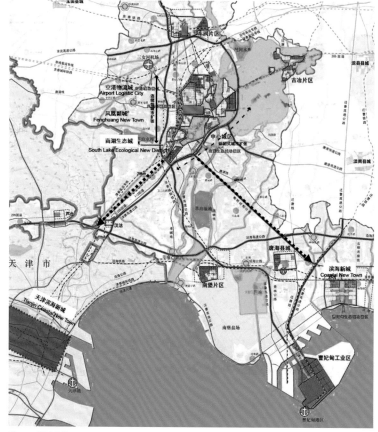

图 02　区域现状
Fig. 02　Status quo of planning area

城市发展 URBAN DEVELOPMENT

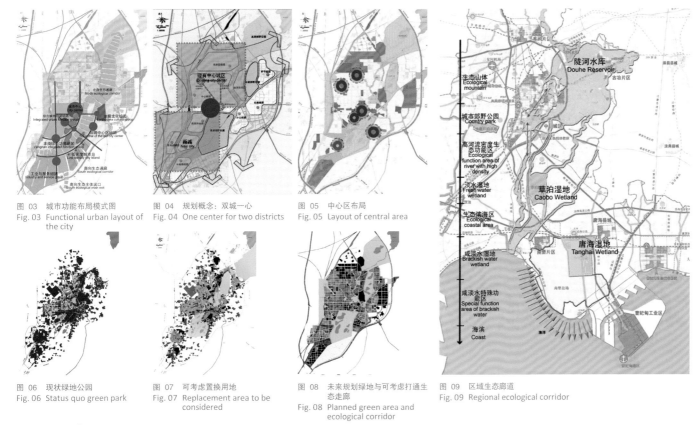

图 03 城市功能布局模式图
Fig. 03 Functional urban layout of the city

图 04 规划概念：双城一心
Fig. 04 One center for two districts

图 05 中心区布局
Fig. 05 Layout of central area

图 06 现状绿地公园
Fig. 06 Status quo green park

图 07 可考虑置换用地
Fig. 07 Replacement area to be considered

图 08 未来规划绿地与可考虑打通生态走廊
Fig. 08 Planned green area and ecological corridor

图 09 区域生态廊道
Fig. 09 Regional ecological corridor

规划理念 Planning concept

生态峡谷

唐山市现有城市建成环境中，包容了多处东北西南走向的生态结构，建议以南湖生态新城为契机，整合现有的生态功能核，打通东北向出口，在城市尺度上形成一道生态峡谷，同时在南侧结合生态与城市建设交织区域，全面接合侧高生态价值区域。

双城一心

南湖生态新城，在北侧将充分补充利用现有的城市结构，继续加强现有核心区功能优势，建立高效稳健的新城发展方案。以此为基础建立用地功能模式，高效组织城市5个功能组团：围绕行政中心的综合功能组团、小山历史文化组团、旅游休闲与生态居住组团、清洁工业与配套居住组团。

生态的南湖，生活的南湖

各个城市组团中心区与市级公共服务功能均位于南湖生态绿心沿线，一条壮美的滨湖林荫大道，串联起各个功能组团，以及城市行政办公商贸综合中心、唐山历史文化副中心、体育休闲中心、旅游中心、农业科研中心。各个中心如同火焰，围绕赋予唐山希望和凝聚力的火焰，形成环湖生命红环。

Ecological Canyon

The existing urban environment of Tangshan covers a number of ecological structures running from northeast to southwest. With the opportunity to create a South Lake ecological district, we suggested integrating all the existing ecological functions and opening the northeast corridor to form an ecological canyon. At the same time, we connected the mixed ecological area and urban development with the area of high ecological value in the south.

One center for two districts

The existing urban structure to the north of the South Lake Ecological District should be enhanced to develop an efficient development program, taking into account the existing functional advantages of the core area. Based on that determination, five functional groups were organized: the comprehensive group around the administrative center; the historical culture group on the hill; the tourist, leisure, and eco-living group; the clean industry group; and the supporting residence group.

Ecological Nan Lake, living Nan Lake

Every municipal public service facility, along with the inner area of each group, was located around the green ecological center. That center is a magnificent lakeside boulevard that connects the various functional groups: the comprehensive center of administrative offices and trade; the sub-center of historical culture; the sports, leisure and tourist center; and the agricultural research center. All these centers are positioned in a concentric ring-like circular pattern, and give Tangshan hope and cohesion.

185

从两种文化中学习　LEARNING FROM TWO CULTURES

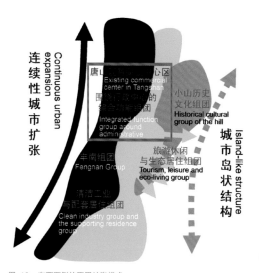

图 10 东西两侧的不同扩张模式
Fig. 10 Different expansion patterns in the west and east

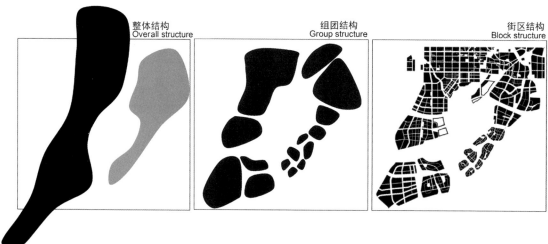

图 11 城市结构
Fig. 11 Urban structure

图 12 空气流动
Fig. 12 Air flow

图 13 水流
Fig. 13 Water flow

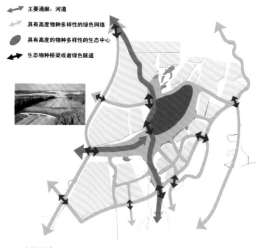

图 14 生物流动
Fig. 14 Bio flow

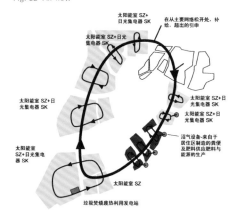

图 15 太阳能利用布局图
Fig. 15 Flow of solar energy utilization

图 16 物质流动
Fig. 16 Material flow

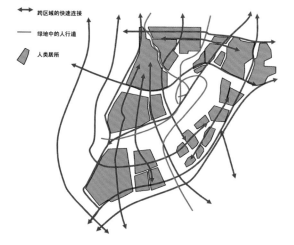

图 17 人力流动
Fig. 17 Human flow

西侧集约化城市，东侧生态化组团

从生态角度出发，通过对生态景观价值和城市建设强度的辩证关系研究，摸索出二者之间的平衡点。结合现有唐山城市结构，在南湖西侧强调连续性城市扩张，东侧强化大型生态布局中的组团式城市岛状结构。

流动城市

生态规划中提出流动城市（Flow City）这一概念，通过南湖绿色心脏的搏动，通过空气流动、水流动、能量流动、材料流动、生物流动、人文社会流动，形成一个活跃的南湖文化生态新城。在南湖生态新城各个区域内培养各种能源的生态试验基地。

Intensive city life in the west, ecological groups in the east

The dialectical relationship between the landscape value and the intensity of urban construction was examined from an ecological point of view. Based on this study, there was an attempt to create balance between the two factors through combining additional construction with the existing formation, guiding the continuous urban expansion to the west of South Lake, and stressing the island-like grouped structure in the large-scale ecological area of the east.

The concept of a "Flowing City"

The concept consists of several steps, specifically planning the area as a green heart offering air flow, water flow, energy flow, material flow, biological flow, humane and social flow. In this way, an active cultural and ecological area could be created at the South Lake. All kinds of ecological foundations were formed throughout every region of the South Lake district.

图 18 滨湖重要城市节点
Fig. 18 Important lakeside city node

公共空间发展模式
Development pattern of public space

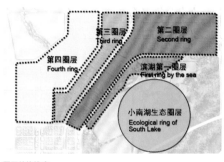

圈层结构模式
Circle-style structure pattern

图 19 总平面图
Fig. 19 Master plan

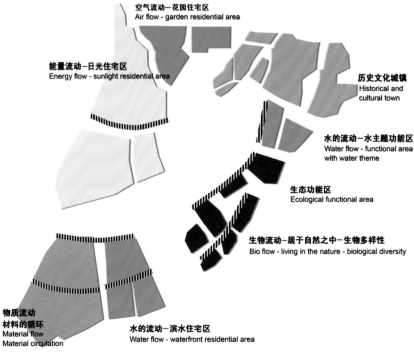

图 20 流动的城市
Fig. 20 Flow city

图 21 鸟瞰图
Fig. 21 Bird's-eye view

起步区城市设计 Urban design

The overall planning
The plan set up the government administrative area, the conference and exhibition center, and parts of the business office area in the heart of the city. These areas all belong to both the ecological new district and the central part, and include the cultural industries of the traditional district east of the city. In this way, a highly integrated public service circle could be created. In the short term, the comprehensive benefits should form a large-scaled all-round functional center that effectively links the South Lake New District to the central part.

Ring-like functional mode
A series of important urban functional rings, like concentric circles, were planned around the lake. The South Lake itself is the regional ecological core. The first ring consists of culture, leisure, commerce and tourism with a high amount of greenery around the Lake. The second ring is for the administrative center, business offices, individual retail stores, culture, leisure and hotels. The third one consists mainly of offices, large-scale shops and residences. The fourth outer ring is mainly residential.

Administrative center
The new government center needed to represent rigorousness, efficiency and delight for life. The planning concept utilized the north-south space axis to create a 'square city.' Four groups of governmental office buildings were arranged, wherein city hall and its adjacent square were positioned symmetrically. Solar panels were installed on the large roof of city hall and the main government office building, covering a large winter garden inside. These were all designed to be landmarks of the new district. A public concert hall and a VIP building were planned, surrounded by a green open space to the south of Lake. Also located in the central area are a water-droplet-shaped square, a large-scale water area, and a greenbelt that is truly a beautiful spectacle.

总体规划
在生态新城与中心城区共有的城市心脏位置，规划集中设置行政中心区、会议会展和部分商务办公区，其东侧传统城区结合文化产业的改造工作，共同形成一个高综合度的公共服务圈，短期内以乘法效应迅速提高综合效益，形成有效联系南湖新城与中心城区的大型综合功能中心。

圈层化功能模式
以小南湖作为区域的生态核心圈层，依次形成由沿湖高绿化率文化休闲、商业、旅游组成的第一圈层，集行政中心、商务办公、商业购物、文化休闲及酒店的第二圈层，以办公、大型商业购物和混合居住为主的第三圈层，以及外侧以居住为主的第四圈层。沿湖在行政中心之外将带状发展出一系列的重要城市功能中心。

行政中心区
以方城为规划理念，结合南北向空间轴线，以对称方式由北向南序列分布四组行政办公建筑群体、市民接待中心、市政广场。市民接待中心及政府办公主楼结合太阳能技术统一采用大型栅栏屋顶形式，覆盖内部大型冬季花园，形成生态新城的标志性中心建筑。小南湖区域市民音乐厅、贵宾楼呈现开放、自由的城市形态，中央区域跌落广场、大型水面、绿岛交相呼应，景色优美壮观。整体营造高效严谨而又充满了生活情趣的新型政府中心区。

城市发展 URBAN DEVELOPMENT

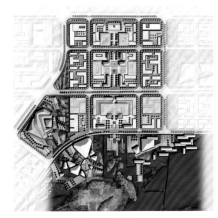

图 22 政府办公及滨水区总平面图
Fig. 22 Site plan of government and waterfront area

图 23 滨水"一站式"商业综合区总平面图
Fig. 23 Site plan and bird's eye view of waterfront comprehensive business area

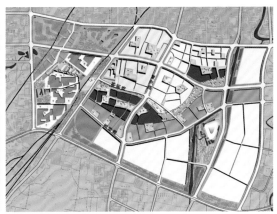

图 24 小山历史文化片区更新概念平面图
Fig. 24 Urban renewal concept of historical and cultural area on the southeast hill

滨湖多功能综合区

水面、绿地环绕的商务办公区，带动两侧SOHO办公居住区和五星级酒店形成起步区办公功能中心。滨湖区以大型城市水景广场为中心，环绕设置群艺馆与影音娱乐中心，大型滨湖主题型商业购物中心，以及七星级度假酒店、大型企业俱乐部等设施。"一站式"商业文化娱乐综合区域弥补了唐山该类商业业态的缺失。两者之间的带状大型地景广场，借助城市展示、景观雕塑群体塑造，形成一个艺术气息浓郁的文化公园，并通过轴向休憩绿地与中心绿地公园，向城市各个区域及南湖渗透。整体上倡导的是一个高品质、高综合度、高度活跃性的城市休闲文化中心。

东南小山历史保护组团

小山历史保护组团地处老唐山发源地，震前繁华商业片区之内，历史文化价值以及土地待开发价值都很高，属于开发敏感区，开发保护规划很重要。在保护面积10平方公里，核心保护面积6平方公里，现状为居住、市场、工业呈无序混合状态的情况下，需保留风貌特征鲜明的区域作为城市记忆的碎片，保证人居环境的可持续发展，并以此为基础制定保护框架结构。

整体概念的创新严谨以及工作团队的细致专业，使本次规划设计工作得到了政府及各位专家的首肯，并继续委托我们进行了唐山东南片区城市设计及控制性详细规划的工作。

Comprehensive multi-functional lakeside area

This area was planned as a business office region surrounded by water and green space. Together with the residential area and five-star hotels, it should serve as the functional hub and would be equipped with the following facilities: a large-scale urban square with a waterscape view, a public art center in the heart of the district, entertainment, a large shopping center with a lakeside theme, a seven-star resort hotel, large clubs and other facilities around the core. This comprehensive area for business, culture and entertainment with a "one-stop" model should complete all the commercial establishments in Tangshan. Between the two areas, a large-scale belt-like landscape park was created as a rich cultural public area featuring many landscape sculptures. This greenbelt will branch out into every area and connect the South Lake with both the axial green spaces and the central park, so that a high quality urban leisure center can be created.

The preservation area of the southeast hill

This area was the old business district before the earthquake, and held a great deal of historical and cultural value in the region. So it had to be developed sensitively in order to preserve the urban heritage.

Despite the existing chaotic circumstances of modern life, the core area of 6 km2, including the market and individual retail shops, needed to be preserved. All the distinctive characteristics had to be maintained as a memory of city; it guarantees sustainable development for the living environment. On the basis of this, a protection framework was created.

Thanks to the rigorousness and innovation of the planning concept as well as the meticulous work of our professional team, the planning concept was approved by both the expert and the government, and was adhered to during the urban design and control planning process.

带状大型地景广场向城市各个区域渗透，形成一个高品质，高综合度的城市休闲文化中心。

A large-scale landscape park should branch out into every area and connect the South Lake with both the axial green spaces and the central park, so that a high quality urban leisure center can be created.

城市发展
URBAN DEVELOPMENT

01 世界文化遗产城市波茨坦城市总体规划
The Urban Development Planning of
World Cultural Heritage City of Potsdam

02 丽江玉龙新城规划
Masterplan of Yulong New Town, Yunnan

03 泉州市城市新区建筑天际轮廓控制规划方法研究
Planning of the Skyline of the City of Quanzhou, Fujian

04 泉州市区和环湾地区空间发展战略研究
Strategic Study on Development of the Urban Districts
and General Area around Quanzhou Bay, Fujian

05 广州市南沙光谷地区城市发展规划
Urban Development of Nansha Guanggu in Guangzhou

06 阳宗海旅游度假区西北部概念规划设计
Conceptual Planning of the Northwestern Part of
Yangzonghai Resort, Yunnan

07 东山蝶岛发展战略规划
Dongshan Butterfly Island in Fujian:
the Strategic Planning of Urban Development

08 十堰市东部新城概念规划及城市设计
Conceptual Planning of the East New Town
in Shiyan, Hubei

09 福州市东部新城中心城市设计
City Planning for the Center of
the Eastern New Town of Fuzhou, Fujian

10 杭州天堂鱼生活：杭州市运河新城概念规划
A Fishing Life Paradise in Hangzhou:
Concept Planning for the Canal New Town
in Hangzhou

11 唐山市南湖生态城概念性总体规划设计
及起步区城市设计
Conceptual Planning and Urban Design
for an Ecological Zone in Tangshan, Hebei

12 大同市御东新区概念性总体规划
及核心区概念性城市设计
Conceptual Planning of Yudong District and
Urban Design of the Core Area, Datong

13 文昌市抱虎角概念规划
Conceptual Planning for Baohujiao Area
in Wenchang, Hainan

14 昆明市绿地系统概念规划
Concept Planning for Kunming Green Space System

12 大同市御东新区概念性总体规划及核心区概念性城市设计

Conceptual Planning of Yudong District and Urban Design of the Core Area, Datong

规划面积 AREA		74.43 km²
人口规模 POPULATION		150000
完成时间 PROJECT DATE		2008.06

城市新貌 A new image for the city

Datong is an industrial city on a mountain, a city of high-tech industries, services, culture, education and tourism, which is required for the development of modern cities. Datong would develop into an ecological green city, which could provide a high quality of public life to meet the rapid growth needs of the Chinese people. In this context a strong city image of high society and urban environment, characterized by ecological landscape, must be developed. The traditional spatial structure of old Datong is the basis of the design concept for the new town Yudong. The Yudong district is to be modeled after traditional building types and space systems such as the combination of vertical axes with a rectangular greenbelt extending from the development axis of Datong city.

大同市是一个在山地上发展起来的工业城市，发展高科技产业、服务业、文化、教育和旅游业成为大同市满足现代大城市发展要求的必经之路。人民对生活质量各方面需求的快速增长，也促使大同必将发展成为一个能够为市民提供高质量生活的园林式生态城市。因此，一方面突出高度文明所表现出来的强烈的城市形象，另一方面塑造以生态园林为特征的城市环境，成为了设计方案的出发点。

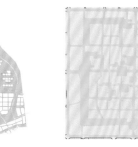

图 02　历史和传统变迁下的城市空间痕迹
Fig. 02　Urban space trace with focus on historical and traditional changes

图 01　继续营造鲜明城市形象的可能性研究
Fig. 01　Feasibility study on continuous creation of a distinctive city image

图 03　东西向城市历史轴线
Fig. 03　East-west urban historical axes

图 04　文瀛湖
Fig. 04　Wenying Lake

图 05　御河及沿岸
Fig. 05　Yu River and the coast

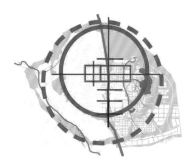
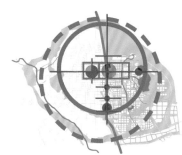

图 06　空间结构构思
Fig. 06　Concept of spatial structures

城市发展　URBAN DEVELOPMENT

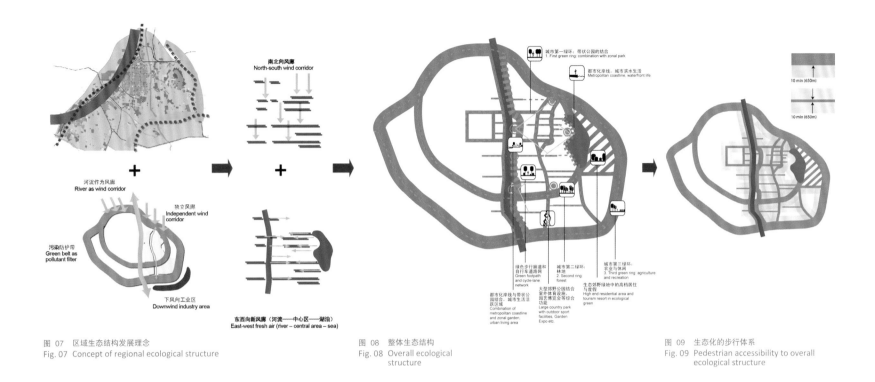

图 07　区域生态结构发展理念
Fig. 07　Concept of regional ecological structure

图 08　整体生态结构
Fig. 08　Overall ecological structure

图 09　生态化的步行体系
Fig. 09　Pedestrian accessibility to overall ecological structure

尊重历史和传统文化的延续性
Respecting the continuity of history and traditional culture

在分析城市发展历史的基础上形成新规划。将城市古老的发展准则应用到现代化新城建设模式当中，实现历史的延续性。

以大同古城的传统空间结构作为新区城市设计的基本概念构思。旧城的基础要素——正方形结构、具有文化传承性的几何模型，定义了老城清晰的边界，并作为新区城市设计的母题。御东新区以矩形环状绿带与中心城区周边两层生态背景相衔接，公共空间结构与纵轴线也遵从同样的原则进行全面接合。新城主体功能与现有中心城区具有明确的差异，但同时又延续了其核心结构——商务中心轴线的扩展方向。东西向的古城历史之轴和南北向的御河生态之轴是两条城市发展轴线，结合用地内部的文瀛湖，共同营造鲜明的城市形象。

Our group made a new program based on analysis of the history of urban development. The old guidelines for city development were applied to modern city construction methods to allow historical continuity. The "old city" is characterized by a clear geometric form; with squares as its basic elements. The two axes of city development are the east-west axis of the historical old city and the north-south ecological axis of the Yu River. The new city plan reproduces the city's main grid structure, extending the east-west axis while allowing the functional axis to develop in stages in the north-south direction. Landscape and culture tie each district together forming a unified entity connected by the city axes and parks, while the Yu River plays the role of connecting the central north-south axis and the east-west axis of the old city.

The center of the new city, which itself consists of several smaller centers, would provide services for Datong City. The development patterns closely match those of the old city. The CBD trade and business center in the north was developed based on the old city. The administrative and cultural center was located on Wenying Lake. And, the exhibition center in the south and industrial groups in the west were closely linked with the industrial functions within the region, transforming it into a modern service center. In this way, the old and new districts would form a new entity.

将传统城市发展准则应用于现代城市建设方式中，实现历史的延续。

The old guidelines for city development were applied to modern city construction methods to allow historical continuity.

历史+生态+功能+公共空间结构+城市形态轴线=新城的整体空间结构。新规划在继续延续东西轴线的同时,在南北方向上衍生出可分期发展的功能轴线,再现十字形城市主结构。城市各区域将与城市轴线和公园相连,在景观与文化联系上形成一个整体,御河起到缝合与联系中心南北轴线与古城东西轴线的作用。各条纵向轴线的绿化结构又在横向上被大同城市发展轴所贯穿,形成富有变化又具有传统风貌的城市空间形态。

新城城市中心以多中心的形式为大同市提供服务,开发格局与旧城紧密联系,北侧CBD商贸中心依附旧城发展,行政与文化中心依托文瀛湖发展,南侧会展中心与西侧工业组团与本区域内工业功能紧密联系,形成了一个现代服务业中心区域。借此,新旧城区形成了新的统一体。

以现状生态基础结合区域生态结构,形成区域内两个方向各种生态廊道的基础,叠加城市职能与其背后的城市生活,形成一个"生态城市"的整体生态结构,它同时借助步行可达性进行检验。

图 10 总平面图
Fig. 10 Master plan

核心轴线上鲜明的城市形象 Creating a distinctive civic image

The new district would attain a distinctive civic image by contrasting its modern urban architecture sharply with its old city elements.

The east-west development axis is the main functional tie between Datong's existing districts and the future district Yudong; but it is also a functional transportation axis of significant historical value, which has borne witness to the development of Datong city. This axis has the central district of the old city at its center. It also crosses the Yu River up to Wenying Lake and links the new CBD core area to the science and education area, cultural and administrative center, as well as many public open and leisure spaces. The interconnected areas along this axis will represent a variety of different landscape interfaces.

通过新区建设提升自身城市形象,创造与古城形成鲜明对比的现代城市文化。

The new district would attain a distinctive civic image by contrasting its modern urban architecture sharply with its old city elements.

In the north-south direction in the new district Yudong, two main urban functional axes were planned parallel to the main north-south axis of Datong old city: a commercial axis along the river with a business core and an administrative axis with cultural facilities in the center of city. In addition, Datong University is surrounded by two ecological axes, which limit its expansion on the east-west axis.

城市东西向发展轴为大同现有城区和未来御东新城区的主要横向联系轴线,同时也是一条见证大同城市发展的极具历史价值和意义的功能与交通轴线。这条轴线以原有古城中心区为源头,跨御河一直延续到文瀛湖。之间串联CBD核心区、高等科教园区、文化行政中心区以及多个开敞型公共空间和休闲空间,沿线将形成各种丰富多样的空间景观界面。

在南北向上,御东新区将主体打造与大同古城南北主轴线平行的两条主要城市功能轴线——分别为以CBD商贸核心区为中心的滨河轴线和以行政文化中心为核心的行政轴线。此外在大同大学边沿两条生态轴加以控制,用以限定其在城市东西轴线上的扩张。

图 11 位于御河与东西向主轴交汇点的城市中心区有力地强调了城市轮廓
Fig. 11 The new city center on the crossing of the river and the new east-west axes shapes the strongest accents of the city skyline

城市发展 URBAN DEVELOPMENT

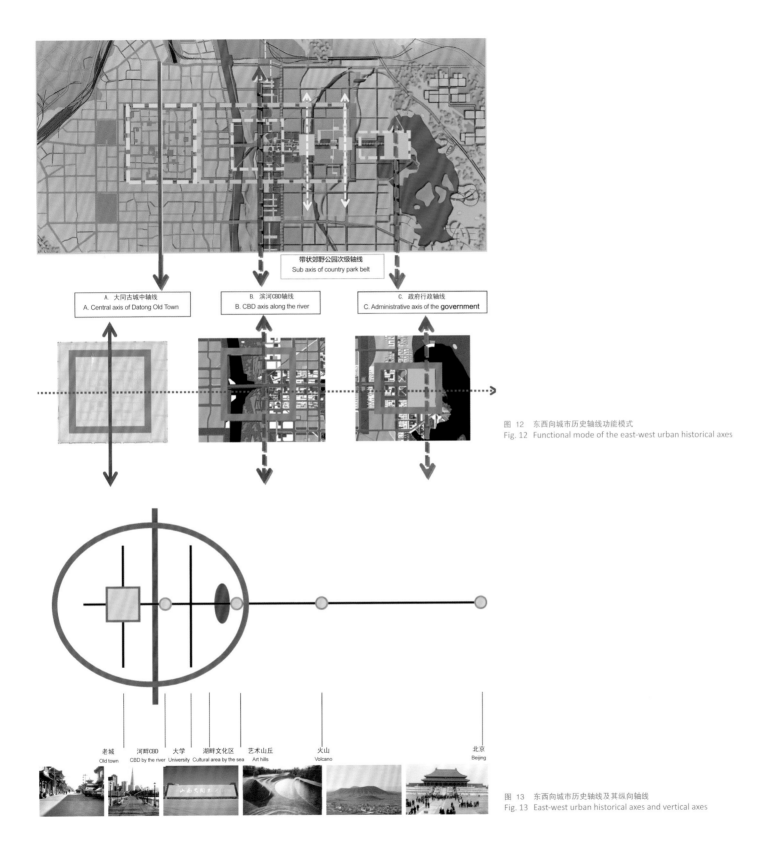

图 12　东西向城市历史轴线功能模式
Fig. 12　Functional mode of the east-west urban historical axes

图 13　东西向城市历史轴线及其纵向轴线
Fig. 13　East-west urban historical axes and vertical axes

195

图 14 城市核心轴线总图
Fig. 14 General layout of the urban core axes

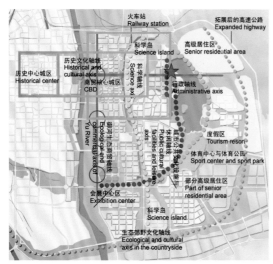

图 15 公共空间及轴线规划
Fig. 15 Public space and axis concept

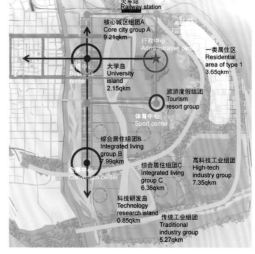

图 16 功能结构规划
Fig. 16 Functional structure concept

图 17 文瀛湖滨水行政文化次轴线
Fig. 17 Waterfront administrative and culture sub-axes of Wenying Lake

城市发展
URBAN DEVELOPMENT

01 世界文化遗产城市波茨坦城市总体规划
    The Urban Development Planning of
    World Cultural Heritage City of Potsdam

02 丽江玉龙新城规划
    Masterplan of Yulong New Town, Yunnan

03 泉州市城市新区建筑天际轮廓控制规划方法研究
    Planning of the Skyline of the City of Quanzhou, Fujian

04 泉州市区和环湾地区空间发展战略研究
    Strategic Study on Development of the Urban Districts
    and General Area around Quanzhou Bay, Fujian

05 广州市南沙光谷地区城市发展规划
    Urban Development of Nansha Guanggu in Guangzhou

06 阳宗海旅游度假区西北部概念规划设计
    Conceptual Planning of the Northwestern Part of
    Yangzonghai Resort, Yunnan

07 东山蝶岛发展战略规划
    Dongshan Butterfly Island in Fujian:
    the Strategic Planning of Urban Development

08 十堰市东部新城概念规划及城市设计
    Conceptual Planning of the East New Town
    in Shiyan, Hubei

09 福州市东部新城中心城市设计
    City Planning for the Center of
    the Eastern New Town of Fuzhou, Fujian

10 杭州天堂鱼生活：杭州市运河新城概念规划
    A Fishing Life Paradise in Hangzhou:
    Concept Planning for the Canal New Town
    in Hangzhou

11 唐山市南湖生态城概念性总体规划设计
    及起步区城市设计
    Conceptual Planning and Urban Design
    for an Ecological Zone in Tangshan, Hebei

12 大同市御东新区概念性总体规划
    及核心区概念性城市设计
    Conceptual Planning of Yudong District and
    Urban Design of the Core Area, Datong

13 **文昌市抱虎角概念规划**
    **Conceptual Planning for Baohujiao Area
    in Wenchang, Hainan**

14 昆明市绿地系统概念规划
    Concept Planning for Kunming Green Space System

13 文昌市抱虎角概念规划
Conceptual Planning for Baohujiao Area in Wenchang, Hainan

规划面积 AREA		43.5 km²
人口规模 POPULATION		100000
完成时间 PROJECT DATE		2009.04

起点 Starting point

Wenchang is located at the northeast corner of the Hainan province, 70 kilometers from the city of Haikou. Wenchang and Haikou enjoy certain advantages regarding regional development that represent a comprehensive tourist zone in the northern ring of the Hainan province. The Baohujiao tourist resort, with a planning area of 43.5 square kilometers including 23 km of coastline, is a prime location for the tourist industry thanks to an abundance of natural resources and good ecological conditions.

文昌市位于海南省东北角，距海口市70公里，与海口市共同构成城市区域联动发展优势，形成海南省北部圈层的旅游综合区。抱虎角旅游度假区规划面积43.5平方公里，拥有23公里海岸线，自然资源丰富，生态情况良好，是天然的旅游产业用地。

图 01　规划区域在文昌市的区位
Fig. 01 Location of planning area in Wenchang

图 02　海岸线长度及规划区面积
Fig. 02 Coastline of planning area

工作方法 Work methods

Planning extracted the three main keywords "climate, sea and beach", to develop an overall concept dubbed "R.M.P.S." was devised, with "R" indicating resources, "M" indicating market, and "P" indicating production; all of which would be integrated into "S" indicating space.

Based on the existing ecological conditions of the Baohujiao area, it was imperative that ecological elements such as the coastline, the mountain farmlands and rivers be rehabilitated and only developed selectively. In accordance with GIS and analysis of the present ecological conditions and landscape protection, a land use concept was created that designated land for protection, land for construction and multi-purpose land. For the land use of building construction, several blocks of a 7.8 km² area were selected for the first development. The Baohu Bay block is fairly close to a sightseeing point, and the Jingxinjiao and the west blocks are close to the Mulangang block with its smooth terrain and landforms. The other eight planning blocks represented 2.2 km² of land.

As for the complex development of the area's 23 kilometers of coastline, our group conducted an analysis and comparison of the internal resources of Baohujiao point by point. Because the development of beaches is a decisive factor for tourism, research methods for assessing the quality of the coastline were created. The criteria included factors affecting tourist

规划通过提取"气候"、"海"和"沙滩"等关键词，以R-资源（Resource）、M-市场（Market）、P-产品（Product）分析为基础，结合进行S-空间规划（Space），形成了连续整体的RMPS工作逻辑。

针对抱虎角地区生态现状，对现有海岸线、山林农田、河流进行了生态保护、更新和选择性开发。根据GIS分析、现状生态及景观保护性分析，确定了整个区域中不可建设用地、不适宜建设用地以及在保持总量不变的前提下可进行调整的用地。整合可建设用地，从中选取距离景观价值点较近的抱虎湾片区、景心角片区及地形地貌较为平整同时距离木兰港片区较近的西侧片区为一期开发用地，用地面积7.8平方公里；其他规划可开发用地共计8块，用地面积2.2平方公里。

针对23公里长的海岸线的开发复杂性，对抱虎角内部资源进行有针对性的逐点分析比较。对于旅游开发起决定因素的海滩因素，开创了海岸线资源品质研究方法：将影响旅游开发的诸多因素分为直接影响要素（如长度、浅滩宽度、坡度、植被类型、沙质、礁石等）、间接影响要素（如潮汐活动的范围、海浪情况、风向及风速

以R-资源(Resource)、M-市场(Market)、P-产品(Product)分析为基础，结合进行S-空间规划(Space)，形成了连续整体的RMPS工作逻辑。

An overall concept dubbed "R.M.P.S." was devised, with "R" indicating resources, "M" indicating market, and "P" indicating production; all of which would be integrated into "S" indicating space.

城市发展 URBAN DEVELOPMENT

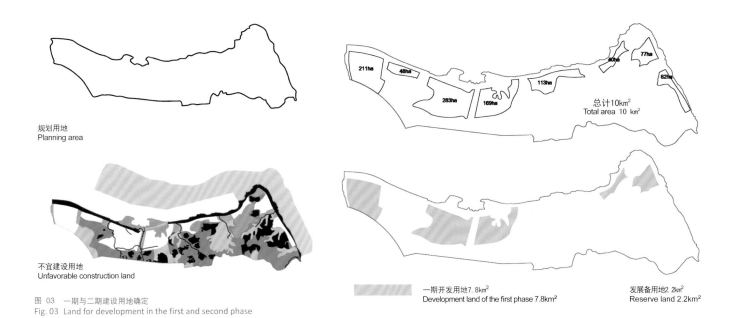

规划用地
Planning area

不宜建设用地
Unfavorable construction land

一期开发用地7.8km²
Development land of the first phase 7.8km²

发展备用地2.2km²
Reserve land 2.2km²

图 03　一期与二期建设用地确定
Fig. 03　Land for development in the first and second phase

等），再根据对于海滩品质的影响程度给出相应的加权系数，进行综合比较分析。在对于选定的六个沙滩进行量化分值比较中，分别对影响程度较大的直接和间接因素赋以一定的加权系数，计算确认景心港（抱虎角）排在第一位，而抱虎港沙滩资源品质最优，并且区位、环境方面综合品质最佳。

development, such as physical elements like length and width of shore, slope, vegetation, quality of sand, rock, etc., as well as climatic factors such as tidal activity range, wave conditions, as well as wind speed and direction. The impact on the quality of beach would give the relevant weighted coefficient to conduct the comprehensive weighted comparative analysis. When comparing the quantified scores of the six selected beaches, we gave a weighted coefficient of 5-10 times to physical factors compared to climatic ones. The conclusion was that Jingxin Harbor (Baohu Corner) took first place in overall ecological quality, but Baohu Harbor had the highest quality in terms of beach resources, location and environment.

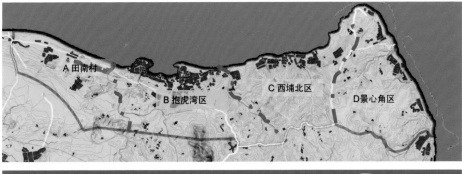

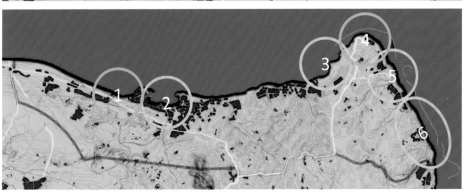
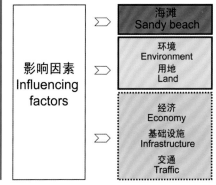

图 04　整体资源评价比较结论图
Fig. 04　Comparative evaluation of sea resources

201

图 05 旅游产品建设指导原则
Fig. 05 Guiding principles for tourism development

规划重点 Planning focus

The Baohujiao District was planned to remodel the character of the cultural landscape with the prerequisite of ecology, establish a unique coastline and develop fashionable sports and leisure activities. It would become a high-grade and middle-grade tourist town for vacation and leisure activities, serving global internationalization with a combination of sports, resorts and high-grade residences.

The three key development blocks were determined. Baohu Bay would function as the tourist center, an urban union of large hotels, restaurants, convention centers, cultural exhibitions, offices and residences. The Jingxin Block would become an area full of high-quality seascape hotels, leisure golf, and would include many tourist facilities for surfing, diving and other sea sports. Hulu Harbor would become a leisure area with a focus on water sports, including a water sports center composed of a multi-sport community club, training centers, a yacht port and other facilities.

The plan also covered tourism, overall space planning, the urban design of Baohu Bay, the Jingxin Block and Hulu Harbor, as well as a series of other projects. Additionally, it conducted research on potential implementation of a Formula 1 Championship and wind power station in Baohujiao.

确定三片重点开发片区：抱虎湾片区以充分保护内侧沙滩为核心，向内扩展景观塑造空间，并因此扩大了景观的影响区域与稳定性，功能上确定为旅游中心区，形成集大型酒店、餐饮、会议中心、文化会展、办公居住于一体的城市综合体；景心角片区为高档海景度假酒店及高尔夫休闲区域，包括以冲浪、潜水等海上运动为主的多种度假旅游设施；葫芦港片区为水上运动主导休闲区，围绕游艇等水上运动，形成包括多种运动社区俱乐部培训中心的水上运动中心与游艇港口。

本次规划的工作内容贯穿了旅游策划、空间整体规划、抱虎湾、景心角及葫芦港片区城市设计等一系列工作，并对F1一级方程式锦标赛及抱虎角片区风力发电站项目进行了可实施性研究，为城市提供了满意的工作成果。

我们规划的是一个优美的旅游城市，而不是旅游地产堆积的空城。

The goal was to plan a vivid and beautiful city, not a city of real estate accumulation.

城市发展 URBAN DEVELOPMENT

图 06 总平面图
Fig. 06 Site plan

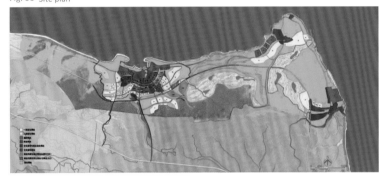

图 07 土地使用规划图
Fig. 07 Land use plan

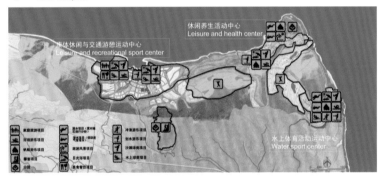

图 08 运动设施规划图
Fig. 08 Plan for sport facilities

图 09 抱虎湾中心区鸟瞰图
Fig. 09 Bird's-eye view of Baohu Bay central area

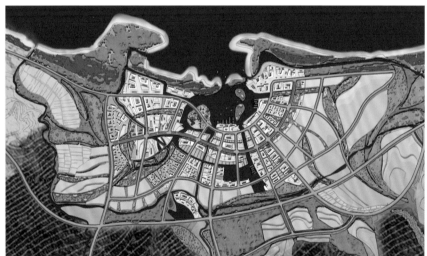

图 10 抱虎湾中心区平面图
Fig. 10 Site plan of Baohu bay Area

城市发展
URBAN DEVELOPMENT

01 世界文化遗产城市波茨坦城市总体规划
The Urban Development Planning of World Cultural Heritage City of Potsdam

02 丽江玉龙新城规划
Masterplan of Yulong New Town, Yunnan

03 泉州市城市新区建筑天际轮廓控制规划方法研究
Planning of the Skyline of the City of Quanzhou, Fujian

04 泉州市区和环湾地区空间发展战略研究
Strategic Study on Development of the Urban Districts and General Area around Quanzhou Bay, Fujian

05 广州市南沙光谷地区城市发展规划
Urban Development of Nansha Guanggu in Guangzhou

06 阳宗海旅游度假区西北部概念规划设计
Conceptual Planning of the Northwestern Part of Yangzonghai Resort, Yunnan

07 东山蝶岛发展战略规划
Dongshan Butterfly Island in Fujian: the Strategic Planning of Urban Development

08 十堰市东部新城概念规划及城市设计
Conceptual Planning of the East New Town in Shiyan, Hubei

09 福州市东部新城中心城市设计
City Planning for the Center of the Eastern New Town of Fuzhou, Fujian

10 杭州天堂鱼生活：杭州市运河新城概念规划
A Fishing Life Paradise in Hangzhou: Concept Planning for the Canal New Town in Hangzhou

11 唐山市南湖生态城概念性总体规划设计及起步区城市设计
Conceptual Planning and Urban Design for an Ecological Zone in Tangshan, Hebei

12 大同市御东新区概念性总体规划及核心区概念性城市设计
Conceptual Planning of Yudong District and Urban Design of the Core Area, Datong

13 文昌市抱虎角概念规划
Conceptual Planning for Baohujiao Area in Wenchang, Hainan

14 昆明市绿地系统概念规划
Concept Planning for Kunming Green Space System

14 昆明市绿地系统概念规划
Concept Planning for Kunming Green Space System

规划面积 AREA		253 km²
人口规模 POPULATION		6432212
完成时间 PROJECT DATE		2006.09

背景提问 Background

In the aerial photograph of Kunming entitled "Spring City", it is readily apparent that over time the buildings and other structures have intermingled with the greenbelts between them to form a kind of jigsaw pattern. Because of the high pressure on development, the natural ecological structure of the city is difficult to define. There is no clear delineation between the city's metropolitan area with its sections of urban enclaves, and the extensive ecological environment.

在把绿地资源作为生态资源的同时,发掘其文化、体育、教育、旅游、经济资源的价值。
The value of ecological resources, particularly green resources must be explored in relation to culture, sports, education, tourism and economy.

In a preliminary weeklong survey, focus was placed on the current features of the urban natural landscape that exhibit the characteristics of an urban 'green space'; i.e. public spaces, proposed parks, etc., with regard to their ecological, cultural and economical values. A German design team conducted constructive analysis, in order to identify what potential green spaces could be secured through supplementation and transformation of existing ecological landscapes.

The analysis highlighted two prominent aspects:

- "Spring City", Kunming is already a very ecologically friendly place and there is little environmental risk. In the center of the city, the ratio of green space is 0.64 square meters per person, but this area is composed mainly of large-scale parks with low accessibility.

- Kunming has a good historical background of structure and texture of urban development, but the overall structure of the urban space is not clear at present. Green space is disjointed throughout the city, dotted across cultural heritage sites and around structures.

"春城"昆明的航拍图上,绿地结构已经与城市建设结构形成了犬牙交错的格局:城市内部受建设压力影响,自然生态结构阅读困难;城市外部片段化的城市飞地楔入自然生态环境,没有明显的近郊森林集中区域与大型生态功能富集区——而本次绿地系统规划虽然是重要法定专项规划之一,但3周的工作时间已经基本控制了这一工作的基调:一个新思路而不是一个新方案。

约1周的前期调研,对现有城市自然景观特色、城市绿化特色、公共空间形态、拟建公园区域进行了重点调研。与此同时,德国设计团队进行了以航拍图为基础的城市可建设性分析,力图寻找以生态地景补充、改造为手段的绿地规划的可发展空间。

现状与航拍图综合呈现两个最为突出的问题:

- 以"春城"的美名闻名于世的昆明市是一个大圈层生态环境品质突出、小尺度城市自然环境品质堪虞的区域。中心区人均绿地0.64平方米,并且以大型公园绿地为主,可达性很低。

- 其城市发展的历史、结构及肌理虽然有良好的基础,但空间的实际感受性并不清晰。绿地结构与文化结构、空间结构、城市生活脱节。

研究内容 Research contents

In our research efforts we strove to plan a combination of ecological living and landscape functions that would support a development framework of green spaces in "Kunming".

The city of Kunming bears the largest burden of ecological responsibility for green areas in the entire region. The main planning tasks lie in restoring the existing branches of ecology, landscape and space; as well as identifying the ecological and green resources with potential value in the areas of culture, sports, education, tourism and economy. Our research emphasized three main types of green areas: countryside, low maintenance suburban parks and greenbelts in the city center with high ecological and living value. This became the basis for creating the hiking system surrounding the Dianchi Lake and its mountains, improving on the overall ecological situation in the area and adding alternatives to the lower quality green areas in central Kunming.

The green areas in Kunming city are the largest carrier of city life, besides being responsible for landscape and ecology. The development goal is to increase the total quantity of green space throughout the city, while simultaneously improving the cohesiveness of the city's overall urban structure. Taking into account the central parks of Frankfurt and Vienna, as well as the general history of urban space development, we have selected the historic Kunming city walls, a 300 meter wide space along the inner ring, as the experimental site for ecological transformation. Through analysis of aerial photography, our inquiry discovered that open spaces, smaller villages within the city, industrial areas

由此,我们试图在本次规划中以"昆明"为背景,建设一个生态功能、生活功能、景观功能的结合体,一个绿化体系的未来发展框架。

昆明区域范围绿化体系为城市担负着最大的生态任务。规划工作核心在于示范性理顺现有生态、景观、空间脉络;在把绿地资源作为生态资源的同时,发掘其文化、体育、教育、旅游、经济资源的价值。在三个层面的绿化体系中,着力强调郊野绿化这一低维护需求、高生态与高生活价值的近郊公园类型,从而形成围绕滇池与周边山脉的城市远足体系,对昆明现有中心城区绿化水准低下的情况形成补充。

昆明城区内绿化体系担负了景观、生态任务,是城市生活的最大载体。发展目标除了总量的提高,还要增强城市结构的可读性。参考法兰克福与维也纳城市中心公园的布局实例,根据城市空间发展的一般历史,我们选择了昆明历史城墙所在位置——城市目前内环沿线400米进深空间作为生态化改造实验区域。在航拍图分析中,果然发现城市空地、城中村、工业用地、低质量结构混乱区域约占据了内环沿线45%以上的面积,此外在这一区

城市发展 URBAN DEVELOPMENT

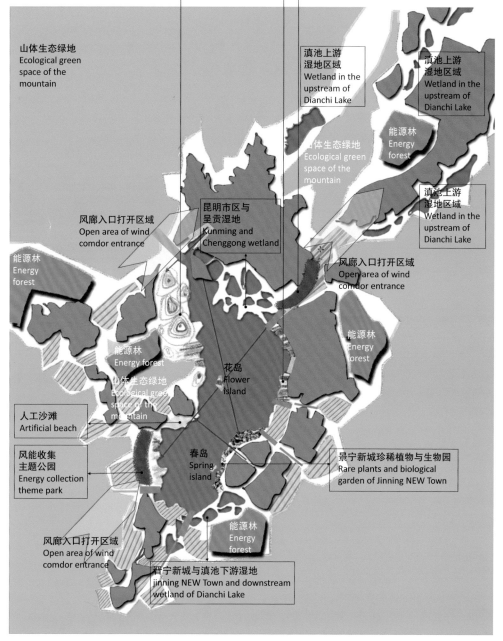

图 01　昆明区域范围内绿化体系
Fig. 01　Greenbelt systems in Kunming

原有城市绿地
Old green areas

原有绿地+城市空地
Old green area + Fallow land

原有绿地+ 城中村
Old green area + Village

原有绿地+城市空地+城中村+工业用地+低质量混合区+推荐拆除区
Old green area + Fallow land + Village + Industry land + Low quality area + Recommended demolition area

原有绿地+低质量结构混乱地区
Old green area + Low quality and structure disordered area

原有绿地+推荐清理地区（中层高度，建筑质量中下）
Old green area + Recommended demolition area (mid-rise building with bad construction quality)

图 02　城市核心区生态改造区域
Fig. 02　The Urban core as ecological transformation region

and regions with low quality and/or disorderly structures occupy roughly 45% or more of the area along the inner ring. Some of these sites may still potentially be repurposed as ecological areas.

整体空间结构通过绿化结构得到凸显：围绕一环的"玉环"环城绿带体系、沿主干道的"绿翼"、环城运河体系沿岸绿地与通过特殊植被色调而形成的代表城市文化核心区域的"红色中轴"。

We have selected the historic Kunming city walls, a 300 meter wide space along the inner ring, as the experimental site for ecological transformation.

The average width of the proposed green space transformation is about 400 meters. If the entire area were transformed it would mean for the city an increased green area of about 5.5 square kilometers. More than 70% of the current transformable areas consist of brown fields, industrial areas and obsolete villages, representing a total structural floorspace of only 165,000 square meters that would require removal.

The areas surrounding transport corridors and along the canal ring were selected for analysis as the space to realize the new modified greenbelt. The quality of the urban structure and public spaces in the current city core of Kunming, which exhibit a lack of green areas, will be enhanced through reorganization. The formal city walls, public spaces as well as the river "Panlong" are the main elements that constitute the city's image. The Confucian Temple, Justice Road and other public spaces also become noteworthy nodes, which represent Kunming's urban cultural and spatial structure. The general spatial structure of the city is accentuated by its green spaces, forming a "Jade Ring" around Kunming's First Ring, greenbelt systems called the "Green Wing" throughout the city, and the core of urban culture or the "red central axis" is established along the main roads including a greenbelt along the canal ring system featuring specially colored plants. Also within the Second Ring of the city, a green area of at least 800 hectares could be created.

域中还存在结构性可更新用地，可以作为生态结构的联系性用地。

内环沿线整体绿地平均改造深度大约400米，如果全部改造完成，新增绿地约为5.5平方公里。现有可改造范围内，空地、城中村、工业用地占了70%以上，容积率平均估算为0.3，拆除面积为165万平方米。

在假设被验证后，我们选择了重要交通干道周边、环城运河沿线进行了同等分析，得到了极其相近的可改造情况分析结构。由此原有昆明中心城区这一城市绿化最缺乏的区域，将借助生态绿地结构，重塑空间规划的整体结构，它同时肩负起城市文化与空间结构的强化作用。原有城墙—十字公共空间核心—盘龙江作为城市意象的结构；文庙正义路等公共空间作为承载的具体节点，基本代表了昆明历史性城市文化与空间结构。这一空间结构也通过绿化结构得到凸显：围绕一环的"玉环"环城绿带体系、沿主要干道的"绿翼"环城运河体系沿岸绿地以及通过特殊植被色调而形成的代表城市文化核心区域的"红色中轴"。除去部分重复计算，在城市二环范围内应当至少形成800公顷绿地面积。

城市发展 URBAN DEVELOPMENT

原有城市绿地
Old green areas

原有绿地+城市空地
Old green area + Fallow land

原有绿地+城中村
Old green area + Village

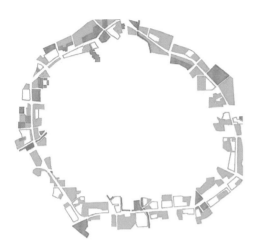

原有绿地+城市空地+城中村+工业用地+
低质量混合区+推荐拆除区
Old green area + Fallow land + Village + Industry land + Low quality area + Recommended demolition area

原有绿地+工业用地
Old green area + Industry land

原有绿地+低质量结构混乱地区
Old green area + Low quality and structiure disordered areas

原有绿地+推荐清理地区
(中层高度，建筑质量中下)
Old green area + Recommended demolition area (mid-rise building with bad construction quality)

图 03　内环沿线400米进深作为生态化改造实验区域
Fig. 03　A 400 meter wide space along the inner ring as an experimental region for ecological transformation

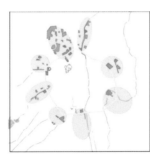

原有城市绿地
Old green areas

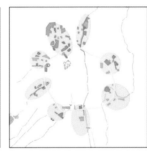

原有绿地+城市空地
Old green area + Fallow land

原有绿地+城中村
Old green area + Village

原有绿地+城市空地+城中村+工业用地+
低质量混合区+推荐拆除区
Old green area + Fallow land + Village + Industry land + Low quality area + Recommended demolition area

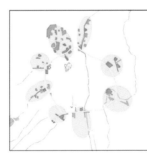

原有绿地+低质量结构混乱地区
Old green area + Low quality and structure disordered area

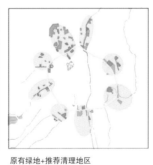

原有绿地+推荐清理地区
(中层高度，建筑质量中下)
Old green area + Recommended demolition area (mid-rise building with bad construction quality)

原有绿地+工业用地
Old green area + Industry land

图 04　重要交通干道周边作为生态化改造实验区域
Fig. 04　The area around essential transport corridors as an experimental region of ecological transformation

209

从两种文化中学习 LEARNING FROM TWO CULTURES

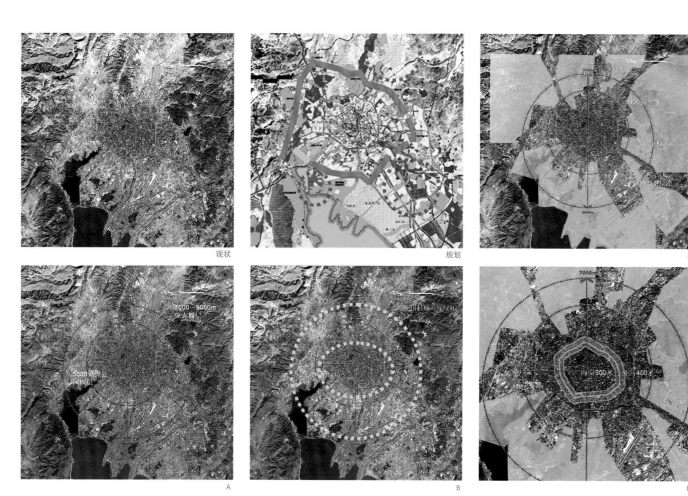

A. 远景规划

城市将在50年之内，通过一系列城市发展手段，达到如下目标：
——国家与世界级园林城市目标，即在绿地率等指标上的一系列提高
——在绿地的可达性上改变绿地结构不均衡的情况，目标是下列绿地可达性情况：
在200米内可以到达城市街头绿地或开放性附属绿地
在500～600米内可以到城市级公园
在5000米范围内可以到达郊野公园，即郊野绿带辐射进入城市区域
在7000-9000米范围内可以到达生态绿地，城市在该半径内60%以上为不同类型的绿地系统
——城市生态管理与生态规划已经成为各个方面工作的基础，包括如下重要内容：
城市土地使用集约化，高强度土地利用与大面积绿地相辅相成
城市雨水收集系统基本成型，通过人造生态系统将污水自我更新能力大规模提高
生态教育与文化观念深入人心，成为社会普遍共识，在私人建筑与绿地建设中得到普遍应用

B. 郊邻绿地结构塑造

绿环体系一：建议在贯彻总体规划的基础上进一步加强沿城市机场一线的绿地界定，围绕城市形成一个较少依托道路系统的绿环。这一绿环以生态功能为主，界定城市发展扩张轮廓。绿带将在城市北、南、东方向形成三个小的组团结构，与城市通过绿地隔离开，结构更加清晰
绿环体系二：在二环沿线应当发展相应的道路用地以形成第二条绿环，它以道路防护绿地为主。进一步确定都市化圈层

环间绿地类型

C. 周边长虫山、西山、世博园、滇池绿地等多条绿地系统辐射进入城市5000米圈层位置以内

5000-7000米圈层以内范围按照郊野公园类型设计，可以包括：体育公园在内的主题公园与生产性绿地以及其他生态绿地，即具有以休闲游憩为中心的相当部分城市公园的功能与活动类型，但以生态原生绿地的培养为主，维护工作方向相对较少
7000米圈层以外范围以生态绿地为主，可以包括：自然保护为前提的主题公园，如森林公园、自然生态公园、湿地公园，以及受到相应控制和管理的农业活动。绿地类型以生态绿地与水面为主，管理工作为主
5000米圈层以内为城市公园类型。
在城市绿地范围之内将以一系列其他措施，将城市继续以绿轴划分，以加强各个区域的可达性。

D. 绿环体系三：围绕内环路：整个城市建成区的中心圈层，不仅仅停留于线性林荫道的规划，而且沿线在一定范围内发展公园带。具体手段如下：
沿一环路继续强调林荫道特性，两侧的步行道绿化继续强化，有意识使用特殊的植物种类与色彩，形成具有识别性的城市骨架
沿一环路内向300米，外向400米范围内，有意识地拆除城市低质建筑、工业建筑、城中村、以及充分利用城市闲置地段，在必要地情况下局部拆除或更新一些尚可使用质量平庸地街区。将其发展为城市的环城公园。古城的历史结构在此重现。同时整个内城区内的公园可达性距离将被平均缩小一半左右
部分难以改造的新建区域应当在指标上提高其绿化率，有意识发展高档办公、居住、商业娱乐功能。整个环带的改造工程所带来的土地价位变化，将对整个城市更新带来重要的作用

图 05　昆明市生态与公共绿地结构优化思路
Fig. 05 Structural concepts for optimizing ecology and maximizing public green spaces in Kunming

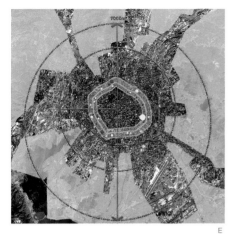
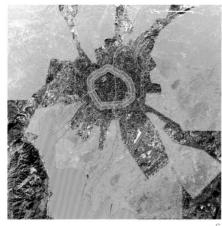
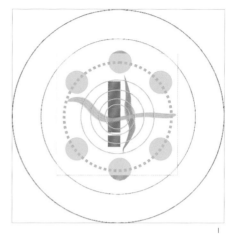

E

G

I

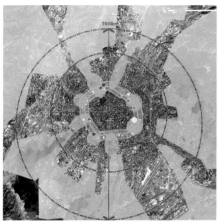

F

H

J

E. 游步道路与公共设施

该区域内部除了环城公园这一基本功能之外，仍然应新建部分建筑，其功能应以文化建筑、体育建筑、城市管理建筑、娱乐休闲服务建筑为主，形成文化轴线。但是所有的公共建筑都应具有相应的开放性，以成为城市的生活轴线。

环城道路中的步行与自行车功能将被转移到绿带之中，提高该区域的使用率，不仅产生了一条长度超过13公里的优美游步道路，而且在安全性上大大提高。环城道路因此甚至有了更高的汽车通行能力。

F. 绿翼

外环范围内将通过从一环上引出的多条"绿翼"将整个周长再次分割。在环内径上距离为1.3公里，外环上平均间距2.3公里，外环内部的可达性将进一步提高。

这些"绿翼"建立在现有有高绿地，或可改造区域相对较多的区域，同时考虑到绿地的均衡性问题。

它包括：
　　原有的绿地、公园、山地
　　高绿地率的公共设施、体育馆、大学等等
　　较高的可改造用地

未来发展中，每个绿翼将包括：
　　至少一个区级公园
　　高绿地率的新城市功能区
　　与城市轴向道路结合的林荫道与散步道

G.现有城市水体体系，将在三个功能上进一步加强

——生态功能
　　滇池与运河的净化
　　流域内污水与雨水的部分生态处理能力
　　活水网络的形成

——景观功能
　　水体网络与绿地网络一起形成城市景观网络的重要组成部分
　　水体形成新的景观特征，重点在城市中心区与滇池方向的一系列景观点。

——生活功能
　　水体再次与城市生活紧密联系起来，水体与公园绿地的结合，水体与城市散步道的结合，水体与重要轴线的结合，将"水"再次纳入城市生活与文化之中。

H.措施

——运河的上游与下游建议有意识地形成湿地体系：
昆明城区已北的湿地体系以金殿风景名胜保护区为基础对盘龙江与金汁河上游加以就地整治，控制来自于滇中北部地区的污染源；昆明城区以南的湿地体系以现有的大面积水产养殖、运河网络为基础，对农业生产加以控制或收缩，有意识地发展湿地系统，水面全部为软岸，种植芦苇等具有清洁污水作用的水陆两生植物，在枯水季节，水面相应缩小，不影响整体景观系统。生态目标为一方面积极净化滇池水质，一方面净化昆明市区产生的雨水与污水。

——以此平衡产生部分建设用地，大观河南侧形成岛状度假中心，可以集中发展相应的设施，其他地区也可以定点发展科研中心，主题公园服务中心，游人栈道，季节性猎场与渔场等等农业设施。

——城市中多条运河相互关联，形成城市景观与生活体系
三条南北向运河在市区内应当在南侧通过部分街道的打开以及人工景观河道的沟通，在北侧通过翠湖与盘龙江的沟通，形成运河网。在局部特殊节点处应当有意识地将水面放大。加强水面与运河的意象。

沿环城公园绿地在有条件的基础上，可以发展环城运河，形成优美的滨河巴士游览线。或者规模较小，水面的放大与收缩形成了公园中的特殊景观体系。

I、J.总体形成的绿地结构空间规划

从两种文化中学习　LEARNING FROM TWO CULTURES

图 06　昆明城市中心区建设状况
Fig. 06　Urban structure of the city center

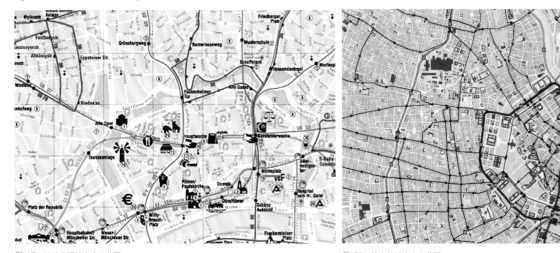

图 07　法兰克福城市中心公园
Fig. 07　The green space system of Frankfurt's downtown

图 08　维也纳城市中心公园
Fig. 08　The green space system of Vienna's downtown

城市发展 URBAN DEVELOPMENT

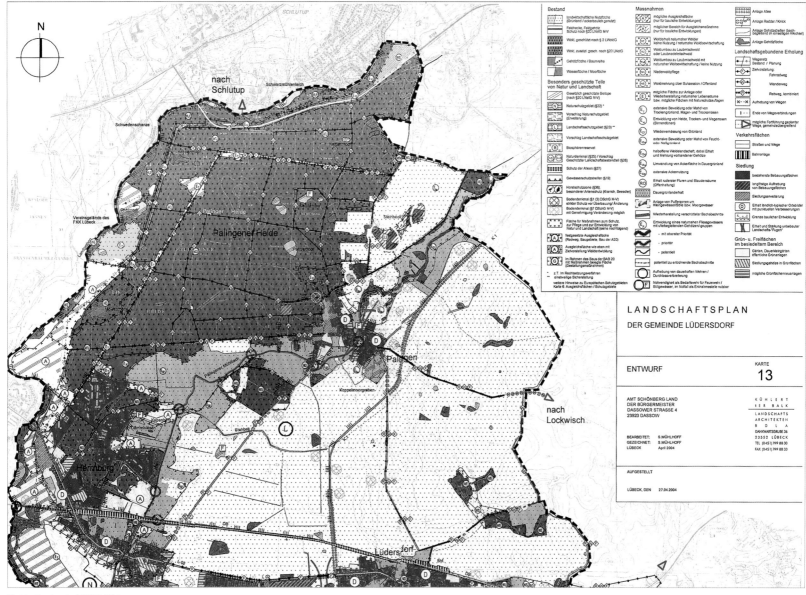

图 09　Luedersdorf 镇绿地规划
Fig. 09　Green space planning in the German village "Luedersdorf"

非典型研究内容：法制化与实施
Contents of atypical research: legalization and implementation

对比德国的Landschaftsplanung(自然生态体系规划)，中国的绿地系统规划更注重为总体土地规划的法规可执行性作铺垫，受到其指定方式与实施主体的限制，其本身的法律严格性较为有限。而德国在生态保护与农林保护甚至一些景观特色突出的空间保护上都有明确的要求——以Luedersdorf景观规划为例，共计形成16类不同类型的

In China, where relevant laws are more relaxed, the planning of green areas is regulated only by laws concerning overall land use planning, and is restricted mainly in its manner and methods of implementation. In Germany however, there are clear requirements for agricultural, forest and ecological conservation and the protection of areas with prominent landscape features. Taking Luedersdorf's landscape planning as an example, 16 different types of areas were designated as protected, including the 3 prerequisite types necessary for the implementation of landscape planning. The drawings here cover a directory of recommended measures and the treatments for different urban structures, traffic and urban green space. Such legal instruments are necessary to protect the natural environment and limit urban development.

In the urban renewal of large areas, overall planning often calls for the large-scale removal of preexisting structures and redevelopment into green spaces. This is difficult for China to reconcile with its current urban construction and development objectives. Integrated with a special index used in German landscape planning, we introduced an index with a high green rate in this conceptual planning. This is of special significance in the aspects of construction intensity and vertical development of Chinese urban construction seen in subsequent projects and discussion with management sections of urban planning. Using this index will create a unique urban image for the city and contribute to economic balance, besides the significant increase of the total quantity of green spaces.

In the Inner Ring for example, the green space makes up 70% of the total transformable area, amounting to 3.85 square kilometers. 1.64 square kilometers can then be rebuilt with high quality high-rise apartment buildings restoring the total floor space of 165,000 square meters. Additionally, property values before and after the transformation should remain constant thanks to improved post-transformation real estate values.

保护区域，其中包括3类推荐保护区域。这些构成了景观规划的执行前提。此外图纸上包括了推荐措施目录与不同城市建设、交通与城市绿地的处理方式——这样一个法律文书势必造成较低的发展弹性——恰合绿地系统规划的目的：保护绝对重要的自然环境，限制城市建设环境的无序发展。

在这样大区域的城市更新中，大规模地拆除改造为绿地，就中国目前的城市建设发展目标而言，是较为困难的。结合德国的土地规划中的一个特殊图例，我们在本次概念规划中推出了高绿化率用地这一指标，并在随后的一系列项目中与城市规划管理部门商讨，看到它对于中国城市建设的水平强度与垂直发展方面确实具有相当特殊的意义。它在绿地总量的显著提高之外，形成了局部地区特殊的城市风貌以及部分经济性平衡。

其中以内环为例：
新建区域中，以绿地率占总数的70％计算，产生绿地3.85平方公里。1.64平方公里建设用地，可以适当建造高层高档公寓建筑。以容积率平均为1计算，建筑面积约165万平方米。两者建设量上基本持平，但建筑质量与价格远远高于改造前，并同时带动了整个沿线的土地价格。整体上应当较好地平衡了改造费用。

非典型研究内容：私人绿化体系 Contents of atypical research: private green space

Quantity, however, is only one part of green planning; another important consideration is quality. In Kunming, its other nicknames "city of the flower", "Spring City" and "green city" refer not just to physical attributes of city itself, but also to the life within.

We also integrated semi-public and private green spaces into the city plan. Based on preexisting models found in other European Cities, the use of private green spaces can be guided through various service facilities, management and recommendations which help to improve the overall quality of life.

城市绿化体系不仅仅包括以城市管理部门为主体维护者的综合生态绿色体系，也包括由私人与机构维护的专用绿地体系，两者分别对生态品质与生活品质有着不同程度的贡献。

The urban green system includes not only large scale green spaces managed by local municipalities, but dedicated green spaces under private and commercial ownership as well.

The urban green system includes not only large scale green spaces managed by local municipalities, but dedicated green spaces under private and commercial ownership as well. The two systems contribute respectively to the overall ecological quality in the region as well as the quality of life on many different levels. Together they form an integrated urban green landscape system.

During the three-week project period, based on aerial photography, on-site surveys and the adopted experience and principles of European planning, we provided a final concept that takes the local issues and the needs of the local community into consideration. The judges unanimously praised this work and it became the basis of the Kunming statutory development planning of green space.

绿化的数量仅仅是城市绿化规划的一个部分，另一个重要的核心点来自于质量。昆明——花城，春城，绿色的城市——不仅仅是城市的形态，也是生活的形态。

本次规划中，我们还引入了一个特殊绿化体系：半公共与私人性专业绿化体系。欧洲城市建设的经验是：私人绿化是可以通过各种服务设施、管理推荐进行相应地指导的，它直接提高了生活品质。城市绿化体系不仅仅包括以城市管理部门为主体维护者的综合生态绿色体系，也包括由私人与机构维护的专用绿地体系，两者分别对生态品质与生活品质有着不同程度的贡献，共同构造成城市绿化景观体系。

在这3周的时间中，依据航拍图与现场调研，我们上交了一个结合欧洲规划经验原则与本土问题的整体性解决思路，这一工作得到了评委的一致好评，最终成为昆明绿地法定规划的发展基础。

城市发展 URBAN DEVELOPMENT

图 10 昆明绿地景观
Fig. 10 Landscape of Kunming

城市更新
URBAN RENEWAL

01 伦茨堡控制性规划
Framework Development Planning of Rendsburg

02 "霍费尔天空":
提升城市品质的范例项目
"Heaven of Hof":
an Initiative-project for Upgrading the City

03 埃尔旺根步行区
Pedestrian Zone of Ellwangen

04 埃斯林根总体规划和公共空间城市设计导则
Master Plan and Design Guidelines
for the Public Spaces of Esslingen am Neckar

05 "粉红岛":韩国首尔方背洞街区广场
"Pink Islands":
a Street Square in Bangbae-dong, Seoul, South Korea

06 Gangseogu城市更新规划
Urban Renewal Gangseogu

07 北京市昌平区奥运自行车训练馆
及周边地区城市设计
Urban Design of the Area Surrounding
Changping Olympic Cycling Training Center, Beijing

伦茨堡控制性规划
Framework Development Planning of Rendsburg

规划面积 AREA		200 hm²
人口规模 POPULATION		10000
完成时间 PROJECT DATE		2009

引言 Introduction

The town of Rendsburg in the federal state of Schleswig-Holstein is a seaport city in the middle of Northern Germany. It lies on the European Suez Canal, the Kiel Canal, which links the North Sea and the East Sea. More than 800 years old, it was once a city of commerce as well as a fortified Danish border town. Today Rendsburg is the secondary center and county seat, in addition to an industrial, port and shipbuilding center, theatre town, university town and hub for polytechnics. It is expected to continue to expand as a service and industrial center for the region.

石勒苏益格—荷尔斯泰因州的伦茨堡是一个内陆港口城市。它位于连接北海和波罗的海的运河（如同欧洲的苏伊士运河）之畔，拥有超过800年的历史，曾经是丹麦边境的堡垒城市和商业城市。今天的伦茨堡是集工业、港口、造船、戏剧城、大学城等功能于一体的中心城镇，并将得到进一步的发展，成为地区的服务业、工业中心。

在几百年的城市发展中，伦茨堡始终保持着不可替代的城市特性，其中建筑的发展始终同城市景观保持着高度的协调。而过去一个半世纪的深刻变化，特别是本次规划前十年间发生的变化，却对城市的未来及个性形成了威胁。

The town of Rendsburg has preserved its historical character until today through its density and closeness. For centuries, the city has been framed like a gemstone within its walls. Any changes were made in accordance with the principles of the city's architecture, which was self-evident for our ancestors. Thus a distinctive character of the city evolved over centuries in which the individual architecture of the buildings was in accordance with the townscape.

直到今天，伦茨堡仍保留了它的历史特点、城市的完整性以及明晰性。几个世纪以来，这个城市好像一块宝石般得到城墙的保护。这座城市建筑的变化理所当然地依然遵守着祖辈定下的原则。在几百年的城市发展中，伦茨堡始终保持着不可替代的城市特性，其中建筑的发展始终同城市景观保持着高度的协调。

A distinctive character of the city evolved over centuries in which the individual architecture of the buildings was in accordance with the townscape. Yet the profound changes of the last few decades in particular, jeopardized the future and individuality of the city.

However, profound changes of the last one-and-a-half centuries, the last few decades in particular, jeopardized the future and individuality of the city. More and more buildings were broken or were in need of repairs. Thus, contradictions in the townscape increased from year to year. Rendsburg was in disastrous shape and the unique urban landscape of a medieval and baroque fortress city began to disappear.

过去一个半世纪的深刻变化，特别是本次规划前十年间发生的变化，对城市的未来及个性形成了威胁。越来越多的建筑物被损坏或需要修复，旧建筑与城市景观的矛盾越来越大。伦茨堡城面临着更新，以及中世纪和巴洛克要塞城市风貌的流逝。

图 01 伦茨堡：一个内陆的港口城市
Fig. 01 Rendsburg : a seaport city in the midland

伦茨堡城市规划和实施战略 The planning and development strategy

This is the reason why Rendsburg developed a planning and development strategy, which has been in place since roughly 1970, wherein all possible planning, legal and financial resources were combined to further develop the city and townscape in a contemporary form. Master plans, building design codes and stylistic statutes were developed, among others. All the possibilities of building legislation, preservation of historic monuments and urban planning promotion of the federal and local government were legally exploited. Last but not least, financial opportunities for grants were continuously investigated and taken out as loans. They were made available on the federal, state and town level in order to finance the urban renewal strategy of the public sector, in addition to supporting private renewal projects which could get two-thirds of the necessary funding – for private costs – through a grant.

基于这个原因，伦茨堡自1970年开始制定规划和实施战略，将所有可能的规划、法律和资金资源相结合，以修复城市面貌并将其发展成与时俱进的更现代的城市。其中包括控制性规划、建设规划、设计导则等类型的规划，以及文保建设法规、联邦及州政府的建设要求等各种法规。资金环节——最后一环但并非最不重要的——需要筹集联邦、州和各城市可支配的资金，其中三分之二是私人资金。

在一个较长时期内，ISA涉猎的工作领域有控制性规划[1]、城市规划[2]、设计导则以及阅兵广场设计等单体设计。ISA所完成的控规、设计导则，虽然已经过了多次的调整，但自生效起迄今的25年仍然发挥着重要的作用[3]，对后来意厦为弗伦斯堡、波茨坦、维尔茨堡、斯图加特、康斯坦茨等地做的规划工作，以及其他许多服务于联邦政府城市政策方面的规划工作来说，具有示范性的意义。下面将对这些工作进行详细介绍。

Over a long period of time, the ISA urban planning studio has been involved to a large extent in preparing the strategy of the master plan [1], the townscape planning[2], the building design statutes and even diverse designs such as the square Paradeplatz. The strategies and statutes developed by the ISA urban planning studio have naturally experienced several revisions since taking effect. But for 25 years, they have retained their fundamental validity[3]. This planning work was a model for further tasks that have been developed by the ISA international urban planning studio for cities such as Flensburg, Potsdam, Wuerzburg, Stuttgart, Constance, and many others in the context of urban policy grants from the federal government, provinces and cities. Therefore it will be worth it to show the work in more detail below.

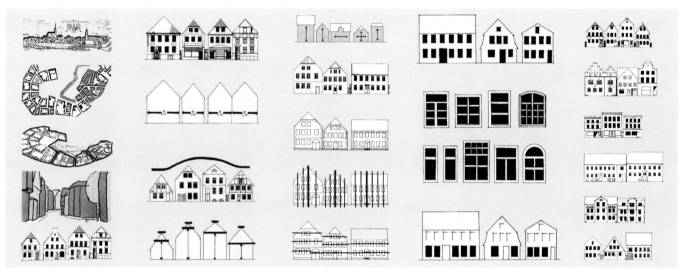

图 02　伦茨堡城市景观规划：城市肌理、城市天际线、城市空间、建筑及立面类型
Fig. 02　Townscape planning of Rendsburg: city layout, skyline, urban space, buildings and facade typology

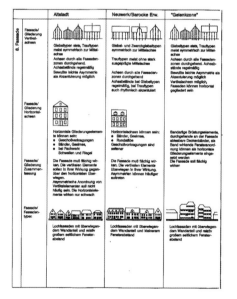

图 03　伦茨堡城市景观规划：城市建筑的基石
Fig. 03　Townscape planning of Rendsburg: building structure

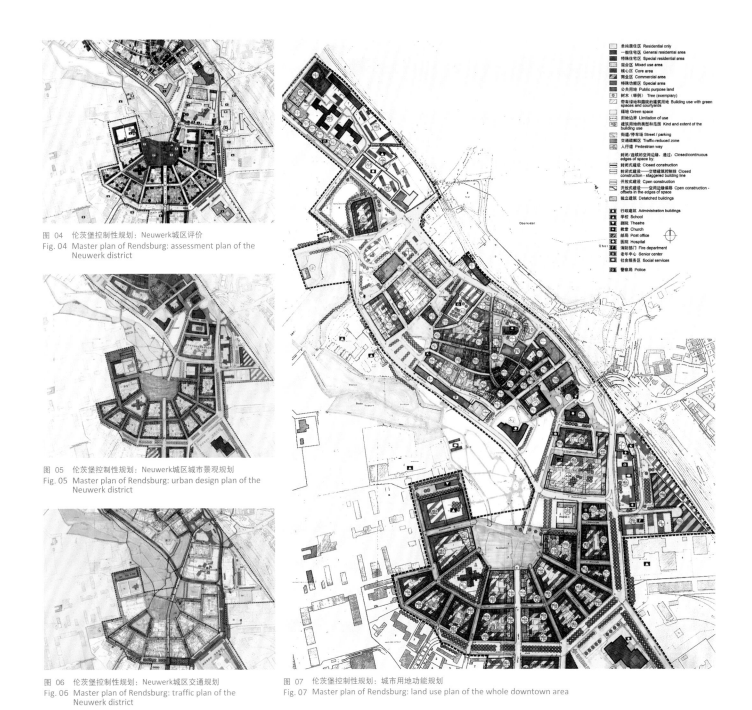

图 04　伦茨堡控制性规划：Neuwerk城区评价
Fig. 04　Master plan of Rendsburg: assessment plan of the Neuwerk district

图 05　伦茨堡控制性规划：Neuwerk城区城市景观规划
Fig. 05　Master plan of Rendsburg: urban design plan of the Neuwerk district

图 06　伦茨堡控制性规划：Neuwerk城区交通规划
Fig. 06　Master plan of Rendsburg: traffic plan of the Neuwerk district

图 07　伦茨堡控制性规划：城市用地功能规划
Fig. 07　Master plan of Rendsburg: land use plan of the whole downtown area

控制性规划：从愿景到实施的指南　The master plan: guide from vision to implementation

Urban development and urban design have gained a great deal of importance in today's way of thinking, which requires all decisions, goal and plans to be justified and understood, in all areas of life. The preservation and development of a city, therefore, requires extensive, comprehensible and clear planning in the urban area. This is especially true for

在现今的时代意识中，城市发展以及城市景观已经包括所有生活领域的评价、目标、规划的依据和可实施性的要求，并且具有重大意义。因此城市的保护与进一步发展以全面、可实施和直观的城市规划为前提。这一点对于伦茨堡复杂的城市结构同样适用——它实际上是一个具有双重特质的城市，一边是中世纪古镇，另一边是巴洛克新城。这两个部分都有自己的历史、文化、社会使命，通过由大型广场和重要设施（城市剧院等）构成的衔接区域进行联系，从社会、经济乃至建筑、城市规划的角度相互补充。

控制性规划是一种规划工具，它一方面介于城市发展规划和土地功能使用规划之间，另一方面介于详细规划和建设法规之间，是以现状评价（SWOT分析）为基础，为城市区域的未来建设发展确定目标并制定规划。它记录和评估用地、交通、城市景观、开放空间，发展相应的规划目标并通过用地、交通和形态理念加以体现，通过措施规划对其实施进行辅助。在欧洲的规划实践中，虽然也有不同的形式，但控制性规划一直对城市规划实践具有重要的意义。对于所有城市规划过程参与者——政府、城市以及市民——来说，控制性规划是非常重要而具体的基础。

伦茨堡控制性规划中最重要的目标有：加强内城的居住功能，实现不同功能用地互不干扰的混合，分流过境交通，设立慢行交通区和步行区。同时包括地块内部居住环境的改善，兼顾功能和审美的平衡发展，以及建筑形态、材料、色彩的和谐关系。

控制性规划的目标和措施

主要的设计目标是维护、修缮城市核心和历史保护区Neuwerk，继续倡导住宅、商业、工业、服务业和文化功能用地的混合发展，将起始—目的地交通与过境交通分离，设立慢行交通区，保持和发展伦茨堡独一无二的城市景观特色。

建立在规划发展目标基础上的具体规划任务的重要措施包括：

功能使用：一方面是保持小规模的、可持续的功能混合和由此产生的包括封闭的街区建筑以及小地块结构的城市形态，通过建筑的现代化和建筑间隙的填充加以实现。另一方面，将所有具有干扰性的企业搬出老城和Neuwerk区，以提高生活品质、居住品质和居住环境的价值，并减轻交通负担。

道路交通：通过过境交通的分流最大可能地实现对机动车交通的限制，将居住区街道建设为行人友好型街道；重新设计所有重要的市区街道和广场空间，拓展街道、广场空间的绿化面积。对交通的梳理及新秩序的建立，不仅可以提高居住品质，还可重建能够容纳多样化功能的街道空间，对伦茨堡人居品质的提升起到了至关重要的作用。

the complex urban structure of Rendsburg that is actually "a double city," consisting of a medieval old town and a baroque new town. Both parts have their own historical, cultural and social function – in addition to a connecting area with a large square and central institutions such as the theatre, etc. – and each complements the other socio-economically and socially speaking, as well as architecturally and in their urban development planning.

A master plan is a tool between the urban development plan, i.e. land use, and the building design code plan, i.e. the design statutes. Objectives and strategies were developed for future urban development on the basis of SWOT analysis. It comprises and evaluates land use, traffic, urban form and open space, and develops relevant planning goals. It would be completed through a series of measures for its implementation.

Within the larger European planning practice, the idea of a local planning practice in its different forms has gained importance. The master plan is a critical and very precise basis for information, involvement and participation from everyone within the urban planning process, the government, the city and its citizens.

The main objectives of the Rendsburg master plan were, for example, the strengthening of the residential function of the inner city, the mixture of diverse functions that are not mutually interfering, the relocation of through traffic and the creation of calm traffic and pedestrian zones. Other projects, such as creating green open space in the building block court, were meant to improve not only the living environment, but also the balance of functional and aesthetic aspects, and moreover, the adjustment of building forms, materials and colors for the urban architecture.

Goals and measures of the master plan for Rendsburg

The essential planning goal was to: preserve and strengthen the city center and the town monument of Neuwerk; further develop the mixture of living, trade, industry, services and culture; decentralize terminating traffic and through traffic creating low traffic zones; as well as develop the unmistakable townscape of Rendsburg.

The main groups of measures based on the development objectives, which resulted in specific planning schemes, included:

Land use: Maintaining a compatible mixed use in the small scale and the resulting urban form, with closed perimeter block development and mostly small-scale plot structures. This was then supported by the modernization of buildings and in-fill development. Conversely, this also implies resettlement of all loud industries from the old town and Neuwerk in order to increase the quality of life, the quality of housing and value of the living environment, as well as to relieve traffic congestion.

Traffic: Reducing traffic to a large extent by taking out transit traffic and creating pedestrian-friendly street designs, along with the restructuring of all major urban roadways and spaces with trees and greenery. Through the traffic reorganization, not only was the quality of residence to be increased, but the restoration of a multifunctional street space was also to be encouraged, which plays a crucial role in the quality of life of Rendsburg.

在现今的时代意识中，城市发展以及城市景观已经包括了对所有生活领域的评价、目标、规划的依据和可实施性的要求，并且具有重大意义。

Urban development and urban design have gained a great deal of importance in today's way of thinking, which requires all decisions, goals and plans to be justified and understood, in all areas of life.

相关的重要设计目标是维护、修缮城市核心和历史保护区Neuwerk，继续倡导住宅、商业、工业、服务业和文化功能用地的混合发展，将起始—目的地交通与过境交通分离，设立慢行交通区，保持和发展伦茨堡独一无二的城市景观特色。

The essential planning goal was to: preserve and strengthen the city center and the town monument of Neuwerk; further develop the mixture of living, trade, industry, services and culture; decentralize terminating traffic and through traffic creating low traffic zones; as well as develop the unmistakable townscape of Rendsburg.

Design: Preserving the typical urban form with its historic perimeter block development and small-scale plot structure by restoring and modernizing existing buildings and monuments as much as possible. Instead of demolition and reconstruction, building renovations were supported to stabilize the building structure and the mixture of functions worth being maintained throughout the historic city. In addition, the ground floor zones of historical building structures were supposed to be restored to the original scale. New buildings are to be integrated into the existing building structure.

城市形态：保留典型的历史临街建筑和小地块结构，通过修缮和改建实现对所有可能的、有价值的历史建筑的保护及更新。代替拆除和新建措施，通过对整个历史旧城区内旧建筑的现代化改建，使原始结构保持稳定，并增强有保留价值的功能混合，此外也有可能使一再被扰乱的楼房首层区域恢复原始的尺度。必要的新建筑要尽可能地同旧的建筑现状相融合。

图 08　伦茨堡控制性规划：用地功能、道路交通和城市景观理念的发展
Fig. 08　Master plan of Rendsburg: the development of land use, traffic and design concept

Development of the master plan: from the inventory of existing conditions to measures

The master plan for Rendsburg had the task of creating a vision for the future development of the city center and to bring that development into a legally and financially viable form. The plan needed to refer to the principle of the federal government's urban policy in order to qualify for their financial promotion program. There was a general paradigm change regarding urban development. For example, one of the ministerial orders of the state government of Schleswig-Holstein was for land use development to be by mixed use instead of a zoning concept.

控制性规划的发展：从现状分析到方案的确定

伦茨堡控制性规划的任务是，根据联邦政府的城市建设政策，结合城市经济发展要求，着眼于城市建设价值观念上的变化——从用地功能分离到功能混合，以州政府提出的要求为背景，对伦茨堡内城的未来发展进行规划，并形成在内容上、法律上和经济上都具有可行性的方案。

城市更新 URBAN RENEWAL

这项工作始于对用地功能、道路交通和城市景观目标的精确制定,并以对所使用的重要法定参数的准确释义为基础,包括用地种类和规模、静态交通面积、慢行交通区、建设方式、高度控制乃至建筑类型和环境保护。

与之相关的是现状图纸,表明了从居住区到军事区等各种用地类型的道路交通、建设方式,对学校、教堂、医院、警察局等公共设施的标注,以及各不同功能地块的面积、容积率和建筑密度。

在现状图的基础上,进行评估规划,包括对用地密度、建筑状况、空地、动态和静态交通以及城市景观的评价。通过评估显示出以下的城市设计层面的弊病:

- 住房及办公地点的灯光、日照和通风不足
- 建筑、住宅和办公地点的建筑状况不佳
- 居住和办公功能的过度混合
- 根据种类、尺度以及现状,已建、未建区没有相宜的功能开发
- 由于用地、企业、一般设施和交通设施的噪声、污染、空气污染和震荡而带来的负面影响
- 土地开发不充分

控制性规划实际上是以评估规划为基础发展而来的。首先进行的是功能规划,确定可能的用地功能类型和规模。

功能规划中,"旧城"将主要用作商业中心及密集的公共用地,形成城市核心;在Neuwerk区,住宅用途应占主导地位,零售和无干扰的小型工业亦被保留。这样,旧城区作为核心区(MK)和混合功能区域(MI),可能的容积率在1.5～2.5之间;Neuwerk区作为以居住为主体的区域(WB),容积率在0.7～1.5之间。

交通规划展示了所有专业规划的合成结果,即私人交通、公共交通、自行车交通、行人交通乃至汽车、自行车停车位规划等各项规划。交通规划首先基于对交通形式的确定,如动态交通、静态交通、慢行交通区内的交通、自行车和行人交通。因此,交通系统规划包括过境交通、区内交通、静态交通、自行车及行人交通系统。

城市景观规划是基于1981年的伦茨堡城市景观和文保建筑名单,因此仅包含与控制性规划直接相关的重要内容,特别是对空间界线(建筑边线和建筑控制线)、建设方式和高度控制。对于其中最重要的内容还单独进行了规划,并纳入城市景观规划中。

该控制性规划以实施措施规划为结束,综合表述了计划、从属关系、程序等,显示出了控制性规划的实施以及向设计导则及管理法规延伸转化的可能。规划的各种措施包括:具有示范意义的措施、街道空间以及交通组织措施、静态交通措施、总体以及分区的更新措施。

The work began with a precise formulation of goals for the land use, traffic, and design concept based on a clear definition of the legally relevant parameters to be applied – such as the type and extent of land use, parking lots, traffic reduction, building structure, building height requirements, building typology, and the ecology.

This was followed by an inventory plan that was used to describe the existing land use from the residential area to the military zone, the traffic system, the building structure, and public infrastructures such as schools, churches, hospitals, police stations, etc. This plan also included the plots and building densities in terms of coverage ratio and floor area ratio (FAR).

On the basis of the inventory plan an assessment plan was developed that specifically analyzed the density, buildings, vacant lots, traffic and parking system, and the quality of urban design. This evaluation revealed the following urban grievances:

- Lack of exposure, sun and ventilation of flats and workplaces
- Poor structural condition of buildings, homes and workplaces
- Detrimental mixture of residential housing and workplaces
- Inadequate land use according to type, dimension and condition
- Negative environmental condition due to noise, pollution and air pollution
- Plots without street access

The master plan was developed out of this assessment plan. For the areas in the historic downtown, the land use plan was developed first and then the type and dimension of the possible uses were determined.

The land use plan provided, for example, that the historic downtown should be mainly used as a business center in order to maintain a high-density town center. In Neuwerk, residential use should be prevalent, but existing retail and non-disruptive industries would remain. Consequently, the historic downtown was defined as a core and mixed-use area according to the legal terminology of the land use plan, which allows 1.5 to 2.5 of the floor area ratio (FAR). The Neuwerk district was categorized as a special residential area allowing for densities of 0.7 to 1.5 FAR.

The traffic plan was a synthesis of all technical planning, i.e. planning for individual transport, public transport, bicycle traffic, pedestrian traffic and parking facilities from car to bicycle. The transport plan was initially based on a determination of the forms of transport such as flowing traffic and resting traffic. Thus, transportation systems for through traffic were developed, as were for access roads and resting traffic, including cycling and pedestrian systems.

The design concept was based on the urban townscape plan of Rendsburg, which was developed in 1981, along with a list of historical building monuments. It therefore contained only those statements that were directly relevant to the master plan. In particular, these were statements referring to building lines, building heights and building structure, which define the 'street walls.' For the most important categories, their own plans were developed and put together in the design concept.

The master plan was completed with the action plan, which represented projects in a multidisciplinary way – dependencies, procedures etc. – that seemed appropriate to enable the implementation of the urban land use plan and its transformation into the building design code and statutes. It was structured into measures with impulse, road construction and infrastructure development, resting traffic, and modernization initiatives referring to specific areas.

> 这项工作始于对用地功能、道路交通和城市景观目标的精确制定,并以对所使用的重要法定参数的准确释义为基础。
> *The work began with a precise formulation of goals for the land use, traffic, and design concept and was based on a clear definition of the legally relevant parameters to be applied.*

图 09 伦茨堡：作为城市历史的一面镜子，反映多元化的城市景观
Fig. 09 Townscape as a mirror of history

规划结果和实施策略 Results of the planning and implementation strategy

The results of the planning and implementation strategy for Rendsburg referred to [4] allotment, infrastructure, fallow land, pollution, ecological quality, technical infrastructure, urban image, attractiveness, housing and commerce. The impacts of all these factors had an affect on the housing market, the commercial structure, real estate, the surrounding and private willingness to invest, and ultimately on actions ensuring the results of urban renewal. The following presentation on the example of urban renewal within the "Neuwerk" district was based on the final report of Rendsburg and the urban renewal corporation BIG Staedtebau. [5]

伦茨堡规划和实施策略的结果[4]涉及开发、基础设施、闲置土地、环境污染、自然品质、市政设施、城市景观、地区吸引力、居住和工业等方面。这些结果将对住房市场、商业结构、地价走势，对周围地带的辐射效应、私人投资意愿，确保城市更新效果的措施产生影响。下面基于伦茨堡市以及城市更新工作的承担者BIG城市建设有限公司关于城市更新的总结报告，以伦茨堡Neuwerk城区的更新为例进行介绍。[5]

城市更新 URBAN RENEWAL

开发、基础设施、闲置土地

今天，部分已完成改造的交通及开发区分布于历史内城的整个重建区内，首先是一系列被改造的街道和广场空间。通过拆除中心区旧建筑、闲置土地再利用、搬迁有干扰的企业以及建设停车楼、停车位等，使居住环境以及城市空间环境品质得到显著提升。

城市自然环境、市政基础设施、环境污染

市政基础设施如电信、电力、煤气、给排水将结合相关地下工程建设措施逐步进行改造更新。内城的自然空间品质通过大量各式的绿化措施在公共空间和地块内部的采用而得到显著提升。噪声、废气等环境污染通过一系列交通整顿、慢行区域建设、污染企业搬迁、自然空间品质提升等措施大幅减少。

城市意象、城市景观、城市吸引力

通过维护和改善城市景观，伦茨堡在周边区域以及石勒苏益格—赫尔斯泰因州的城市意象得到巨大提升。做到这一点归功于非常多互相关联的修缮及现代化措施，包括对建筑间的空地进行建设，拆除中心区的老旧建筑，以及通过对街道和广场进行大力改造——使之与地方风貌相协调、慢行化并有利于居住环境的改善。

住房、就业、贸易、文化、旅游

通过大量由联邦政府资助的修缮和现代化改造的措施，以及许多私人自发进行的改造措施，住房状况已大为改善，面积也得以增加，内城的居住功能稳定并加强。

运营环境大幅升值，零售贸易得到了促进，使得这里成为一个有吸引力的购物场所，与建设区域中心的要求相适应。

干扰企业外迁，内城生产型企业数量必然减少，且发展的重点将从制造业转向文化服务业，形成了Neuwerk区在居住以外的第二个主导产业。

通过公共空间和建筑的现代化，特别是对重要古迹的保护更新，伦茨堡内城的旅游功能也得到了加强。

25年的规划和实践对整个城市的影响是显著的，而且一直在起积极的作用。更新区内大量经过修缮和改造的住房（从数量及质量上）大大提高了整个城市住房市场的品质。随着历史内城价值被重新发掘其功能升级，手工业、贸易及服务业等的环境和区位条件也得到了优化；这些也改善了整个城市的产业结构。相比更新之前，地价也得以摆脱价格约束而提高。由于功能上及景观上的价值提升，内城重新焕发神采，并使邻近地区也可感受到它的魅力，这使得伦茨堡内城对周边区域乃至整个城市起到了明显的辐射带动作用。

Allotment, infrastructure, fallow land

Step by step, the roads and streets have been improved in the whole urban renewal area of the historic downtown. That the network of streets and squares are in excellent condition today is the result of that improvement. Through the creation of green open space in the interior blocks, conversion of fallow land, relocation of disruptive industries and construction of car parks and parking lots, the quality of the living environment was substantially increased.

Nature in the city, technical infrastructure, environmental pollution

Technical infrastructures such as telecommunications, electricity, gas supply and sewage were renewed and upgraded in the course of each incurring underground engineering service. The ecological qualities of the downtown area have been significantly increased through diverse planting actions in public spaces and the block interiors. The pollution caused by noise and exhaust fumes has been significantly reduced through numerous traffic reducing and regulating measures, as well as the removal of disruptive industries and the improvement of green space.

City image, townscape, attractiveness

The image of the city of Rendsburg in the region and the state of Schleswig-Holstein has been improved to a large extent through the preservation and improvement of the townscape. This was achieved by diverse repair and modernization measures: in-fill development, green courtyards within block interiors, traffic reducing measures appropriate for the townscape, improvement of residential environments through impressive street and square redesign.

Housing, employment, trade, culture, tourism

Thanks to numerous repairs and modernization measures subsidized by the government and private actions, the housing stock has substantially improved and enlarged and the residential function of downtown has been stabilized and strengthened. Likewise, due to the significant improvements of the urban business environment, conditions for retail trade promotion have been strengthened and the city's reputation as a main regional center now encompasses an attractive shopping destination. The manufacturing business in the city was reduced due to the outsourcing of disruptive factories. The focus shifted from industrial to cultural service sectors, which particularly in the district of Neuwerk were the second focus for land use after housing. Through the modernization of public space and buildings, like important monuments, the role of tourism in downtown Rendsburg has also been strengthened.

The results of twenty-five years of planning and implementation for the whole city were significant and positive. Due to the high proportion of repaired and modern equipped apartments within the redevelopment areas, the housing market of the whole city was stabilized and improved in respect to quality and quantity. With the functional improvements of the historic town center, the location and conditions for crafts, trade and service facilities were improved, which in turn had a positive impact on the business structure for the whole city. Yet the land costs did not rise after the release of fixed prices during the urban renewal proceedings. The impact of downtown Rendsburg's urban renewal was very clear to see on the surrounding areas and the whole city. Because the downtown became attractive again due to improved function and design, the attractiveness of adjacent territories was increased significantly.

25年的规划和实践对整个城市的影响是显著的。

The results of twenty-five years of planning and implementation for the whole city were significant and positive.

图 10　伦茨堡控制性规划：用地功能规划及实施措施
Fig. 10　Master plan for Rendsburg: concept of land use type and dimension including measures

城市更新：以控制性规划为发动机，以城市建设需求为动力
Urban renewal Rendsburg: master plan as motor, urban renewal promotion program as fuel

Urban renewal was not a single action, but a process usually taking place in diverse, small steps, often lasting decades. The sewer system is to be repaired, street surfaces and pavement are to be renewed, lighting is to be modernized, buildings are to be renovated, demolished and replaced by new ones, lands are to be converted. The list of urban renewal activities is endless.

Without a common, long-term planning concept, such small-scale urban renewal would only lead to chaos and even sometimes lead to a mutual interference. In order to organize such various activities in urban renewal, each renewal effort needs an instrument of coordination – that is the task of the master plan.

Urban renewal is a short-, medium- and long-term process that encompasses all aspects of urban planning in terms of location, type and dimension. The volumes and shapes of those aspects must be thought of ahead of time and coordinated. This is why the master plan is something like the two- and three-dimensional controlling tool in urban planning.

As a city is something like a living organism, old problems may disappear and new tasks may arise during the renewal of a city. It is the principle, then, to "extend" or "update" the master plan, as has been done here several times over the course of 40 years of urban renewal.

城市更新不是一次性的行为，而是一个过程，通常需要多方面地一小步一小步实施，而且往往要持续几十年。污水系统修复、路面更新、人行道和照明的现代化、翻修房屋、拆除和修建新建筑物、用地功能更新——市区重建活动的清单还远不止于此。

如果没有一个共同的、长远的规划概念，小规模、无计划的城市更新只会带来混乱，甚至有时会相互干扰。为了协调多样化的城市更新活动，每个更新活动都需要协调的工具——这正是控制性规划的任务。

城市更新也是一个短期、中期和长期的过程，其中所涉及的城市规划的各个方面（位置、种类、尺度、规模以及景观）必须提前考虑好并相互协调。基于这个原因，控制性规划如同二维及三维的城市规划管理工具。

由于城市如同一个有生命的有机体，在城市更新的过程中，旧问题可能会消失，然而也会不断出现新的挑战，这使得控制性规划要不断随之"扩展"或"调整"，在过去40年的城市更新中正是这样。

城市更新对于公民个人、城市和政府来说也是经济发展的共同任务。如果没有公共部门的调控，这种形式的城市更新将不可能存在。如果国家、城市和市民在过去40年中对一个城区的城市更新已投资超过1亿欧元的话——在Neuwerk区就是如此，那么这个行为不是出于无私，而是为了改善城市的吸引力、功能以及人居品质。

一个深层原因也在于对私人投资的刺激——国家投资1欧元可以带来7欧元的私人投资，并且提高城市的生产力，增加税收，同时创造了规划和建筑行业的就业机会；这样，城市建设发展在欧洲就成为了一个有效的刺激经济的方式。

Urban renewal is also a common economic task of the individual citizens, cities and governments. Without the commitment of the public sector, urban renewal would not be possible in this form. If the state, city and citizens invested about 100 million euros in the urban renewal of a region over the course of 40 years, like here in the Neuwerk district, they would be doing so not for altruistic reasons, but to improve the attractiveness, functionality, and quality of life for the sake of that city.

A more profound reason, however, lies in the encouragement of private investment. Per 1 euro invested by the state, there may be an investment of up to 7 private euros. In return, the productivity of a city – i.e. tax revenues – would be maintained and increased, as well as safeguarding or creating jobs within the field of planning and construction. Therefore supporting urban renewal has become one of the most effective economic programs in Europe.

城市如同一个有生命的有机体，在城市更新的过程中，旧问题可能会消失，然而也会不断出现新的挑战。

As a city is something like a living organism, old problems may disappear and new tasks may arise during the redevelopment of a city.

注释
1) 伦茨堡市控制性规划图解和斯图加特1984年解说报告
2) 伦茨堡市1981年城市景观规划
3) 伦茨堡市长，2008年7月17日新闻发布会，和受托承担Neuwerk区更新工作的BIG的总结报告:城市规划，基尔
4) "38年的更新，1亿欧元的投资:伦茨堡市生机勃勃的历史保护区Neuwerk的持续性的保护"
5) "38年的更新，1亿欧元的投资"

References
1) Master plan of the city of Rendsburg, construction documents and explanatory report, Stuttgart in 1984
2) Townscape planning of Rendsburg in 1981
3) Mayor of the city of Rendsburg, Press release from 17 July 2008 and final report on Neuwerk carried out by the reorganization in trust of the support organization for redevelopment BIG - town planning, Kiel
4) "being renovated for 38 years, it is investing 100 million euros - in the urban living memorial of Rendsburg and of Neuwerk has been permanently preserved"
5) "being renovated for 38 years by investing 100 million euros"

城市更新
URBAN RENEWAL

01 伦茨堡控制性规划
   Framework Development Planning of Rendsburg

02 "霍费尔天空":
   提升城市品质的范例项目
   "Heaven of Hof":
   an Initiative-project for Upgrading the City

03 埃尔旺根步行区
   Pedestrian Zone of Ellwangen

04 埃斯林根总体规划和公共空间城市设计导则
   Master Plan and Design Guidelines
   for the Public Spaces of Esslingen am Neckar

05 "粉红岛": 韩国首尔方背洞街区广场
   "Pink Islands":
   a Street Square in Bangbae-dong, Seoul, South Korea

06 Gangseogu城市更新规划
   Urban Renewal Gangseogu

07 北京市昌平区奥运自行车训练馆
   及周边地区城市设计
   Urban Design of the Area Surrounding
   Changping Olympic Cycling Training Center, Beijing

02 "霍费尔天空"：提升城市品质的范例项目
"Heaven of Hof": an Initiative-project for Upgrading the City

规划面积 AREA		1.2 hm²
人口规模 POPULATION		47300
完成时间 PROJECT DATE		2009.01

从两种文化中学习 LEARNING FROM TWO CULTURES

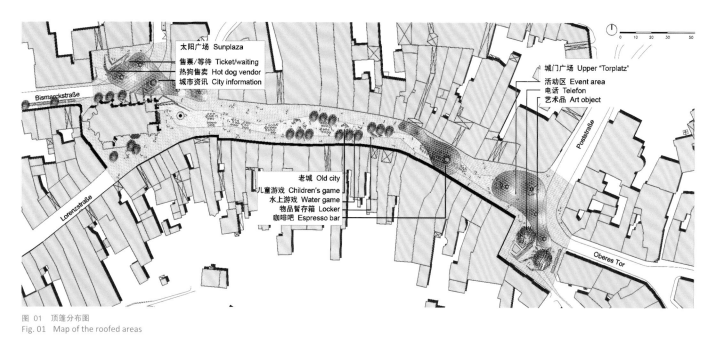

图 01 顶篷分布图
Fig. 01 Map of the roofed areas

规划现状与规划任务 Background and planning

Every city has its own unique character manifested in the physicality of its townscape. The shape of a city is influenced by the surrounding landscape, along with its social, economic and cultural histories. A city takes form through its architectural interactions with the landscape, urban layout, skyline, streets and squares, building types and facades.

For decades little attention was paid to the overall image of a city. Other issues such as reconstruction, building maintenance and infrastructure improvements were deemed to be of higher priority.

But today townscapes are taking on increasingly important roles. Citizens now consider the city's image an essential and highly visible symbol of their hometown. It is becoming clear that how attractive a community is depends not just on employment opportunities, shops, building sites and recreational facilities, but also on the quality of life, cultural heritage and the overall townscape.

In times of economic difficulty, German city centers suffer from decreasing purchasing power, widespread vacancies and fierce competition from large shopping malls on the city outskirts. Cities must react by developing a sustainable plan to secure and revive the future of shopping streets and public areas in urban centers.

Based on urban development concepts and concepts for retail centers, the city of Hof has established several 'Initiative-Projects', one of which is called "Sky of Hof - Upgrading the City". The concept of this downtown revival is a courageous step in the fight against "bleeding" inner cities. The street name "Altstadt" means 'old town', but the project differentiates itself from other cities and shopping malls by reconsidering light itself, using transparent roofing over the public spaces in the pedestrian zone. Together with retailers and concerned citizens, a unique contemporary selling point was reached for the pedestrian thoroughfare. As a next step, in an effort to generate alternative solutions for

在经济困难的时期，德国必须考虑内城的购物街和公共空间在未来的发展前景。
In difficult economic times, Germany must consider the future of its shopping streets and public areas in inner cities.

任何一个城市都有它特有的城市性格，并表现为城市景观。城市景观受其所处的地理环境影响，被城市的社会、经济和文化层面的历史所塑造，通过城市的布局和城市中的建筑体现出来。城市景观的内涵包括自然景观、城市空间格局、城市天际线、道路和广场空间、建筑类型和立面特征等多项因素。

城市在其发展过程中，对于城市景观塑造并不总是能投入充足的力量。其他的任务，比如城市重建、更新维护和基础设施改善，总是在城市建设中被优先考虑。

今天，城市景观正变得日益重要。市民将城市景观看作他们家乡的视觉特征。商人和企业家们意识到，消费者不只看重店铺内的部分，企业雇员也不只把目光局限在工作位置上。人们可以越来越明显地观察到，一个城市的吸引力不仅取决于就业机会、商业设施、可建设用地和娱乐设施，也取决于其人居品质，其中文化遗产保护以及当下和未来的城市景观品质也是重要的层面。

目前，德国城市的内城正面临着由郊区购物中心的兴起所带来的购买力流失和店面空置的问题。因此，城市的管理者要对此作出恰当反应，通过规划和建设手段，维护内城城市公共空间的质量，给内城购物街区注入新的活力。

德国的霍尔费市在城市发展规划和零售业中心区方案的基础上，决定启动一批相关项目，其中"霍费尔天空——提升城市品质的范例项目"便是近期打算实施的第一步。振兴内城计划是针对内城的"衰竭问题"迈出的勇敢的一步，与零售商和对此关注的市民一起讨论，

城市更新 URBAN RENEWAL

为"老城"的步行区寻找一种有别于其他城市购物中心的特征。讨论的结果是要为步行区的公共空间加盖一种轻盈透明的顶篷。

霍费尔市政府接着选拔了一组设计师(包括结构工程师、建筑师和城市规划师),与零售商和土地所有者一起,于2007年12月进行了一次研讨,为加强老城区的公共空间品质寻找提案。

规划的任务是为"老城"的步行区改造和覆顶工程提供一个富有吸引力的"抛砖引玉"的方案。相关目标一方面是将霍费尔市建设成具有竞争力的购物型城市,另一方面要吸引周边地区的来访者和消费者。老城的品质提升工作应巩固霍费尔市的区域中心地位,并使之成为吸引市民和周边访客的亮点。

德国ISA意厦国际设计集团和SBP工程事务所的方案荣获了一等奖,并被委托进行进一步的工作。

improving the public spaces of Altstadt street, the city of Hof conducted a design workshop with an exclusive group of structural engineers, architects, urban planners, retailers and property owners in December of 2007.

The task was to develop an attractive concept for the reorganization and roofing of the pedestrian street Altstadt. The initiative was intended to boost shopping in Hof, which is in direct competition with other nearby rapidly developing urban centers, and to serve as an attractor for visitors and consumers from surrounding areas. The upgrade is intended to strengthen the town of Hof's long-term standing as a regional center, and increase the attractiveness of its downtown for the benefit of its citizens and visitors.

ISA - International Stadtbauatelier together with the engineering office Schlaich, Berger and Partner won the competition and was authorized to develop the work.

> 壳形格栅屋顶应该通过它富于空间及雕塑感的形式发挥长期的作用,从而形成霍费内城新的标志,并且能直接为老城所用。
>
> *The design concept envisions the creation of a central zone in an exceptionally wide Altstadt street as a three-dimensional 'landmark' having partial roofing and street furniture.*

开放空间设计 Open space design

方案在老城区宽阔的街道上规划出一个中央区域,在这个三维尺度的标志性区域中,以局部覆顶的城市小品形式强化老城区与周边绿地的关系,并连接了附近的购物街区。同时通过绿化措施,改善区域内居民和来访者的空间体验和休闲感受,给房地产市场带来推力。通过绿化元素进行适度的城市现代化,通过城市小品创造合乎时代要求的驻留空间,并将商店和餐厅的面貌与覆顶结

The design concept envisions the creation of a central zone in an exceptionally wide Altstadt street as a three-dimensional 'landmark' having partial roofing and street furniture. It is thought that this will strengthen the relationships between the street, the surrounding green spaces and the river Saale, and in doing so help to establish a stronger connection to the neighboring shopping streets. By planting trees and other greenery and introducing leisure facilities, the overall quality of life for residents and visitors can be improved, while at the same time relieving some of the stresses on the real estate market.

图 02 加顶篷后的老城入口效果
Fig. 02 Entrance to Altstadt street with new roofing

图 03　夜晚的氛围
Fig. 03　The atmosphere at night

图 04　街道小品的模块式系统及可能的组合方式
Fig. 04　Modular construction system for the street fittings and possible combinations

A modern and appropriate urban renewal project with green, contemporary recreational spaces, street furniture and the coordinated appearance of retail shops and restaurants, combined with the form and function of roof construction, offer heightened visual recognition for the visitor, and opportunities for community events and performances will help to accentuate the feeling of community for residents.

This design solution was reached through analysis of the existing situation and the desired downtown development, taking into account administration, political issues and the needs of local retailers. The lattice structures are meant to achieve long-term effects through their spatial design and sculptural form, becoming an identifying feature of downtown Hof and a utilitarian addition to the Altstadt street. The partial roofing improves weather protection on the street, and variable roof height meshes with a city skyline that includes buildings of varying heights and multiple uses along the street. The maximum transparency of 'the sky of Hof' ensures that the sun's rays continue reaching pedestrians while reference to the existing buildings is maintained. And, at night the street is bathed in special artificial lighting to create a very special atmosphere.

The partial roofing and design of the public spaces form a complete ensemble. The basic concept of the open space design is to create a continuous strip in the pavement which stretches from Sonnenplatz square through the pedestrian street Altstadt to Oberer Torplatz square and from there leads foot traffic down to the river Saale. The important 'nodes' of the city are connected in this way and the individual roofs become part of an overall concept that can be expanded into other areas of the downtown as needed. By way of this strip, the street space is divided into two zones: a living zone with running paths and delivery and emergency routes along the shop windows, and a quiet zone used for rest, recreation, children's play areas and restaurants. Located in the latter zone are strategically placed "islands of peace", which consist of multi-functional elements specially developed for the city of Hof. The movable street furniture can be reconfigured to accommodate different functions including ticket offices, lockers and toys for children.

构的形式和功能结合进行考虑，这些在为来访者提供视觉识别性的同时，也提供了结合大型活动的组织形成霍费尔市城市景观特色的可能。

开放空间和顶篷的设计建立在对内城现状和未来发展分析的基础上，同时结合了当地政府、政党和企业的意愿。壳形格栅顶篷应该通过它富于空间及雕塑感的形式发挥长期的作用，从而形成霍费尔内城新的标志，并且直接为老城所用，通过局部覆顶使得老城区的全天候环境得以改善。顶篷的高度配合不同的建筑功能和建筑高度而变化。透明的材料可以使阳光更好地透入顶篷下的空间。夜晚的老城在顶篷灯光的照射下将展示出别样的风情。

顶篷设计和开放空间的设计共同组成一个整体。开放空间设计的基本思路是要创造一个有着良好铺装的带状开放空间，其从太阳广场穿过老城，延伸至城门广场，并且一直延续到萨勒河。通过这种方式，内城中被称为"关节"的几个重要城市广场也被联系在一起。单个顶篷的设计是在整个方案理念的指导之下进行的，在必要的情况下，这种顶篷也可以放入内城的其他区域。这条带状开放空间将老城分为两个部分：紧邻橱窗两侧的充满活力的步行区，以及位于中央的休憩区（用作孩童嬉戏和餐饮的场地）。这个休憩区中重要的位置上设计了"休憩岛"，由特别为霍费尔市设计的独特的街道小品布置而成。这些充满现代感的多功能街道小品可坐可躺，也可在上面游戏，在街道空间中占有重要位置。它们是可以移动和重新组合的，可按照需要进行改装，也可以充当自动售票机、物品暂存箱和儿童游戏设施。

城市更新 URBAN RENEWAL

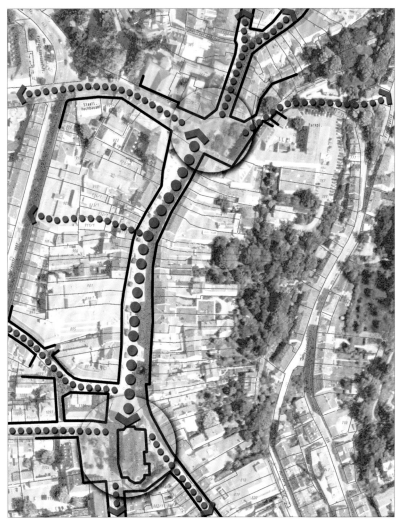 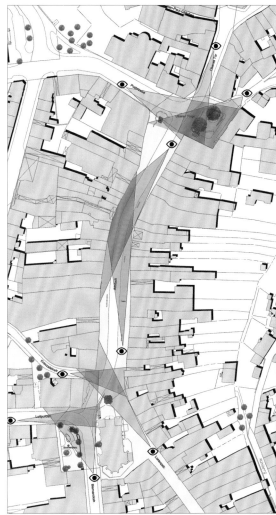

图 05 引出顶篷的街道网和视觉廊道通向老城
Fig. 05 The relation of paths and visual connections to Altstadt as a derivation of the roof positions

顶篷的位置与形式

步行街上方顶篷的位置与形式一方面由于防火、私产保护和结构要求，因而不允许与建筑立面直接相连；另一方面，通过有意识地将玻璃顶篷与建筑立面保持距离，可以使得现有的历史建筑立面连续流畅地呈现在人们面前。依据这个设计构思，重要的历史保护建筑圣玛利亚教堂首先被塑造为老城区引人注目的视觉终点。并且在玻璃顶篷的设计过程中充分考虑与教堂的间距问题，以免影响现有的视觉效果。高低错落的建筑檐口以及凸凹的建筑立面则展现出一幅老城区所特有的活跃气氛。由此，未来可以把步行街的玻璃顶篷设计成为一个安静优雅的建筑形体，与现有立面形成对比。

Position and shape of the roofs

In part, the position and shape of the partial roofs are because of fire safety regulations, along with private and structural requirements that do not allow them to be directly attached to the facades. However, the roofs are also deliberately distanced from the facades in order to protect the public view of historical buildings. This is especially important in the case of the Church "Marienkirche" - an important historical building and natural point of focus, at the end of the pedestrian walkway. The building facades on the street with their heterogeneous eaves and varying ground floor heights create an uneven, almost bumpy appearance. It makes sense then that the future roof to cover the path in front of these structures should have a subdued, slightly curvilinear form to them that contrasts with the varied heights of the building facades.

柔和、略带曲线的屋顶形状，与不同高度的建筑立面形成对比。作为从各个方向、各个角度看都明显易见的形态特征，将游客迅速吸引到购物街。

The roofs have a subdued, slightly curvilinear form to them that contrasts with the varied heights of the building facades. They serve as an alluring feature that is highly visible from all directions and angles, quickly drawing visitors into the shopping street.

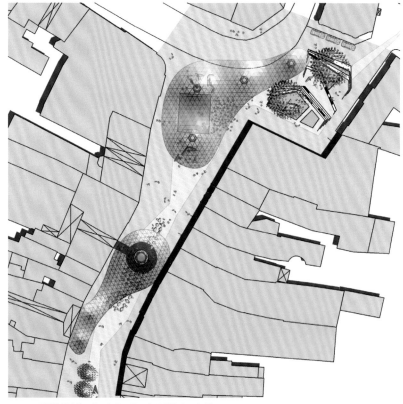

图 06 顶篷覆盖了行人区，老城与教堂间视线通畅
Fig. 06 The roof covering the pedestrian zone maintains clear views of the historic church

图 07 不同使用功能：节日舞台，国际电影节的"红地毯"
Fig. 07 etailed map with different variations of usage: a stage for events, "Red Carpet" at the International Film Festival

Another goal was to develop the added architectural elements to serve as an alluring feature that is highly visible from all directions and angles, quickly drawing visitors into the shopping street. They simultaneously function as both nodes and connectors that link the adjacent shopping streets to Altstadt. The partial roofs are not only visible from outside the street, but also within the street, overlooking connections that exist from one element to the next. The two ends of the street are thus visually connected. Thanks to their expressive appearance, they become a new visual icon for the town center of Hof.

The sculptural shape of the roof structure sets itself apart from the structures of the existing buildings in order to preserve their individual character. The result is three separate roofs, but they speak a common design language and are seen over the entire street as a harmonious whole. Each component of the roof should appear transparent and delicate. The aim is to achieve a generously high and airy atmosphere. The sculptural columns create space and at the same time appear transparent, as they are comprised of very delicate individual components: stainless steel pipes connected to castings, intended to reduce maintenance and to maximize durability.

The first roof is located on Oberer Torplatz square where the existing green spaces are to be upgraded. Together with the curved roof, they will create an upbeat atmosphere and serve as a visual gateway into the street. In this way, a square with green elements and a fountain is created, which invites visitors to linger and attend events such as public viewings, concerts and markets.

另一个目标是把玻璃顶篷设计成一个具有视觉标识作用的"吸引点"，藉此把游客目光都锁定到这条购物街上来。同时也通过这个设计把这条街道与整个老城联系起来。不仅从"外部"具有连续可见性，在老城内部也使其作为一种建筑符号把人们的视觉轨迹延伸到老城的其他地方。老城的两端就这样联系起来，最终赋予霍费尔这座老城一个崭新的视觉印象。

通过这一具有雕塑性的结构突出了现有建筑的个性，三片顶篷既独立存在，拥有各自的形式，又和谐地采用了统一的设计语言，并延展到整个老城区。每个顶篷都尽量给人一种通透轻盈的视觉感觉，而且尽最大可能实现通风。顶篷支柱作为支撑结构，通过使用细致精巧的零部件，力求做到视线通透，给人以轻巧的感觉。使用不锈钢材料，以铸件连接，以尽可能地减少维修，延长使用年限。

第一个顶篷建在城门广场。把现有绿地加以改造，与弧形的屋顶一起在老城的入口处形成一道靓丽的景观。这样就形成绿树掩映下的休憩地带，其间有喷泉点缀，可以举办公共集会、音乐会或集市等活动。周围的街道，特别是路德维希大街，将与老城形成通路，因为远处的人们从四面八方都可以看见这些屋顶。

城市更新　URBAN RENEWAL

第二个顶篷建在老城中心区与玛丽大街之间的步行街入口处。这样街道入口在视觉上得以突出，并为附近的电影院留出一个适当的入口。出于结构和空间考虑的支撑脚在必要时可以改装为小酒吧或者小型咖啡屋。而这个屋顶下方的区域则为蜚声海外的霍费尔电影节提供了合适的活动场地。

第三个顶篷坐落在太阳广场上，这里位于老城区的入口处，有着空间门户的效应。顶篷覆盖了公交候车区，荫庇了街道，并一直延伸到了老城里，直到老市政厅前。通过交通静化措施和对广场铺装的设计使得太阳和老城能够更好地联系在一起。顶篷位于圣玛利亚教堂的一翼，配合重新设计的台阶将使得教堂更加有吸引力，老城和教堂间的视线通畅也得到了保障。教堂边老市政厅的阳台上加盖了顶篷，这样可以延长其使用寿命。

由于毗邻的街区都可以看到顶篷，内城区和老城区的步行街也得以联系起来。太阳广场顶篷下新设置的公交站能使等公车的旅客或者要去老城的人们免受风吹雨淋之苦。

The second roof in the middle of the pathway connects pedestrians directly to the passage between Altstadt and Marienstrasse streets. This makes the entrance of the passage visually clear and gives the adjacent cinema an appropriate entrance. The sculptural and spatial column base can function, as needed, as a small champagne or espresso bar and the area under the roof provides an adequate framework for events in Hof during its internationally renowned film festival.

The third roof on Sonnenplatz square at the city gate strengthens the gate's character and connects it to the redesigned plaza. It covers the area of the bus stop and the street space, and stretches into Altstadt street. This, aided by traffic control measures and the design of the street pavement, will link Sonnenplatz square to Altstadt street. Marienkirche church is flanked by the roof and gains in appeal in connection with the large stairs defining the space, but the main focal point from Altstadt street remains unobstructed.

由优美柱体支撑的自由曲面覆顶将吸引人们进入老城区，并在空间节点处吸引着行人从一个顶篷走向另一个顶篷。

The free-curved grid shells, with their sculptural columns lead visitors from one roof to the next and even integrate the adjacent squares.

图 08　作为地标和公交站的顶篷
Fig. 08　The new landmark roof functions and bus stop

枢纽广场

两个"枢纽广场"——太阳广场和城门广场将借助一直延伸到行车道的铺装来提升其品质,从而使得历史的空间界面得以展现。下沉的阶梯广场利用并强调了地形的特征,形成一个"公共生活的舞台",令人流连忘返。这些露台和看台也可以在各类活动中用作观众席。在太阳广场,通过公交线路的迁移,在教堂前赢得了一块开敞开间,使得教堂更加伟岸,改变了目前教堂前只有狭窄过道的状况。新的巴士站、舒适的候车区和城市问询处赋予此处富有活力的城市特质。与此相反,城门广场则绿荫密布,并有一条通向萨勒河滨的绿地。

公共空间的功能

配合开放空间的设计,三个顶篷对老城进行了重新组织,并给空间带来功能上的多样性。由优美柱体支撑的自由曲面覆顶将吸引人们进入老城区,并在空间节点处吸引行人从一个顶篷走向另一个顶篷。这将形成灵活、宽敞且富于变化的城市空间,可以满足不同居民群体的需求。无论是举办每周户外活动、时装表演、节日庆典,还是进行其他各类活动,比如撑竿跳高节和霍费夫国际电影节,"霍费尔天空"项目都为各个季节的活动提供了不受天气影响的活动空间。

在日常生活中,这种多功能的空间给大家提供了位于中心地带安静的露天餐饮和休憩区,在这里人们可以尽情欣赏缤纷的橱窗。绿树荫蔽下的中央休憩区可以设置嬉戏、运动和服务设施,如提供照看儿童、城市信息查询和物品寄存之类的服务。

Urban hinge places

The two 'urban axes', Sonnenplatz square and Oberer Torplatz square in front of the city gates are upgraded through new paving, which crosses over the car lanes. The topography is put to good use, accentuated through spacious terraces and inviting seating that encourages people to linger and sets the stage for public life. These seats can also be effectively used as spectator seating for events. At Sonnenplatz square, the church is highlighted by shifting traffic lanes out of the way to create an adequate pedestrian pathway; a major change considering the narrow sidewalks of the past. The area takes on a vibrant urban character, shaped by the new bus stop, the comfortable waiting area and the city information area. In contrast, Oberer Torplatz square is dominated by existing trees and greenery, thereby creating a transition to the green area of the river Saale.

Functions in public spaces

In conjunction with the open space design, the three roofs reorganize the pedestrian street Altstadt and provide a new multifunctional space. The free-curved grid shells, with their sculptural columns lead visitors from one roof to the next and even integrate the adjacent squares. The result is a flexible, large-scale staged urban space with interconnecting squares and open spaces that allow for a multitude of uses depending on what groups populate the space. From the weekly market, open-air events and fashion shows, to festivals such as the Hof International Film Festival; the "Sky of Hof" provides a year-round space for events regardless of weather conditions.

In everyday life, this multifunctional area serves as a quiet yet central public space where one can dine outdoors or just to watch the urban bustle of the shopping street. The central zone with trees and seating areas provides play areas, spaces for sporting events and service facilities, such as childcare and information on the city and lockers.

图 09　城门广场上作为空间门户的顶篷
Fig. 09　Roof as gate : Oberer Torplatz square

图 10 透明的、雕塑性的顶篷结构细节图
Fig. 10 Details of the transparent and sculptural roof structure

城市更新
URBAN RENEWAL

01 伦茨堡控制性规划
Framework Development Planning of Rendsburg

02 "霍费尔天空":
提升城市品质的范例项目
"Heaven of Hof":
an Initiative-project for Upgrading the City

03 埃尔旺根步行区
Pedestrian Zone of Ellwangen

04 埃斯林根总体规划和公共空间城市设计导则
Master Plan and Design Guidelines
for the Public Spaces of Esslingen am Neckar

05 "粉红岛": 韩国首尔方背洞街区广场
"Pink Islands":
a Street Square in Bangbae-dong, Seoul, South Korea

06 Gangseogu城市更新规划
Urban Renewal Gangseogu

07 北京市昌平区奥运自行车训练馆
及周边地区城市设计
Urban Design of the Area Surrounding
Changping Olympic Cycling Training Center, Beijing

03 埃尔旺根步行区
Pedestrian Zone of Ellwangen

规划面积 AREA		1 hm²
人口规模 POPULATION		24600
完成时间 PROJECT DATE		2000.11

简介 Initial position

Ellwangen is a small town with 25,000 inhabitants located in southern Germany about 80 kilometers east of Stuttgart. It has the status of a large town and forms a secondary center for the surrounding villages and towns. The surrounding picturesque landscape is typical of South Germany, characterized by gently rippling topography with numerous small streams and meadows. Many parts of the surrounding countryside are designated nature preserves or protected landscapes. West of the city center, the Jagst River bisects the city north to south.

Whereas cities like Heidelberg or Tubingen attract many international tourists from America and Asia; Ellwangen is more popular with cycling tourists and weekenders from the surrounding countryside. Most visitors don't come just to see the city of Ellwangen, but also for the well preserved and unusual combination of small villages, old towns and scenic landscapes that the region has to offer. In partnership with neighboring communities, Ellwangen strives to protect this heritage. Encroaching cities are avoided by maintaining a clear delineation between rural and urban areas, with outdoor areas preserved and intended to encourage tourists to go on hiking and cycling trips. Area towns by contrast, are intended to attract visitors with their lively urban character and variety of restaurants and souvenir shops.

Residents, along with their city councils assess the importance of a city in terms of its political, cultural and economic values and do not display exaggerated or false ambitions concerning urban development policy. Instead, cities are limited to identifying only the qualities that will maintain the everyday life of their inhabitants and for visitors from surrounding areas. In this harmonious dialogue with residents about the city they live in, a longstanding urban tradition becomes obvious, and citizens begin to take pride in the historic buildings that make up their enduring city.

埃尔旺根位于德国南部斯图加特以东80公里处，是一座拥有25000人口的小城市。它在形态上呈环抱状，是周围村镇的中心。四周的景观风光优美，有平缓的草场和众多的溪流，展现了南德的典型景观，其中大部分为自然或者景观保护区。中心区西边的雅格斯特河由北向南穿城而过。

海德堡、图宾根等城市吸引的是来自美洲和亚洲的国际游客，埃尔旺根则得到周末近郊游和骑自行车旅行的游客的青睐。游客们前来的目的不仅仅是因为埃尔旺根这座城市，也是因为通过旅游可以完整而深刻地了解欣赏该地区的小城镇及古城景观。埃尔旺根同其相邻城镇一起致力于当地文化的保护，尽量避免城区的扩张，以实现乡村同城区的清晰区分。外部区域则力争保留农业景观，吸引步行游人和自行车游人。与此相反，城市则要展现其生动活泼的城市特性，例如用特色餐饮和纪念品商店来吸引眼球。

这座城市的居民和城市管理部门很珍惜城市在政治、经济、文化方面的现实意义，在城市发展的策略中不刻意强化或改变城市本来的面目。城市管理者将城市的品质界定为要同时满足市民日常生活及游客观光的需求。埃尔旺根市民自信地向世人展示了这座城市的悠久传统，并对城市所拥有的源远流长的历史深感自豪。

图 01 修道院与集市广场，2008年
Fig. 01 Monastery with canon's houses and market place in 2008

图 02 埃尔旺根的短途旅游
Fig. 02 Day-trippers in Ellwangen

城市更新　URBAN RENEWAL

图 03　埃尔旺根城以及老城中心区
Fig. 03　Ellwangen city's historic old town

图 04　从南部看埃尔旺根老城区
Fig. 04　Historical view of Ellwangen from the south

历史背景 Historical background

历史上埃尔旺根的宗教、政治、行政和公共生活是紧密交织在一起的，并从中衍生出一种在今天也能感受到的精神和建筑学的文化。这也是埃尔旺根从城市规模上看同许多亚洲城市相比只能算乡村，但却拥有宏伟的建筑和浓郁的城市化氛围的原因。

埃尔旺根今天的城市结构可以追溯到一座修建于8世纪的修道院，该修道院成立伊始即成为法兰克帝国卡尔大帝的皇家修道院。修道士的地位逐步提高，其势力范围也逐渐扩张。12世纪初的一场大火之后修道院进行重修，修道士住在北边，南边则是普通信徒的房子。

12世纪前这个修道院的辖地还是一片荒野，后来发展成一个居住区，这就是埃尔旺根城市的起源。

经过另一场大火的洗礼之后，在从1182年到1233年间的整修中，这座修道院教堂被修成了后罗马时代的风格，它至今仍是埃尔旺根市的标志。这座教堂被视为该时期南德最重要的建筑之一。

在城市经历了一次来势汹涌的经济危机后，修道士阶层愈发贫困，并失去了影响力。1460年它变成了国际性的唱诗班修道院，隶属一位亲王。时至今日，该修道院（包括南边普通信徒的住房）依然令内城景观焕发出别样的宗教色彩。

这座城市在农民战争和三十年战争期间接受到的狂风骤雨般的洗礼和变革，直到17世纪才结束。随着基督徒的迁入，城市发展进入了一个新的阶段，Schoenenberg 山上的朝圣教堂给人以鲜明的印象。之后很短的时间内一批出身自旧贵族家庭的新王侯开始在城市中兴建宏伟宫殿，如大量巴洛克式府邸。这些建筑大部分由当地的手工业者建设完成，显示了高度发展的、至今在埃尔旺根仍占统治地位的建筑文化。

Religion, politics, administration and public life are very closely integrated into the history of the city. All of these factors have helped to develop the city's spiritual, architectural and urban culture, which it still enjoys today. This helps to explain why Ellwangen, although barely the size of a village when compared to many Asian cities, could develop an outstandingly urban architecture and atmosphere.

The present structure of the city dates back to an eighth century monastery, which shortly after it's founding become a royal monastery under the Frankish King Charlemagne. This is how the rise of the monastery began and the extent of its realm. Following a fire at the beginning of the twelfth century, the monastery was rebuilt, while living quarters for the monks' were designated in the north and houses of the laity in the south.

This southern section grew into a smaller settlement under the jurisdiction of the monastery, which was to be the beginnings of the city of Ellwangen. The area was referred to in twelfth century parlance as 'civitas'.

After another fire sometime between 1182 and 1233, the monastery church was rebuilt in late Romanesque style and continues to this day to be one of the city's major landmarks. It is considered to be one of the most important buildings of the period in southern Germany.

Whereas the city witnessed an economic boom in the following period, the monastery became poor and increasingly lost its influence. In 1460 it was converted into a secular choir of monks and placed under a lord provost. The canon's houses, which emerged in this context, south of the monastery complex, still dominate the city's downtown area.

During the time of the Peasant's War, the Reformation and the Thirty Year War, the city endured tumultuous times right up to the end of the 17th century. A new phase of development began for the city with the settlement of the Jesuits, which received its most visible outward expression through the pilgrimage church on Schoenenberg. Shortly thereafter, beginning with a series of new lord provosts descending from old aristocratic families, the city was rebuilt into a Baroque residence with numerous magnificent city palaces. The majority of the buildings were constructed by local craftsmen, a clear sign of the advanced architectural culture that prevailed at that time in Ellwangen.

At the beginning of the 19th century, the Fuerstpropsttum - having been independent until that time - was joined with the House of Wuerttemberg during the secularization of the city. It was then rebuilt as a residence for the county Wuerttemberg and equipped with the relevant authorities. Ellwangen even received a university, but it was soon subordinated to the University of Tuebingen.

In the following period, the importance of Ellwangen continued to wane until the end of World War II, when refugees began pouring into the city, which had been left largely unscathed by the war. During this time the population saw an increase of nearly 50 percent. Ellwangen grew to its current size and reached the status of a large district town in the early 1970s by incorporating surrounding villages.

In 1987, after the construction of the A7 motorway to the east of the city, new industrial areas could be designated thanks to improved infrastructural connections, and this encouraged to the city's continued growth.

Increasing traffic loads, however, endangered the historic old town and overall quality of life as well as historical substance. For this reason, the city council sought ways to develop a new transport system that would solve the city's traffic problems and preserve its historic image.

19世纪初的世俗文化运动中独立的王侯将房屋毗邻符腾堡而建，作为符腾堡的属地，并设置相应的机构，埃尔旺根不久以后成为一座隶属图灵根大学的大学城。

后来该城的重要性日趋下降，直至"二战"时它由于少有战乱波及，大批逃亡者涌入，才又显得重要起来。此间，居民数量增加了50%。通过相邻村庄的加盟，埃尔旺根才形成今天的规模，并于1970年初发展成为一个颇具规模的城市。

1987年城市东部A7快速路建设以后，得到改善的基础设施网络促进了新工业区的形成，加快了城市的发展。

然而日益增加的交通压力给历史老城造成负担且损害了历史构筑与人居品质。出于这个原因，市政府开始寻求发展新交通体系的可能性，以解决交通问题并维护古城的形象。

交通状况 Traffic situation

A major source of Ellwangen's traffic problems was the B290 federal road built in the 1950s that led right through the center of Ellwangen and constituted the area's main north-south route. Thanks to the new federal route, the city surged with traffic and the historic old town was bisected right down the middle. An earlier break in the city was the railway line dating from 1866, which separated the mainland from the western suburbs on the western edge of the old town.

The two traffic lanes of the B290 are the main cause of environmental and noise pollution in the downtown area. In response to the problem, in 1962 the city administration began researching alternative routes for the B290. This decision by the regional and federal governments was seen as an opportunity in Ellwangen to create a plan for the A7 motorway based in the east of the city, which would become a new transport system for the region. With high citizen turnout, many alternatives were discussed, one of which resulted in two main potential solutions: a transfer of the B290 as a bypass to the west, known as the "West Ring Road", or a tunnel under the federal road.

The first solution was very controversial, since it would mean cutting through and impairing the Jagst river – a major fixture of the local natural environment. But unfortunately, the tunnel could not be realized either due to construction cost. The discussion of the two solutions, however, led to the realization that the protection of the city's cultural heritage and natural landscape were important to the overall townscape and these issues were given high priority in preparation for the forthcoming West Ring Road. The speed limit on the new roadway was restricted to 60 kilometers per hour instead of the typical 100, so that the roadways could be kept much narrower than usual in hopes that this would minimize the adverse effects to the landscape. The restriction of traffic was acceptable, since construction of the A7 motorway had already been decided, and significant relief to existing highway congestion could be expected.

现有的一个交通问题的根源是1950年修建的联邦大街B290，它从埃尔旺根的中心区穿过，一直延伸到高速公路处，是最重要的南北贯穿线。联邦大街的接入极大地增加了城市的交通量，历史老城也因此被分隔开来。城市发展另一个大的转折是1866年老城西部边缘铁路线的贯通，这条线将核心区同郊区分开。

这两段交通线是内城环境及噪声污染的主要原因。因此1962年政府开始寻求替代B290的路线。联邦和地方政府决定，在城市东部修建A7高速路，这一决定为埃尔旺根带来新的交通体系。市民的积极参与形成了各种不同的替代方案，其中有两种方案是可行的。一种是将B290移位用作西部的分流道路，也就是形成所谓的西部直路。第二种方案就是在这条大街下开挖隧道。

针对第一个方案存在很多分歧，因为它分裂和影响了那里的自然空间及雅格斯特河，但是隧道方案由于费用问题又难以实现。针对两种解决方案的讨论旨在将景观保护作为文化遗产以及城市景观重要的组成部分，要能够一次性高效地解决问题。新街的车辆运行速度被限定为60公里/小时，而不是通常的100公里/小时，这样7.5米宽的车道就明显比普通车道窄，对景观的影响也大为减轻。市民可以容忍对交通的限制，因为A7高速路的建设提案已经一致通过，因此为B290大街的交通减压是指日可待的。

图 05　改建前arien大街的交通负荷
Fig. 05 Traffic load in the Marientrasse before reorganization

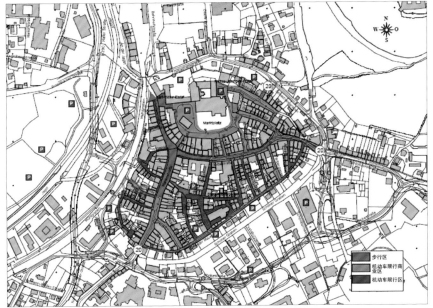

图 06　交通方案：红色为步行区，绿色为受交通管制的商业街，蓝色为受交通管制的住宅区巷道
Fig. 06 Traffic concept: pedestrian zones in red, commercial streets as restricted traffic zones in green, residential alleys as restricted traffic zones in blue

1983年城市开始对分流道路的建设进行规划，一年以后政府通过了该建设项目并于1987年立项。鉴于市民的参与和商榷，整个过程由设想到实践需耗时25年。

西部直路减少了内城南北向交通的负担，然而东西向的交通负荷依然过大。因此在内城南部再修建一条分流道路的项目也被提上日程，也就是准备建设南部直路，将路网补充建设成为环形系统。1994年该项目获批，1999年立项。为了避免南部接邻的城区因新道路而与内城相分隔，南部直路被建成绿地下的一条隧道。

同西部直路委托项目的目的一样，这些措施实现了内城片区的交通减负。这样整个历史古城核心就适宜自行车及行人步行通过，打下了古城城市更新和品质提升的基础。这也意味着，主要的购物街Marien大街、Schmied大街和Spital大街被设计为步行区，这些街区除公交车外一律不允许其他机动车穿行。老城其余的街道则被设定为限速的交通缓行区。

引入这些措施后的1998年，也就是南部直路立项的前一年，Spital大街和Marien大街的竞赛设计项目向公众开放。ISA的设计荣获一等奖，并继而承接该项目规划设计。步行区的规划于2000年完成。

In 1983, construction planning for the bypass began; the project was approved by the national government a year later and completed in 1987. Because of the high level of civic participation and extended deliberation processes, it took a total of 25 years from concept to realization to complete the entire project.

Since its completion, the West Ring has relieved the city of its problems with north-south traffic congestion, but the east-west link was still heavily congested. To remedy this, a second project began with the aim of constructing another bypass road just south of the downtown. This would become the so-called "South Ring Road", completing the transport network's ring system. This project was approved in 1994 and completed in 1999. In order to avoid separating the adjacent neighborhoods to the south from the downtown are with this new road, the Southern Ring Road was constructed in a tunnel beneath a green area.

These projects have succeeded in a two-directional reduction of downtown traffic, realizing the aims of the original "West Ring". Thanks to their success, the entire historic town center was redesigned for pedestrians and bicycles and provided the basis for upgrading and renovating the old town. This allowed the main shopping malls of Marienstrasse, Schmied- and Spitalstrasse to be designed as pedestrian zones with the exception of bus service. The remaining streets of the historic city are to be designated as restricted traffic zones, in which driving is only allowed at walking speed.

In the course of these measures initiated in 1998, a year before the completion of the Southern Ring Road, a competition was announced to design the two streets Spitalstrasse and Marienstrasse. The ISA urban planning studio's design won first prize and they were charged with the planning and the pedestrian zone was finished in 2000.

图 07 改建后的Marien大街
Fig. 07 Marienstrasse after implementation

图 08 Marien大街的建筑要素
Fig. 08 Street furniture in Marienstrasse

图 09 Marien大街效果图
Fig. 09 Perspective of Marienstrasse

步行区规划 Planning of pedestrian zone

The concept of the ISA urban planning studio was selected for three main reasons. First, the road surface is designed as a flexible, usable area with a uniform character. Second, dividing the two streets in sequences was praised for taking into account the spatial characteristics and topographical situation. The third important point was a lighting concept that emphasized the topography, spatial quality and above all the extraordinary architecture of mansions and other simpler buildings in the area at night. The lighting concept helped emphasize the City's rich history, making its particular qualities more visible at night.

Meanwhile, nine years have passed since completion. According to the former chief officer of the organization, the design of the attractive city center of Ellwangen continues to meet all expectations. One aspect that needs to be viewed critically, however, is the concept of mobile landscaping and greening, which had been suggested by the ISA urban planning studio.

The plan included using trees only sparingly so as not to obstruct views of the valuable and beautiful facades. Trees are therefore placed only in selected locations so as to divide the series of roads or highlight visual links in the landscape. Nevertheless, in order to create a green impression of the road space, plant containers were selected and purchased by retailers and local residents. The idea of involving citizens in the design process of the road in this way turned out to be a problem in retrospect because it had the effect of fragmenting the image of the space along the road as a whole. The municipality is therefore currently considering whether or not to standardize the appearance through the use of style guidelines.

埃尔旺根步行区例子证明，公共空间的升级可以成为促进经济和结构发展的一个重要手段。

Ellwangen's example of pedestrian zones illustrates that the upgrading of public spaces may in fact be an important means of economic and structural support.

According to the chief officer, one interesting effect of these projects is that thanks to the upgrades of the roads, increased sales have been reported and new work places created. This is where the city of Ellwangen proves the success of a way of thinking that is different from many local authorities in Asia. This seems to contradict the widespread opinion that employment must first be provided before an upgrading of roads can be undertaken. Ellwangen's example of pedestrian zones illustrates that the upgrading of public spaces may in fact be an important means of economic and structural support.

ISA的方案由于以下三个原因被选中：第一个原因是街道上部平面被设计为具有统一特性、功能灵活的平面。第二个原因是，两条街道的设计都充分考虑了空间特征和地形状况，并据此加以划分。第三个重要的原因在于，规划方案中的照明设计，无论是夜间的地势形态、空间品质、非凡的城市宫殿建筑，还是普通的房屋都得到了充分的强调。这样的设计使得城市的历史和它的特殊品质有了清晰的夜间视点。

该项目自立项至此已过了9年。正如当时的官员对该设计寄予的希望那样："埃尔旺根再次以其独有的魅力和地方特色获取世人瞩目。"如今这一愿望已得到实现。另一方面，也有批评的观点指向ISA提出的灵活的绿化方案。

方案中，为了不挡住立面上有观赏价值的视线，绿化的树木稀疏布局。所以树木被种植在预先设定的位置上，这样街道显得井然有序，景观视角也能通透无碍。街道空间的另一个绿化方式在于，绿化盆栽由街上的零售商人和居民自行选择栽种。这种市民大量参与设计过程的想法，后来被证明存在一些问题：所谓众口难调，大家自行选择盆栽设计出来的街道空间不具有统一性。市政府现在正考虑，是否通过制定景观设计导则来统一街道空间的形象。

一个有趣的观点是，行政部门有关提升街道商业营业额的声明带来了一些新的就业机会。这展示了埃尔旺根市政府同亚洲行政管理部门截然不同的执政思路。亚洲的观点是，首先要创造就业机会，才能提升街道的品质。埃尔旺根步行区的例子有力地说明，公共空间的升级是促进经济和结构发展的一个重要手段。

城市更新 URBAN RENEWAL

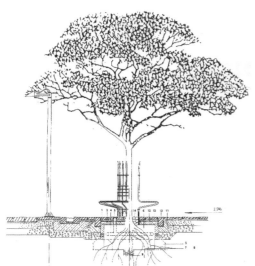

图 10 规划细部
Fig. 10 Detail plans

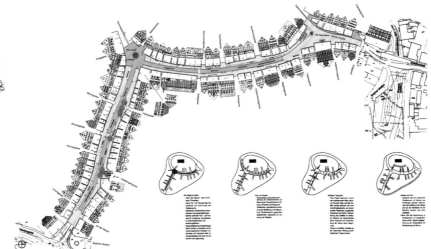

图 11 规划方案
Fig. 11 Design concept

图 12 Adelmann宫殿前的广场
Fig. 12 Square in front of the Palais Adelmann

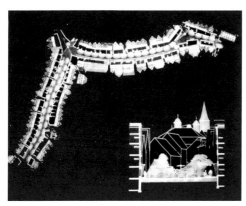

图 13 照明设计
Fig. 13 Lighting concept

图 14 夜间Adelmann宫殿前的广场
Fig. 14 Square in front of the Palais Adelmann at night

城市更新
URBAN RENEWAL

01 伦茨堡控制性规划
Framework Development Planning of Rendsburg

02 "霍费尔天空":
提升城市品质的范例项目
"Heaven of Hof":
an Initiative-project for Upgrading the City

03 埃尔旺根步行区
Pedestrian Zone of Ellwangen

04 **埃斯林根总体规划和公共空间城市设计导则**
Master Plan and Design Guidelines
for the Public Spaces of Esslingen am Neckar

05 "粉红岛": 韩国首尔方背洞街区广场
"Pink Islands":
a Street Square in Bangbae-dong, Seoul, South Korea

06 Gangseogu城市更新规划
Urban Renewal Gangseogu

07 北京市昌平区奥运自行车训练馆
及周边地区城市设计
Urban Design of the Area Surrounding
Changping Olympic Cycling Training Center, Beijing

04 埃斯林根总体规划和公共空间城市设计导则
Master Plan and Design Guidelines for the Public Spaces of Esslingen am Neckar

规划面积 AREA		72 hm²
人口规模 POPULATION		91450
完成时间 PROJECT DATE		2008.08

从两种文化中学习 LEARNING FROM TWO CULTURES

"城市无处不在" "The city is everywhere": the leitmotif of an urban network

In today's cities, the lack of interest in urban public space can be observed everywhere. As a result of this – as well as a similar dearth of interest in design aspects and a sense of urban community – the quality of life and identity of the city and its inhabitants are neglected.

The reality of medieval cities was that urban life occurred everywhere in the whole town, not just in the pedestrian zone. Over time, this quality disappeared in many cities. The city as an overall network is often neglected and efforts to improve public space are limited mostly to the main axis or important squares. Urban life happens only in these representative places, or is shifted to the mega-shopping malls and recreation centres which are usually built on the outskirts of the city. In Esslingen am Neckar the main public space was the pedestrian area and some new urban magnets integrated into the overall network of the city, however something was still missing. The relation to water, the river Neckar, is neglected in many places, and the city gates could not be recognized as such because automobile traffic took over the dominant role on the street.

城市不应该由发生城市生活的点状或者线状中心组成,而应当遵循"城市无处不在"的原则。
The city shall not consist only of a centre in certain places or main pedestrian zone where urban life takes place, but follow the basic principle, "the city is everywhere."

The medieval idea of the city as a network of public spaces is used as an alternative model to the mall on the outskirts of the city. The city shall not consist only of a centre or a main pedestrian zone where urban life takes place but follow the basic principle, "the city is everywhere." The residents need not travel a great distance to participate in urban life; urban life begins in front of one's own door. As for public space, man acquires a new quality, skipping his individuality and becomes urban. He is not in the performance of a specified role, nor in a staged room. With this behaviour, he can leave his given constraints at home or the workplace, and make the freedoms of urban life his own. From this perspective, the design of public space gains a new dimension: design towards freedom and diversity, a new kind of utopia, which could be provided by the city of Esslingen.

The entire public space of downtown is to be included in this concept via networking. This enhances the experience of urban life in the context of representative centres and other special areas as well as the familiarity of the living environment. As a special feature, a network of public spaces offers the opportunity to explore the city going forward and then return to the starting point. The network is an ordered form of developing the medieval labyrinth of streets, alleys, roads and squares. The historic character of public space is hereby brought into the current time.

如今,城市公共空间的乏味随处可见。所以空间应不只是从设计角度出发,也从城市的人居品质、意象以及市民生活的共同角度来思考。

体现中世纪城市品质的原则不仅存在于步行街区,也贯穿于这座城市的各个方面,这种品质在其他的很多城市里已然消失了。将城市作为一个公共空间网络来构建的观点被忽略,通常只是局限于主要公共空间轴线的塑造或者广场公共空间的改善。这使得城市生活展开的地点也只是局限于这些地点,或者是人为建造的中心,如休闲活动中心,以及城市边缘坐落于草地上的购物中心等。

内卡河畔埃斯林根的公共空间建设现状也是将注意力集中在步行区上,缺少与整个城市公共空间网络的联系。很多地方都不再能感受到城市与水体的联系,城市入口无法辨识,私人机动交通仍占据城市空间的主导地位。

将城市作为由公共空间构成的网络,早在中世纪时期就有这一基本思想,而今却被坐落于绿地中的购物中心所遗忘。城市不应该由发生城市生活的点状或者线状中心组成,而应当遵循"城市无处不在"的原则,即:居民无需长途跋涉,就能参与到城市生活中,城市生活应始于每个市民的家门口。在公共空间中人们可以获取新的品质,超越其个体性,人们不再被赋予某种角色,也不是处于一个"设定的场景"内。在这里人们可以遗忘工作或生活带来的巨大压力,享受都市生活的自由。从这个视角来看,公共空间的塑造获得了一个全新的尺度:更加自由和多样化,正如埃斯林根内卡将为市民提供的,一种新的乌托邦式的公共空间。

这种自由不像在其他许多城市那样,只局限于步行区,而是将埃斯林根内城的公共空间都通过网络的构建赋予了这种品质。这样,城市生活就可以与有代表性的中心区、特殊区域以及熟悉的居住环境紧密关联在一起。

网络化的公共空间使得内城在发展中能够形成闭合的网络,重新获得最初的品质。这个网络好似由街道、小巷、道路和广场组成的规整的中世纪迷宫,以此使得公共空间的历史特征在现代得以传承发展。

内卡河畔埃斯林根的网络化建设措施
Measures for the creation of networks in the city of Esslingen am Neckar

The concept of the urban network of Esslingen is structured into spatial, temporal and functional components: the physical network consists of city squares, public buildings and facilities, etc., which are connected by deliberately designed street spaces. The route

"埃斯林根城市网"的概念包括空间、功能和时间三个元素。城市网在空间上包括广场、公共建筑、公共设施等,并由经过设计的街道空间相连。从一个地方到另一个地方的路径不仅仅是运动的空间,也代表一个城市的状况,以及城市生活给人的感受,并由此奠定了城市化的基础。除不同类别交通网之外,功能网还包括连接、划分埃斯林根的购物、休闲、文化、服务和居住功能的网络。时间元素将历史和现代联系起来,构成一个整体的网络。

网络化的原则始于对两条竞争性街道的同等对待(Bahnhof大街和历史街道Pliensau)。自此这一原则陆续被推广到内城其他区域。该网络融合了所有重要的区域和城市入口,此外内城网络还扩展至水体景观和葡萄园周边区域。由此开始了功能的网络化——空间网围绕功能网进一步延伸,包括购物、服务、滨水休闲城市、文化、历史区域及居住。

以历史文化遗产为基础的网络化思路并不意味着要将现代化的购物及休闲中心排除在外,相反它们已经成为内城边缘的重要吸引力点。在重要的三个城市入口处设置单一功能(娱乐、服饰、食品、旅游)的中心,通过历史内城中具有区位优势的小型公共设施形成上述中心的连接轴线,可以使这些中心归于整体网络中。

规划中对内城网络已经形成的部分都进行了分析,以便逐步将其纳入整体网络中。新的联系也被建立起来,曾经看起来不重要的街道成为整体网络中的重要联系要素。内城边缘的单一功能中心将通过内城网络联系起来并融入其中。

from one place to another should not only be a space of movement, but should also represent the urban situation and thus become a space for an experience. In this way, the basis for re-urbanization is created. In addition to different types of traffic, the functional network aims the connection and distribution of different land uses in Esslingen: shopping, leisure time, culture, services and housing. The temporal component links the tradition of historical sites and the modern trend to a common network.

The principle of networking begins with the equal treatment of two important streets: the Bahnhofstrasse running from the main railroad station to the medieval city gate, and the main pedestrian street Pliensaustrasse in the medieval city core. This principle is then bestowed onto the rest of downtown. All key areas and city entrances will be interwoven. In addition, the network will be expanded to the urban river, stream landscape and the vineyards. This is the start of the functional aspects of networking. Gradually, the spatial network is extended into those functional aspects: shopping, services, leisure areas on the river, cultural and historical sites and housing. The idea of the network, which is based on historical heritage, does not exclude modern shopping and leisure centres. They currently appear as magnets on the edge of downtown. Facilities of entertainment, clothing, food and tourism are located mono-functionally in three major city entrance areas and provide corridors linking the historic town centre with small-scale retail shops and business. Thus, they are integrated into the overall network. Already existing parts of the urban network have been analyzed and are to be likewise integrated step by step into the larger network. New connections were created, and seemingly unimportant streets have become important links in the overall scheme. Mono-functional magnets on the edge of downtown are supposed to be integrated and connected via the new network system.

一个城市的形象特征不仅取决于艺术层面,也与城市居民有很大程度的关系,市民对城市规划各阶段的参与程度、对经济的管理以及对街道空间的维护都发挥着重要的作用。

The identity of a city depends not only on creative aspects, but also highly on its residents. The participation of citizens is important even beyond the planning phase in terms of economic management and maintenance of street spaces.

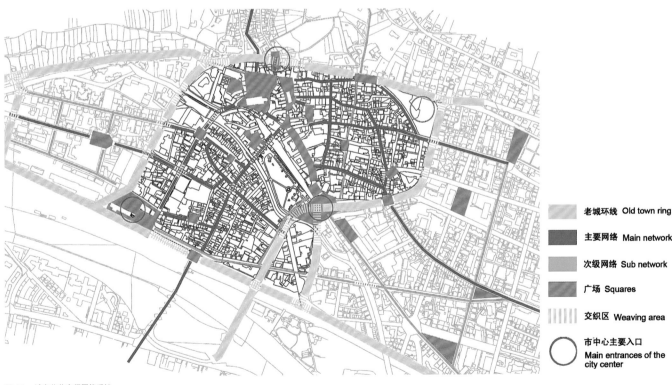

图 01 城市公共空间网络系统
Fig. 01 Hierarchy of the network system

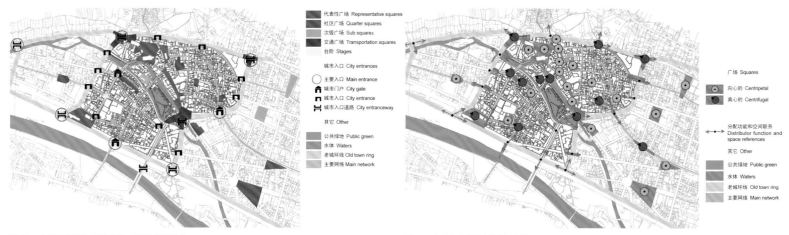

图 02 内城规划导则：城市入口、广场和重要设施
Fig. 02 Masterplan city centre: Entrances, squares and important facilities

图 03 与人流有关的空间特征分类
Fig. 03 Motion-related spatial characteristics

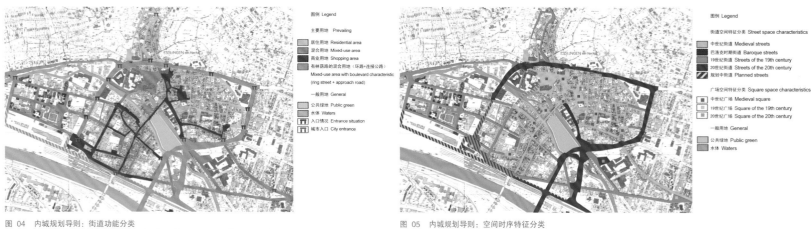

图 04 内城规划导则：街道功能分类
Fig. 04 Masterplan city centre: Organization of streets regarding to main uses

图 05 内城规划导则：空间时序特征分类
Fig. 05 Masterplan city centre: Temporal spatial characteristics

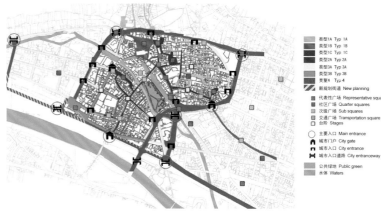

图 06 "广场及城市入口"规划导则
Fig. 06 Master plan "Typology of squares and entrances"

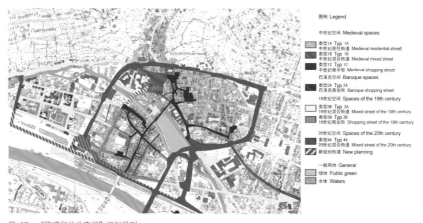

图 07 "街道和公共空间"规划导则
Fig. 07 Master plan "Typology of streets and open spaces"

一个城市的形象特征不仅取决于艺术层面，也与城市居民有很大程度的关系，市民对城市规划各阶段的参与程度、对经济的管理以及对街道空间的维护都发挥着重要的作用。

在深入分析的基础上，从空间、功能、时间层面上对城市网络进行完善补充，形成了针对街道空间、广场、城市入口及城市照明的有实践意义的总体控制规划。从个体层面来看，埃斯林根的Baufibel和街道小品目录为城市更新以及私有道路的景观建设提供了依据。这一整体性的规划能够一步步地与城市"更新周期"各阶段的要求相适应，并最终形成城市特有的个性和独一性。

The identity of a city depends not only on creative aspects, but also highly on its residents. This is why the participation of the citizens is considered of great importance even beyond the planning phase, especially in terms of economic management and maintenance of public spaces.

After analyzing and planning the urban network on the spatial, temporal and functional level, the master plan for street spaces, plazas, city entrances and city nightscape was developed for practical implementation. As for the private sector, design guidelines for the construction, renovation and refurbishing of private buildings, along with recommendations of outdoor furniture for restaurants and retail shops were published. This holistic approach of a planning concept allows for a gradual implementation within the normal regeneration cycle of a city and ultimately leads to its own identity and uniqueness.

街道和广场规划导则 Design guidelines for streets and squares

规划目标

公共空间、街道和广场的设计，是埃斯林根内城一项长期的任务。随着街道空间更新的重点从中世纪老城区逐步转移到邻近地区，不同的地点需要因地制宜采取不同的建设措施，因而有必要为埃斯林根未来的街道、广场空间设计制定一个整体方案。其任务是为每个类型的街道空间类型形成塑造的主题，判断出各街道、广场的不同之处，同时将埃斯林根内城公共空间塑造成一个可感知的整体。

规划的目标是为内城的每种街道空间类型及其限定界面制定指导原则和设计规则。就是在历史分级的基础上，形成指导原则，描绘理想的空间个性，标明主要功能，定义交通区域划分，阐释铺地的样式、色彩和材料，照明的类型和形式等重要设计元素。上述重要设计元素为各个街道空间的设计勾勒出基本的设计纲领并具有不可取代性，并在任何情况下都制约着街道或广场的具体设计。

导则的构建

根据每个街道广场空间形成的时间展示其空间时代特性，是导则设计的第一步，分别划分为中世纪、巴洛克、19世纪、20世纪和当代这几个时间段。

除了对形成时间分类以外，根据广场形态可以将其分成向心和发散两种：遵循向心原则的广场可以起到汇集和联系的作用；城市入口、火车站前广场和交通性广场则遵循发散原则。

而后根据街道的主要功能和使用类型对它们进行划分，如居住性街道、混合性街道、购物性街道以及具有林荫大道特征的混合性街道等。与此同时，内城的出、入口显示出了更为重要的意义，除了功能含义外，在形态上也意味着对城市入口的标识。这些都需要结合地面铺装、植被、照明等重要设计理念进行考虑。

The aim of a design master plan with guidelines

The design of public spaces, streets and squares in the city of Esslingen will take a long time. Since the focus of public space renovation has been shifting from the medieval old town to the surrounding areas with individual construction projects on different levels, it will become indispensible to develop a comprehensive design concept for streets and squares in both downtown Esslingen and the outskirts. The task is to develop an overall design concept and a set of design guidelines, a master plan, for all individual public spaces, which take into account the diversity of the different streets and squares but at the same time let them be experienced as a whole.

The concept and design guidelines, based on the urban development history, describe the envisioned spatial characters and the predominant functions, as well as define the diverse traffic zones within the street space. They will also contain design ideas for street pavement regarding form, colour and material and the type of street lighting. They represent the basic design "attitude," in each case giving the specific rules for individual streets and squares.

Structure of the design master plan with guidelines

As a first step, a master plan had been prepared which shows the spatial characteristics of each street and square, divided into four chronological categories depending on their date of origin: medieval, baroque, the 19th to 20th centuries, and today.

Apart from the classification of the date of their origin, they can be categorized into introverted or outward squares: according to the centripetal principle, places can collect and bundle movements. City entrances, plazas in front of railroad stations and traffic squares arrange movements and orientation lines according to the centrifugal principle.

Afterwards, the streets were classified according to their predominant types of functions that have different relevancies, e.g. predominantly residential use, multi-functional use, shopping character and multi-functional use along with boulevard character. Thereby important access points to and from downtown and even minor entrances requiring a design scheme are classified as "city gates." Design concepts for these gates were elaborated with regard to pavement, planning and lighting.

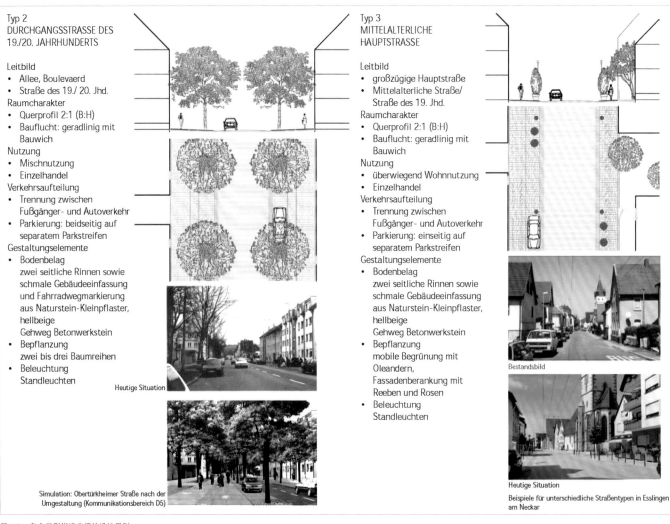

图 08 各个类型街道空间的设计导则
Fig. 08 Guiding principles for individual types of street space

The master plan "typology of streets and plazas" finally summarizes the previous studies and describes the street spaces in a typological overview with subgroups concerning the design of these streets. The description of each type of street starts with the formulation of a guiding principle for future design that defines the predominant use and most important elements of the spatial character, such as the running form of building lines. Typical cross-sections describing the diverse functional zoning within the street surface would be developed, e.g. sidewalks, parking strips and lanes for cars. Design guidelines for the pavement would be elaborated regarding materials, colour and patterns.

To achieve a holistic design effect, a homogeneous design principle is used as a basis for the master plan, which will be diversified and worked out in detail into individual design guidelines according to the space typology; hence the typical local characteristics are taken into account.

"街道和广场空间规划导则"几乎涵盖所有此前提到的相关研究，并在街道空间塑造分类及子类的综述中对每个街道空间进行表述，而后具体形成各种街道类型的设计目标。每个街道类型设计导则都以未来街道空间的形象为开端，定义出主要功能和体现空间特征的重要元素，例如建筑基准控制线和典型剖面，描述基于各种交通功能区（如步行、公园和车行区）进行的街道空间的划分，并形成包括分区、材料、形状、颜色、表层结构等的地面铺装设计准则。

为达到一个整体的设计效果，规划导则为街道、广场空间制定的是一个统一性的设计方针，并以此为基础具体深化形成各个街道空间的设计方针，以保证地方特色的形成。

城市更新 URBAN RENEWAL

图 09 埃斯林根城市夜景效果
Fig. 09 Simulation of the city of Esslingen at night

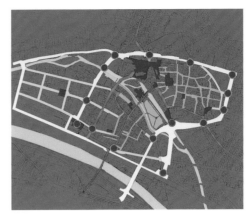

图 10 照明规划导则
Fig. 10 Lighting master plan

照明规划导则

与当今的城市功能要求相适应，城市景观与城市公共空间的夜晚效果与日间效果同样是重要的规划任务。埃斯林根内城及其具有极高价值的历史建筑对街道及城市景观照明提出了特殊要求。

照明规划导则作为街道及广场规划导则和广场及城市入口规划导则的补充，形成城市夜景照明设计的基本原则。

埃斯林根内城照明规划导则的任务在于，构思一个整体照明方案，并为各区域构思设计主题。其中自然景观和历史遗产应该在夜间也可被清晰地辨识。整体城市形象，以及重要的内城区，例如作为城市的心脏地带的市场广场和市政厅广场、与之相连的步行区、城市入口、重要的街道和广场空间、漂亮的立面、内卡河沿岸、桥梁和公园设施，都将被照亮，特别是代表城市历史特征的具有标示性的城堡。这些不同的元素使埃斯林根内城的公共空间形成一个可感知的整体。

工作的目标是发展形成指导原则，并由此为内城各区域延伸形成照明设计建议和设计导则。

作为街道及广场规划导则的补充，照明规划导则对街道、广场和城市入口等处的照明设计及实施提出了详细的要求，针对每个区域都提出了规划目标和设计计划，并推动了进一步的设计及建设。

Master plan for nightscape

For today's city, the nightscape and public spaces represent just as important a planning task as the cityscape during the day. The city centre of Esslingen, with its historic buildings, required many special demands on the lighting of its streets and its entire townscape. The master plan for the nightscape was developed to complete "the master plan and design guidelines for the public spaces" as a fundamental design program at night.

The master plan for the nightscape of Esslingen contained an overall lighting concept and design guidelines for individual areas. The natural layout and historic heritage were supposed to be clearly readable after sunset. The important urban areas such as the heart of the city would be distinguished by night, significant urban elements such as market square, town hall square with its main arteries, pedestrian zones, city entrances, major streets and other types of public spaces like the beautiful facades, Neckar canals, bridges and parks would all be illuminated. The castle as the main feature of the city should be particularly emphasized as well as historic features that serve to provide orientation. All the various elements being put together would allow the public space of Esslingen to be experienced as a whole.

As a complement to the general master plan and public space design guidelines, the demands on the nightscape master plan refer, once implemented, to the practical design of the streets, squares and city entrances, etc. For each case, the nightscape master plan thus describes the design goals and the draft programs, and makes suggestions for the implementation of both the design and its execution.

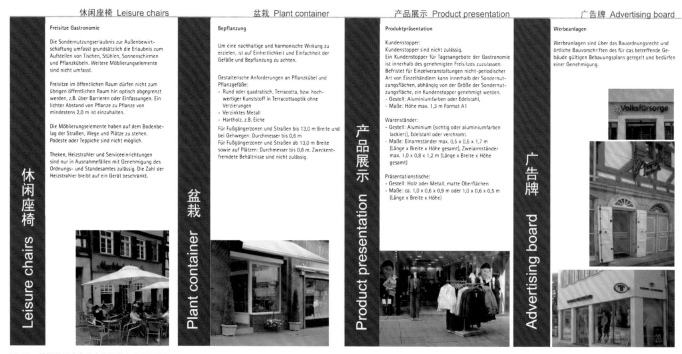

图 11 埃斯林根火车站大街街道小品宣传手册
Fig. 11 Flyer for furnishing elements in the street Bahnhofstrasse in Esslingen

小品目录

公共及私人城市小品对公共空间有着决定性的影响，对一个城市的个性及自我形象的展示贡献巨大。然而在街道更新中常常将公共城市小品当做孤立的一部分来看待，并不顾及相邻街道和区域。餐饮及零售业等设立私属城市小品时更是如此。

随着埃斯林根内城景观的逐步更新，零售商、市民及城市管理部门也提出了将公共空间内的城市小品纳入整体方案中的愿望。结合他们的意见，最终拟定出了一个包括街道家具、广告设施以及植物种植等的城市小品控制目录。该目录的内容作为法规得到了市民组织的采纳。

为了保证城市的协调感，应当使公共及私人城市小品在形式、色彩及材料方面互相协调，不能同历史旧城的色彩和形态产生冲突。除此以外，城市小品的尺寸及形状应建立在对现状视线关系、标识的分析基础上，避免遮阳伞等较大的小品对其的影响。设立小品目录的好处在于，一方面根据街道类型来排列顺序，可以调节各种小品设施的尺度及数量；另一方面根据小品类型（如桌子、椅子……）来确定色彩、形状和材料。此外需要确定一个限定个性化空间的控制框架。

Catalogue of recommended outdoor furniture

Public and private street furnishings affect the quality of public space and make a significant contribution to the identity and presentation of a city. For street maintenance, public street furnishing is often considered in isolation, without regard to adjacent streets and urban areas. The outdoor furniture of private restaurants and retail shops turns out to be even more heterogeneous.

In connection with the gradual urban renewal of Esslingen, the desire to integrate street furniture into the overall concept was expressed by retailers, citizens and city administration. Produced with help from the retailers and city government, a catalogue containing design rules for the special street elements such as furnishings, advertising and planting was developed. The municipal council has embraced the contents as statutes.

In order to create a homogeneous structure in a city, the different public and private furnishing elements have to be coordinated regarding form, colour and materials so as not to compete against the abundance of colours and forms of the historical old town. Additionally, the size and shape of the elements are defined based on an analysis of the existing view axes and city landmarks. The result is a furniture catalogue, classified by the type of streets. It regulates the size of special street zones and the number of special elements and type of outdoor furniture like tables and chairs in regards to colour, shape and material. It lays down a framework, within which individual design is possible.

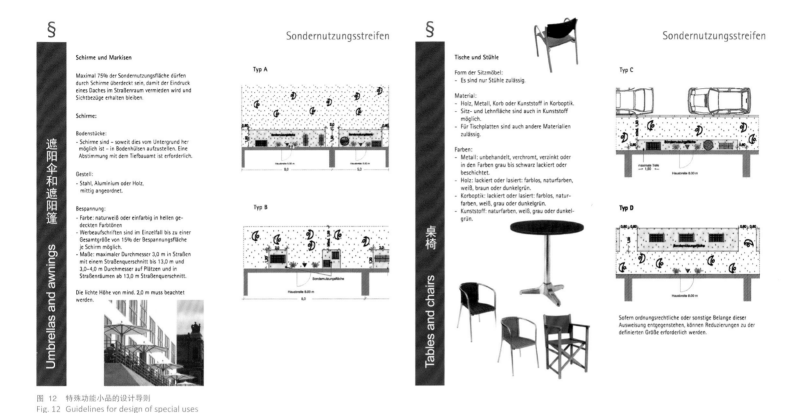

图 12 特殊功能小品的设计导则
Fig. 12 Guidelines for design of special uses

埃斯林根范围内完成的实例 Examples being carried out in Esslingen

在"街道和广场规划导则"指导下,不同的街道、广场设计被逐步实施,构成了公共空间网络中的一个个基石。

火车站大街:新的设计理念、设计导则、照明设计和实施方案

火车站大街,这条连接火车站及旧城的通路自19世纪以来就人来人往、富丽堂皇。作为城市入口,其主要功能是购物街。汽车交通是这条路的主导交通。过去十年间战后重建及相关项目的吸引力呈下降态势,因此未来应当将这条大街塑造为具有开敞的城市空间、灵活的多样性用地、引人驻足的品质及丰富的环境感受的独具特色的购物街,并同周围的城市空间相关联,提升城市品质,作为公共空间的展示舞台,提高人们对城市生活的感受程度。房屋所有者和商人们于1992年在"火车站大街倡议"中联合在一起,与市政府以及德国ISA意厦国际设计集团共同寻求火车站大街功能及设计品质的提升。

15个门型的多功能框架构成城市空间内有趣的感受空间——一个通透的次级空间。它们将这片街道空间装点成个性十足的购物区。它们是城市照明和临时装饰(例如旗帜)的载体。在这种统筹全局的设计理念之下,商

Based on the master plan and design guidelines for public spaces, different streets and spaces have been constructed step by step, which makes up the entire network of individual public spaces.

The street Bahnhofstrasse: design concept and implementation

The street Bahnhofstrasse in Esslingen originally dates back to the 19th century, linking the main railroad station and the old town through a direct route. It serves as an important shopping street and a city entrance, dominated by heavy automobile traffic. In recent decades, the attractiveness of the post-war market buildings had decreased noticeably. Consequently, the Bahnhofstrasse and its many post-war structures needed to be redesigned in order to remain an attractive shopping area; something with a special identity and uniqueness that would be inviting to stroll and linger along the open mall. The intended objective would need to have an urban quality, something with a spacious open area, flexible uses, amenity values, a diversity of experiences and a connection to the surrounding areas. The street as a stage of life had to allow for the presentation of public space and enhance the city experience.

Homeowners and business people joined together in 1992 in the Initiative Bahnhofstrasse and sought cooperation with the City Council and the ISA Stadtbauatelier to search for a vision for a functional and visual upgrade.

图 13 过去的火车站大街
Fig. 13 Bahnhofstrasse before the renewal

图 14 现在的火车站大街
Fig. 14 Bahnhofstrasse after the renewal

Fifteen strictly defined gate-like multifunction carriers form an atmospheric experiential space in the existing urban area, a transparent subspace. They not only display the street lighting, but also advertising signs and temporary decorations such as banners. They transformed the unattractive street into an attractive shopping lane with its own identity. An overall design concept was developed, and joint actions of trade and commerce were organized.

The private outdoor furnishings on the Bahnhofstrasse, such as tables, chairs, umbrellas, display stands and commercial signboards are subject to the guidelines of the outdoor downtown furniture catalogue: items may only be placed within the designated special use strip, in which the city clerk's office must approve the scope of these elements. Deviations from the guidelines are only permitted with the express consent of the city government urban planning and land survey department.

The design concept for the street lighting in the Bahnhofstrasse is equivalent to a spacious urban shopping street for pedestrians. The multifunctional "carriers" provide the ambient lighting. Additional sources of illumination include spotlights imbedded in the street ground and the lights of the shop windows.

Town centre of Esslingen Zell: public space design and implementation

Giving a new heart to the suburbs: many suburbs of Esslingen have lost their own character. Former villages have been merged into an anonymous suburban landscape, their centres having largely fallen victim to car traffic, while their amenity value became only rudimentary and their identities disappeared. Even in Esslingen Zell, the functional centre of a former wine-growing village has lost its identity. Cut by two roads, the remaining open spaces have been turned into parking lots. It offers hardly any living quality, even though it is the shopping and service centre of the village and the central transportation hub of the daily lives of citizens living in Zell – e.g. the local rapid transit station, bus stops, the main circulation of individual transport in Esslingen Zell.

The design goal was to give the centre another identity, to transform it into an urban reference point, to enhance the quality for a pleasant stay and to create an exhilarating and relaxing atmosphere. A strong design concept for public spaces was developed so that an attractive atmosphere could be created in the heterogeneous architecture of the perimeter. Especially the main commercial street, the roundabout and the main railroad station square which were all conceived as an "urban back born." Form and function

业和小企业共同行动起来，如同一个购物综合体，使设计原则更为完整。

火车站大街上的私人街道小品，被小品目录中的桌子、椅子、遮阳伞、商品陈列、广告三角旗等所取代，且按导则要求，小品元素只允许在得到法规和街道办许可的特定空间内设置。为此楼房商店入口、橱窗必须保持开放。同导则要求不一致的情况，必须得到规划局与测绘局许可。

火车站大街的照明设计，是与精心设计的城市步行购物街的形象相适应的。多功能框架承担了基础照明、其他的光源包括点状地面射灯和橱窗照明灯。

埃斯林根采尔中心区：广场设计及实施方案

在郊区创造一个新的中心：埃斯林根的许多地方已经丧失其个性。从前的一个个村庄已经消融于模糊的郊区景观中，它们的中心也成了机动车交通的牺牲品，驻留品质不断退化，独特性也逐渐消失。埃斯林根采尔地区的功能中心也失去了其早先作为葡萄园区中心的特性。尽管它还是采尔地区的购物、服务中心，以及市民日常生活的中央交通节点（快轨站、公交站、埃斯林根采尔地区的主要机动车道路汇聚于此），由于两条交通道路从中穿过，剩下的区域也被建成停车位，使得它基本毫无逗留品质可言。

因此，此次设计任务是，为这个中心区再次塑造独特意象，使其成为一个标志性的节点，并具有适当的驻留品质及轻松愉快氛围。目标是要在周边各异的建筑之间塑造一个给人深刻印象的公共空间，一个具有强烈自我风格的感受空间，并把各个部分包容相连（街道、环形交通和火车站前地区）。通过功能和形态的结合共同

城市更新 URBAN RENEWAL

形成标志性特征,并为中心区所有部分(溪流、灯柱、平台)发展完整的指导原则。规划中的环形道路应当同独具特色的水体灯光雕塑相结合。不同特性的逗留品质为中心区的不同区域提供了多种可能性,例如等待、聚会、聊天、休息、静坐、品尝咖啡、游玩等。各种设计的元素都要符合人们的需求——同绿化的关系(树木、可移动的绿化)、同土地的关系(高品质材质)、同水体的关系(水道和水体雕塑)。对机动与非机动交通进行了重新安排,同时在保持停车位数量不变且预留了发展可能的前提下进行停车位布局。

were merged together to form a memorable urban feature. Common guidelines were developed for all design elements such as creeks, light pillars, benches and plates. The visual and functional powers of intended roundabouts were enhanced through a memorable water light sculpture in its centre. Amenity values of various spatial characteristics provided multiple use options in different parts of the town centre, such as waiting, meeting, talking, resting, sitting, drinking coffee and playing. The concept was elaborated by using design elements that met human needs - the relation to greenery including trees and bushes, to the earth and to the water. Traffic flow was rearranged so that the current parking spaces were relocated, but the number of parking spots will still remain the same and can be increased later on.

图 15 过去的埃斯林根采尔中心区
Fig. 15 Center of Esslingen Zell in the past

图 16 现在的埃斯林根采尔中心区
Fig. 16 Center of Esslingen Zell today

图 17 埃斯林根梅廷根的教堂广场:过去和现在
Fig. 17 Kirchplatz in Esslingen Mettingen: then and now

图 18 过去的埃斯林根梅廷根中心区
Fig. 18 Centre of Esslingen Mettingen in the past

图 19 现在的埃斯林根梅廷根中心区，Hermann-Sohn广场
Fig. 19 Centre of Esslingen Mettingen today, the Hermann-Sohn-Square

The town centre of Esslingen Mettingen: public space design and implementation of the square Kirchplatz and Hermann-Sohn-Platz along with the road Obertuerckheimerstrasse

埃斯林根梅廷根中心区：街道空间设计及教堂广场、Hermann-Sohn广场及Obertuerckheimer大街设施方案

By upgrading its design, the square along the street Schenkenbergstrasse, in front of the Lutheran church, is supposed to become the city centre and a prominent reflection of the Mettingen city identity. The place is intended to provide both space for a market as well as space for the wine festival, where one is invited to linger, offering a meeting point or an attractive setting for ceremonies like confirmations or weddings. The square gradually expands in the direction of the historic bakery. This particular spatial characteristic is supposed to be highlighted in the first phase of construction. The background for the concept was dictated by the master plan for the public spaces of Mettingen. A continuous stone pavement creates a homogenous spatial effect; space-defining trees articulate the square and improve the experience of being in the space. The one-sided, flat drainage channel optically enlarges the square area around the church. A fountain with a tree provides the functional, aesthetic centre and acts as an urban reference point. The church is particularly highlighted at night through the illumination of facades and towers, following the nightscape master plan.

新教教堂前的Schenkenberg大街广场区应当通过设计品质的提升再次成为梅廷根的标志性中心。这样的一个广场应当有给葡萄酒节专用的集市和空场，并形成吸引人们前来聚集活动的，可以举办弥撒、婚礼等仪式的，外观漂亮的场所。该广场的空间来源于Schenkenberg大街的轻微扩展。这些空间特质在第一期建设中应予以突出。梅廷根的规划导则是设计的基础。贯通的石材铺地形成了统一的空间效果，通过树木对广场空间进行了界定并提升了逗留品质。设于一侧的排水槽将广场空间从视觉上扩展至环绕教堂的区域。一口井和一棵树，形成功能和设计的中心点与基准点。通过对立面及钟塔的照明设计，使得教堂在夜间更是熠熠生辉，正如在照明规划导则中提出的，通过夜景照明强化重要建筑的夜景效果。

整体规划设计导则的制定为有步骤的实现"城市更新循环"提供了可能，也保证了在不丧失现存城市特征的前提下，与新的精神需求、功能需求相适应，满足城市发展提出的各项新要求。

This holistic concept for planning allows for a gradual implementation into a normal modernization process for the city and enables adaptation for new developments.

The street Schenkenbergstrasse continues and eventually intersects the other main street, Obertuerkheimerstrasse, creating a linking point between the new centre of Mettingen and the historic church square, Kirchplatz. Within the second construction phase, the far too chaotic church square area, which provided only little amenity value, was reorganized in coordination with the public space master plan. Given the limited budget, a new centre was created by making use of common means, improving the quality and enjoyment of one's stay in the area of shops and restaurants located there.

By removing a lane along the street Obertuerkheimerstrasse, the existing sidewalks could be enlarged to a cobbled square area. In the area of the intersection, the newly created square Hermann-Sohn-Platz featured three

Schenkenberg大街一直延伸到Oberturkheim大街与Schenkenberg大街的交叉口区域，这个交叉口被作为梅廷根新中心和教堂广场历史中心之间的联系节点进行了塑造。在第二期建设当中，目前这个仍显混乱、缺乏逗留品质的区域也将被重新设计，以便规划导则的实施顺利进行。鉴于预算有限，应当通过重点塑造商业、餐饮业所在区域的逗留品质来形成一个新的中心。

通过减少Oberturkheim大街的一条机动车道，人行道得以被拓宽成为广场。在交叉口，形成了以定居于当地的艺术家Hermann-Sohn的名字命名的新广场，通过三个特别为这个地区设计的小品，将视线导向了梅廷根特有的重要建筑或城市元素：教堂、酿酒厂和葡萄园。混凝土石柱形态的小品上的开孔，形成了观赏这些景致的"画框"，从另一种独特的视角展示了这座城市。夜间通过采用暖白色LED灯照明，形

图 20　埃斯林根梅廷根，过去的Schenkenberg大街
Fig. 20 Esslingen Mettingen, Schenkenbergstrasse in the past

图 21　埃斯林根梅廷根，现在的Schenkenberg大街
Fig. 21 Esslingen Mettingen, Schenkenbergstrasse today

成了舒适的氛围。

更进一步的设计是对Obertuerkheim大街的改造。此规划的目标是创造一条具有城市入口标志性的、个性鲜明的街道，通过用收缩机动车道宽度、将自行车道同机动车道相分离等方法，提升逗留品质及周围地带的品质，并改善单调的街道景观。林荫道形态的绿化形成空间序列，并结合照明设计，使Obertuerkheim大街成为有广场性质的多功能用地。

以上及其他已经实现的各项措施是构建埃斯林根公共空间网络的第一步。重要的不仅是单独的广场或城市空间的设计，深化设计及建设实践也是规划导则实施中的重要基石。公共空间网络的构建和塑造、埃斯林根城市意象的形成和发展都是长期任务，而规划设计导则是其基础。整体规划设计导则的制定为有步骤地实现"城市更新循环"提供了可能，也保证了在不丧失现存城市特征的前提下，与新的精神需求、功能需求相适应，满足城市发展提出的各项新要求。

public furniture elements that were especially designed to guide one's view toward notable buildings or elements in Mettingen: the church, the wine press and the vineyards on the hill. The view corridors were created by loopholes within the concrete pillars, and show the city from another perspective. At night they are illuminated by warm white LED guide rails and create a pleasant atmosphere.

Additionally a reorganizing of the Obertuerckheimerstrasse was carried out, giving the street a memorable identity corresponding to the entrance situation. The qualities of life increased by narrowing the lanes and separating the bike from the car lanes. The environment has been upgraded and the monotonous streetscape has been improved. Supported by a similar lighting concept, avenue-like tree plantings created space sequences and a square-like multifunctional space was developed.

These and other projects that have been implemented so far are the first steps to network the public spaces of Esslingen. The individual square and street design is not the sole objective, but the concrete design and implementation of these spaces represents a component of the overall concept. The design and networking of public spaces, and as a result the creation and development of the city identity, is a long-term task for Esslingen. The master plan for the public space works as the basis to achieve this. This holistic concept for planning allows for a gradual implementation into a normal modernization process for the city and also enables adaptation for new developments, the current spirit, and any requirements of the user, without losing track of the elements that define the city identity.

城市更新
URBAN RENEWAL

01 伦茨堡控制性规划
   Framework Development Planning of Rendsburg

02 "霍费尔天空":
   提升城市品质的范例项目
   "Heaven of Hof":
   an Initiative-project for Upgrading the City

03 埃尔旺根步行区
   Pedestrian Zone of Ellwangen

04 埃斯林根总体规划和公共空间城市设计导则
   Master Plan and Design Guidelines
   for the Public Spaces of Esslingen am Neckar

05 "粉红岛":韩国首尔方背洞街区广场
   "Pink Islands":
   a Street Square in Bangbae-dong, Seoul, South Korea

06 Gangseogu城市更新规划
   Urban Renewal Gangseogu

07 北京市昌平区奥运自行车训练馆
   及周边地区城市设计
   Urban Design of the Area Surrounding
   Changping Olympic Cycling Training Center, Beijing

05 "粉红岛"：韩国首尔方背洞街区广场
"Pink Islands": a Street Square in Bangbae-dong, Seoul, South Korea

规划面积 AREA	1.5 hm²
人口规模 POPULATION	118375
完成时间 PROJECT DATE	2008

从两种文化中学习 LEARNING FROM TWO CULTURES

现状情况 Existing situation

The residential areas of Bangbae-Dong and Seocho-Gu district were created in the 1970's in the South Korean capital city of Seoul as a residential area with the small-scale construction of single-family houses. Today, most of theses homes have been replaced with town houses, but the small-scale structure has been preserved. The residential area is very popular among young people with higher incomes as well as European foreigners. The area also features a French school, as well as many French and Italian restaurants.

Real estate prices rose sharply due to the popularity of the urban quarter and the subsequent development pressure on the district increased significantly. Normally, it would be likely for the district to be purchased by major investors who would then replace the existing structures with high-rise apartments, as has been the case with many Seoul neighborhoods before it. However, in this case the municipal government has been trying for several years now to develop a new residential and urban culture oriented around the existing small-scale structures. The goal is to maintain the character and variety of area buildings and to develop public spaces such as streets and squares as living environment and recreational areas. In keeping with this plan, the mayor of the Seocho-Gu district decided to redevelop the headquarters of Bangbae-Dong, while preserving its small-scale structure.

方背洞街区（Bangbae-Dong）位于韩国首都首尔市南部的瑞草区（Seocho-Gu），它建成于20世纪70年代，是由独栋别墅组成的小尺度街区。今天大多数的独栋别墅都被联排别墅代替，然而小尺度的街区结构却被保留了下来。作为居住区，该地区深得高收入年轻人以及欧洲人的喜爱。居住区内有一所法语中学和许多法国、意大利餐厅。

该区的知名度以及随之高涨的房价使得该区的发展压力明显增加，进而面临着同首尔的许多其他城区一样的危险——被大财团收购并用于建设高层建筑。但自从在清溪川（Chonggyecheon）项目中采用了新的内城发展模式后，几年来首尔政府开始倡导一种新的居住与城市文化，即保留带来多样性品质的小尺度街区结构，并将街道、广场等公共空间作为居住环境和社会交往空间来进行建设。

以这一目标为导向，瑞草区官员作出决策，在保留方背洞居住区小尺度街区结构的前提下进行整理。系列项目中的第一步是要重新塑造已经被覆盖的溪流和沟渠地块。由于覆盖溪流或沟渠的盖板的承载能力极其有限，而且也不能满足植被栽培所需的土壤厚度，因此这些地块既不能用于建设，也无法进行绿化。因此，这些地块在当时完全被当做超大尺度的街道或停车场使用，只有几个临时的小型市场，城市形象因为这种杂乱无章受到了很大影响。

该区的知名度使得该区的发展压力明显增加，进而面临着同首尔的许多其他城区一样的危险——被大财团收购并用于建设高层建筑。

The popularity of the urban quarter causes increased development pressure on the district. In many cases, it would be purchased by major investors and replaced with high-rise apartments.

Many urban zones gained by cementing over a former creek and sewage canal, are to be reorganized and redesigned. These covered zones cannot be used as building sites or green spaces however, because the concrete coverage is too weak to hold structural loads and too thin for planting. As a result, these areas are currently used primarily as oversized traffic streets, parking areas and storage space. These areas affect the overall impression of the city because of their temporary and disordered appearance, becoming home primarily to small, temporary street markets.

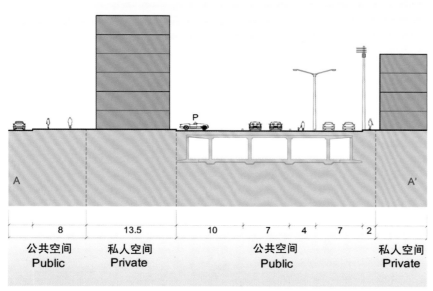

图 01 剖面图现状
Fig. 01 Section of the planning zone

图 02 具有后院特性的典型城市景观
Fig. 02 A typical streetscape with backyard character

城市更新 URBAN RENEWAL

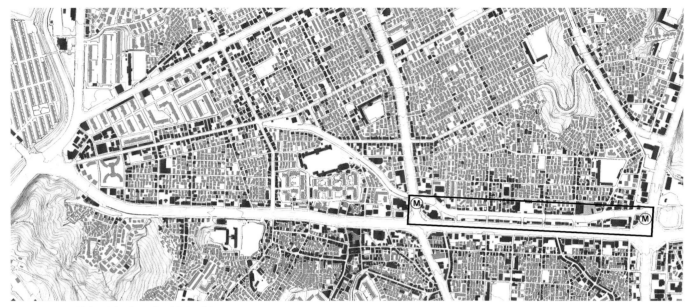

图 03　总平面图（灰色：住宅用地，红色：非住宅用地）
Fig. 03 Site plan, with construction and usage structures (gray: residential use, red: non-residential use)

图 04　现状
Fig. 04 Existing conditions

263

The first section of the southern end of this zone would be the target for restructuring. It is an undeveloped area approximately 30 meters wide and one kilometer long with metro stations at each end. The surrounding building structure is extremely heterogeneous. On one side, the area is flanked by three and four-story buildings, while on the other by skyscrapers. Directly behind the skyscrapers is a main thoroughfare through the city, which most buildings' main facades and entrances face. Consequently the rear side of the skyscrapers with their service zones for delivery and garbage collection is what faces the planning zone, giving the entire area a character more akin to a disorderly backyard.

The function of these two different building types also differs largely. There are simple furniture and other shops on the ground floors of many of the skyscrapers with office space on the upper floors. In the three to four story buildings, there are mainly retail stores, as well as numerous pubs, bars and small restaurants.

首先进行规划的是南部的一个约30米宽、1公里长的未建区，两端各有一个轻轨站。用地两侧的建筑形态迥异，一侧为三至四层的低层建筑，另一侧则是高层建筑。高层建筑的入口和主立面都朝向另一面的城市干道，而朝向规划用地的这一面则被用作货运交通和垃圾收集等，由此使得规划区域呈现出一种"杂乱后院"的特征。

用地两侧的功能差别则更大：高层建筑的底层是单一的家具商店，上层是办公空间。小尺度街区一侧则是零售店和各种餐饮业，主要是小酒馆、酒吧和小餐馆。

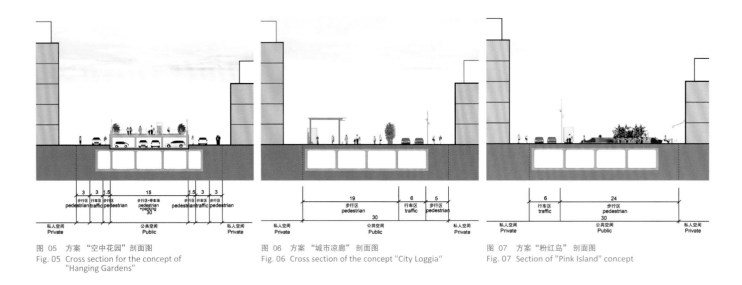

图 05　方案"空中花园"剖面图
Fig. 05　Cross section for the concept of "Hanging Gardens"

图 06　方案"城市凉廊"剖面图
Fig. 06　Cross section of the concept "City Loggia"

图 07　方案"粉红岛"剖面图
Fig. 07　Section of "Pink Island" concept

规划 Concept

The basic overarching concept was to transform a large, chaotic, sometimes depressing space into a brighter, more cheerful urban street environment that would provide space for farmer's markets, concerts and restaurants with ambient street furniture. In addition, a variety of leisure and social activities would be possible in the space such as inline skating, cycling, strolling, sitting and gatherings of various kinds.

基础设想是：把大面积的现有开敞空间规划成一个清新、愉悦和文雅的街道广场，这个广场可以用作周末集市和大型活动用地，也可以用做露天的餐饮业用地，此外这里还可以进行多样的业余活动，比如滑旱冰、骑单车、散步、闲坐或约会、沟通和社会生活。

基础设想是：把大面积的现有开敞空间规划成一个清新、愉悦和文雅的街道广场，这个广场可以用作周末集市和大型活动用地，也可以用做露天的餐饮业用地，此外这里还可以进行多样的业余活动。

The basic overarching concept was to transform a large, chaotic, sometimes depressing space into a brighter, more cheerful urban street environment providing space for farmer's markets, concerts and restaurants with ambient street furniture.

To this end, the traffic concept envisioned reducing the existing oversized access road to two lanes. Large parking areas would be dissolved and replaced in part by curbside parking along the access road, or parking structures in the surrounding areas.

The aim of the design concept was to create a more uniform and orderly appearance for the entire space, which was divided by three square accents into sequences in the middle and at the two entrance positions with the metro stations. Three variations were developed to achieve these aims in different ways.

交通规划的构思是：把现存过大尺度的道路用地减少为两条车道的道路。大型停车场被取消，其中一部分改建成沿街停车位；另一部分必要的车位通过车库的形式得到满足。

在城市设计方面的目标是：使整个空间呈现一种统一有序的外观。整体空间结构通过一个中央广场及两端的轻轨站广场来进行组织并且形成空间序列。这个设计目标下，发展出了三个方案，以不同方式来实现这个目标。

城市更新 URBAN RENEWAL

序列 1　　　　　　　　　　　　　　　　　序列 2　　　　　　　　　　　　　　　　　序列 3
Sequence 1　　　　　　　　　　　　　　　Sequence 2　　　　　　　　　　　　　　　Sequence 3

图 08　空间序列构思
Fig. 08　oncept sequence

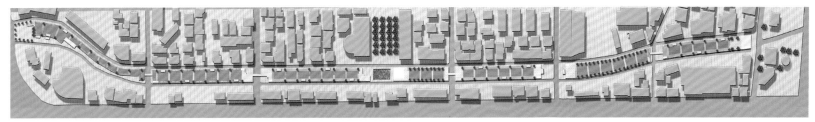

图 09　方案"空中花园"总平面图
Fig. 09　Site plan of "Hanging Gardens"

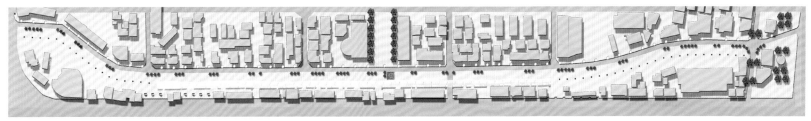

图 10　方案"城市凉廊"总平面图
Fig. 10　Site plan of "City Loggia"

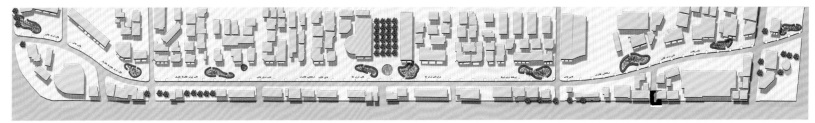

图 11　方案"粉红岛"总平面图
Fig. 11　Site plan of "Pink Island" concept

方案1：空中花园

方案1的构思是，在规划区的中央、位于两条车道中间的开敞空间上，形成带状布局的单层建筑，建筑朝向街道的两侧是杂货店或商店，中央则是车库。真正的公共空间位于建筑的屋顶。这样将形成一个公园，作为一个消除纷扰的自有世界，出现在让人意想不到的地方，在熙熙攘攘的都市中为人们提供一个安静的休闲之所。

Variant 1: Hanging Gardens

The first variant provided for a single story structure in the middle of the planning area between the two car lanes. The inside of this structure would be used for parking, with kiosks and small shops flanking it on either side. Meanwhile, a public space located on the roof of the structure would become a raised park, offering visitors to the area an unexpected break from the chaotic environment below. This raised area is intended to be a joyful oasis of rest and relaxation in the middle of a mega city.

265

从两种文化中学习　LEARNING FROM TWO CULTURES

This alternative offers an expanded parking solution without the need to provide additional parking garages on site. The construction efforts necessary for strengthening the underlying channel, however, lead to constraints in terms of financing. Moreover, later exposure of the channel for medium-term development would be hampered by the proposed structure's position over the channel.

这个方案在不占用规划区外用地的情况下解决了停车位的问题。但是这个方案需要加固河渠盖板的结构，这将带来额外的财政支出。此外，在被覆盖的河渠上建造建筑也使得将来水体更难以重见天日。

方案2：城市凉廊

Variant 2: City Loggia

凉廊可以联系商店和杂货店，并形成一种全新的空间界面，从而消除所谓的"后院"特性。

The 'disorderly backyard' character would be transformed - creating a new front to the building by creating a small architectural structure, a city loggia, housing built-in shops and kiosks.

The establishment of a city loggia along the rear of the skyscrapers represents the second variant. The 'disorderly backyard' character would be transformed - creating a new 'front' to the buildings through a small architectural structure, a city loggia, housing built-in shops and kiosks. The access road would be laid out along the skyscrapers, so that along the three to four story buildings, generous areas could be built and used as a space for restaurants, markets and events.

This alternative considers the possible necessity of exposing the channel below for future construction. However, the plan would require a high level of design and construction quality in order to improve the situation overall.

方案2意图沿着高层建筑的背面建设一条城市凉廊。凉廊可以联系商店和杂货店，并形成一种全新的空间界面，从而消除所谓的"后院"特性。道路将集中到高层建筑一侧，沿着小地块一侧的区域便产生了大块可以用于餐饮、集市及大型活动的用地。

该方案的优势在于，有朝一日可以使被覆盖的水体重见天日。但是它对凉廊的设计品质提出了相当高的要求，以确保这个区域的品质能够被提升。

方案3：粉红岛

Variant 3: Pink Island

The third variant is similar to the second in how it handles the access road along the skyscrapers. In this third variant, however, the focus of design would be on the center of the planning area so that the facades of those skyscrapers with the 'disorderly backyard' character take a back seat. Through improvements to the street square, this variant would create an incentive for the building owners of these skyscrapers to make the necessary upgrades on their own to provide easier access to the now more attractive square.

The uncertain future of the sewage canal was deliberately incorporated into the design. The possibility of exposing it should not be restricted but rather encouraged. Therefore, a provisional and flexible solution was suggested, which could be realized with limited resources leaving the area open to a variety of scenarios. The surface of the street square would be unified with a single continuous paving, and seating in the form of pink rubber snakes would be allocated. These structures would form the edge of a mobile "island", the interior of which would provide the required height for planting and space for playgrounds or necessary infrastructure. The edges serve as benches for sitting and reclining, and the remaining street square space would be kept open to encourage various kinds of urban life. At night, the snakes would be illuminated from the inside.

方案3与方案2的用地结构有些相似，并把道路转移到了高层建筑一侧。这个方案的重点放在了规划区的中心一带，相对方案2而言，规划区一侧的高层后院退入了背景中。街道广场也将给高层的后院空间带来改良，改善其背面的品质，联系后院和新的开敞空间。

河渠上方的盖板在将来可能会被拆除，这种不确定性在规划中已经被考虑进去了。这种可能性不应成为规划的限制而应该成为其助力。因此这里提出一种暂时性并比较灵活的解决方案，它投入最小物力，并保证了将来能够适应各种改造的可能性。广场用连续的铺装加以统一，用硬质塑料制成的粉红色条形椅对空间进行了点缀和划分。这些座椅被组织成岛式的座椅群。"岛"中央既保证了培育植物必需的土壤层厚度，又为儿童游戏及必要设施提供了场地。"岛"的边缘可以坐也可以躺，广场中其他的空间则为开敞空间，为各种形式的休闲生活提供开放性场所。夜晚这些座椅会从内部发出光亮。

这样就形成了一个充满Pop元素的令人愉悦的城市景观。它符合这个街区年轻和富有活力的特点，同时也强调了装置的临时性。

This variant would create a joyous urban 'Pink Island' landscape created with elements of pop art, which is in keeping with the young and dynamic character of the neighborhood.

This variant would create a joyous urban 'Pink Island' landscape created with elements of pop art, which is in keeping with the young and dynamic character of the neighborhood. Since the islands are mobile, they can be regrouped and used for events or to create new streetscapes. This solution also emphasizes the temporary nature of the installation, which is intended only as an interim solution until the underground waterways are uncovered. Along the street square, glass portholes would also be installed allowing visitors to see the covered canal that once dominated the landscape, helping to keep it fresh in people's minds.

这样就形成了一个充满Pop元素的令人愉悦的城市景观。它符合这个街区年轻和富有活力的特点，同时也强调了装置的临时性。因为"岛"是可移动的，所以在有活动或进行新的景观设计时，可以对其进行移动和重组。这些装置是拆除河渠覆盖前的过渡装置，因此必须强调这些装置的暂时性。此外，河道上方的地板上设置了玻璃孔，透过玻璃可以看到下面被覆盖住的水体，这样就可以提示居民下方河流的存在。

城市更新　URBAN RENEWAL

图 12　方案"粉红岛"夜间效果图
Fig. 12　Night simulation of "Pink Island" concept

图 13　规划方案：粉红岛
Fig. 13　Favorite design: mobile Pink Island with greenery

城市更新
URBAN RENEWAL

01 伦茨堡控制性规划
Framework Development Planning of Rendsburg

02 "霍费尔天空":
提升城市品质的范例项目
"Heaven of Hof":
an Initiative-project for Upgrading the City

03 埃尔旺根步行区
Pedestrian Zone of Ellwangen

04 埃斯林根总体规划和公共空间城市设计导则
Master Plan and Design Guidelines
for the Public Spaces of Esslingen am Neckar

05 "粉红岛": 韩国首尔方背洞街区广场
"Pink Islands":
a Street Square in Bangbae-dong, Seoul, South Korea

06 Gangseogu城市更新规划
Urban Renewal Gangseogu

07 北京市昌平区奥运自行车训练馆
及周边地区城市设计
Urban Design of the Area Surrounding
Changping Olympic Cycling Training Center, Beijing

06 Gangseogu
城市更新规划
Urban Renewal Gangseogu

规划面积 AREA		41 km²
人口规模 POPULATION		570000
完成时间 PROJECT DATE		2008.06

从两种文化中学习 LEARNING FROM TWO CULTURES

工作开始状况和规划任务的确定 Initial situation and planning objective

The city of Seoul has currently begun two major projects, the Hangang Renaissance and the Magok Waterfront City. The first one aims to landscape the bank of the river Hangang, which traverses the city Seoul from east to west. This project will directly affect the district Gangseogu, as it extends all the way up in the north to the Hangang River. The latter project is the planning of a new town on the site of Magok, which is the last major constructible site in Gangseogu.

In order to get the district interested and involved in the planning, the district mayor ordered the creation of an overarching plan in which the two projects are integrated into an overall vision for the region. This was to ensure that the two projects did not remain isolated, and that the potential inherent in the plans was fully exploited to improve the entire Gangseogu area.

The planning was awarded to the ISA international urban planning studio in April 2007. The ISA suggested a solid medium- and long-term development scheme for the district, in the form of a master plan with particular references to the two projects planned by the city of Seoul. The master plan, similar to a 'Rahmenplan' in Germany, is designed to mediate between the abstract, existing Land Use Plan and the detailed Building Design Code Plan. It should define spatial structure in terms of design and land use, such as the three dimensional development of building mass.

首尔目前城市建设中两个重点项目均与Gangseogu区有密切关系，一个项目是"复兴汉江"，旨在振兴贯穿东西的汉江，将其建设成生机勃勃的景观空间。由于北边Gangseogu区直达汉江，因此对Gangseogu的整治也被纳入这个项目。第二个大项目是"滨水城市麻谷(Magok)"，目的是在Gangseogu区最大的未建区——山地区域麻谷规划出一个新城区。

为符合该区的规划利益，该区行政总署委托将两个项目的规划置于一个总体规划之下，代表形成区域发展的同一个愿景，融合并举。可以肯定的是，这样可以使得两个项目不是各自为政，而是通过整体规划的协调，最大程度地发挥其潜质，从而提升整个城区的品质。

该项目于2007年4月由德国ISA意厦国际设计集团承接。ISA建议，结合首尔市的实际情况，以总体规划的形式制定出一个稳定的中长期发展规划，同时考虑到两个重点项目的需求。规划形式应该与德国的分区控制性规划（Rahmenplanung）类似，介于抽象的用地规划与详细的修建性规划之间，根据形态与功能确定空间结构，包括给出三维的建设尺度框架。

规划形式应该与德国的分区控制性规划（Rahmenplanung）类似，介于抽象的用地规划与详细的修建性规划之间，根据形态与功能确定空间结构，包括给出三维的建设尺度框架。

The master plan – similar to a common reference master plan in Germany – is to mediate between the abstract existing zoning and detailed development plans in terms of the design and use, thus defining the spatial structure, such as the three-dimensional development of cubic capacity.

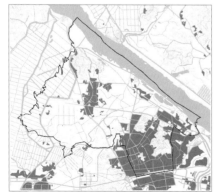

图 01 城市发展阶段，1966年，1976年，1986年，1996年
Fig. 01 Phases of urban development, in 1966, 1976, 1986 and 1996

城市更新 URBAN RENEWAL

图 02　两种对立的建筑结构：wagokdong 的小尺度街区结构、高层居住区的超级街区结构
Fig. 02　Two contrasting building structures in Hwagokdong with small-scale structures and high-rise housing estates

特征所在、机遇与问题
Characteristics, opportunities and problems

城区意象

Gangseogu区位于首尔市西部边缘，与汉江的北部相邻。该区面积约为20多平方公里，人口为57万，规模相当于一个典型的欧洲中等城市。

伴随着金浦（Gimpo）机场的开放，机场所在的城区从1960年开始发展起来。第一批30万个居住单元于1970年完工。沿河岸一条建于1986年的快速路将整体区域与水面之间的联系分割开来。1990年，Gangseogu区在国家城市居住建设政策指导下进一步扩张——通过拆除现状住宅，建设了大量的单调的大型高层板式住区。因此，今日的Gangseogu区有两种面貌，一种是高楼林立的大尺度住区结构，另一种是20世纪六七十年代的小型街区结构。

最近一个阶段以来，Gangseogu区政府开始着手保护失去的历史面貌，建设了一系列博物馆，里面陈列了知名人士——例如画家Jeong Sun和18世纪医学家Hu Jun的遗物。尽管政府花了大力气进行城市整治，Gangseogu区由于其过密的居住区、被私家车停放而显得拥挤不堪的巷道以及充斥着二手车商的街道仍然给人以不愉快和被忽视的印象。这个区域唯一的强烈意象承载者目前仍然是金浦机场。新国际机场仁川机场开放以后，一些韩国国内和东亚航线的航班不再使用金浦机场，原有机场功能的退化造成了大量建筑面临功能更新，其中一部分如今已经形成了一座大型购物中心。

Image of the district

The Gangseogu district lies on the western edge of Seoul and is bordered to the north by the bank of the river Hangang. It is just over 40 square kilometers in size, and with around 570,000 inhabitants it equals the average size of a typical European city.

Development began in the 1960s with the opening of the Gimpo International Airport, which is also in the district territory. The first 300,000 housing units were completed in the 1970s. A highway built in 1986 along the southern bank of the river Hangang separates the district from the river today and makes water access difficult. In the 1990s, the district continued to grow in the wake of a public housing policy that produced monotonous, massive prefabricated housing, usually through demolishing existing neighborhoods with small-scale buildings. Thus, today the district of Gangseogu has two faces: one of very large-scale residential skyscrapers, and one of the small-scale structures leftover from the 1960s and 70s.

For some time, the government of Gangseogu has been trying to revive the history of the district by building museums dedicated to nationally famous figures like the great 18th century painter Jeong Sun or the great doctor Hu Jun. But despite such efforts, the densely built neighborhoods seem miserable and neglected. Automobiles dominate the whole district and even narrow streets are totally filled with parked cars. The only strong image-maker in the district is the now former International Airport of Gimpo. Since the construction of the new Incheon International Airport, only domestic and some East Asian flights are handled there.

今日的Gangseogu区有两种面貌，一种是高楼林立的大尺度住区结构，另一种是20世纪六七十年代的小型街区结构。

Today the district of Gangseogu has two faces, on the one hand that of very large-scale structures of residential skyscrapers, on the other hand, the small-scale structure from the 1960's and 1970's.

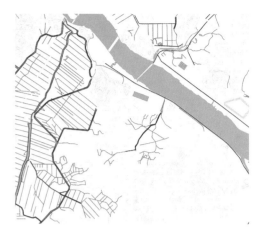

图 03 农业用水道
Fig. 03 Agriculturally used streams

Topography, greenery and water

The Gangseogu district used to be a marshland and is mostly flat. Only a few green and largely undeveloped hills rise from this level and serve as natural orientation points. Two of them, the Gaehwasan and Gungsan hills, offer unique panoramas of the historical city center of Seoul north of the Hangang River.

In order to use the marshes for agriculture, numerous small drainage streams were created that are still found in the area today. There is a chance to use the streams as ecological and design elements in order to keep the history of the landscape in the city alive.

A significant disadvantage of the district is the lack of accessibility to green and water areas. The hills and the ecological park on the banks of the Hangang are too remote for everyday life. Convenient residential recreational space is missing, as is shopping and other regular activities that improve the quality of normal life.

Use and zoning structure

The Gangseogu district is primarily dominated by residential use. Important exceptions are the Gimpo airport and related services and logistics companies, as well as groups of used car dealerships and other small commercial facilities, particularly along Yangcheon Road on the east-west axis.

Another exception is the current agricultural terrain region called Magok, upon which the planning of a new district has already begun. The site can be developed as an incentive for new territory with a mixture of residential uses, services, retail and other commercial facilities. Due to its location in the center of the Gangseogu district, adjacent to the Gimpo Airport in the southwest and the Hangang River in the north, this area offers an opportunity to connect the river with the airport by a central channel.

There is great potential for the development of the district because of its location near the Gimpo airport, the Hangang River and the historic center of Seoul. In addition, the small-scale building structure has been preserved in large parts. They could be developed into a vivid residential area but on a human scale.

地形、绿化与水体

Gangseogu区在过去是一大片沼泽地，整体上用地平整。只有一些绿色的未建的山丘耸立在平原上，形成自然的地表识别标志景观。从其中的两座山（Gaehwasan山和Gungsan山）上，人们可以观赏到汉江北部首尔市历史中心独一无二的整体风貌。

为开发这片沼泽地为农业用地，在Gangseogu修建了大量的排水溪，现在这些排水溪仍然存在。规划中可以将这些溪流作为设计和生态元素来利用，这样就能生动地维持地区的历史景观。

唯一缺憾是，这里绿化和水体区的可达性不佳。对于日常生活来说，河岸边绿色小丘和生态公园过于偏僻。居住区附近缺乏日常的、在住所与工作地点及学校之间的、在购物等日常活动中能被感受到的休闲设施，以便能够实际地益于民众，从而提升地区日常人居品质。

用地与建设结构

Gangseogu区的首要用地类别就是居住用地。此外，就是金浦机场用地、机场相关服务业、物流公司用地、二手车商贾聚集区和特殊的小型贸易用地。这些功能尤以东西向轴线——Yangcheon大街沿线表现得最为突出。

另一种特殊的用地是现有农业用地麻谷片区，该地计划修建成一个新城区。这个地区被规划为融居住、服务、零售业和其他商业用地为一体的混合功能区，旨在形成地区经济的一个起飞点。麻谷穿过Gangseogu区的中部，与西南方向的金浦机场和北部的汉江为邻。鉴于这样的地理条件，麻谷区域具有修建一条中心运河将汉江和机场片区连在一起的条件。

Gangseogu区的潜在发展优势是临近金浦机场、汉江与首尔历史中心的区位。此外其小街块的建设结构及其人性化尺度是值得保留的。如果在此基础上能够进一步提升此地居住环境的品质，该地区将具备理想的条件，建设成为一个城市理想居住地。

威胁这个发展方向的因素来自投资者，他们可能会想要破坏原有的小体量建筑，修建密集的高投资回报率的高楼大厦。这个方案要设法排除这一威胁。还有一个问题，如何对待Gangseogu区的现状已建高楼——它们是在小型街区结构中的严重异体，破坏了空间的连续性。

在项目开展过程中，还会遇到更多必须克服的困难，如密集居住区中夹杂的小型商务岛对居住功能的干扰，以及建筑密度的不断加大而使原有小尺度街区的基础设施日益不堪重负。

Yet, this development is threatened by the demolition of the small-scale structures by the investors, who promise higher returns out of high-rise estates. A concept must be developed that terminates this process. Another question is how to deal with the existing high-rise developments that appear as incongruous objects in the small-scale landscape and disrupt the spatial continuity. Another difficulty the concept needs to solve is the disruption to residential function by small businesses, which are scattered like islands throughout the residential areas and by an ever-increasing building density that overloads the infrastructure in small-scale structure communities.

图 04　小尺度街区的道路
Fig. 04 Residential street in areas with small-scale structures

图 05　Hwagokdong 的建设结构
Fig. 05 Building structures of Hwagokdong

交通系统与空间结构

Gangseogu区的路网由两个不同的体系组成。东北部Gonghangro街区，包括连接金浦机场的主要通道以及1990年建设的高层住宅区，形成了有着宽阔路面的以汽车交通为主导的道路网络。这些街道沿线几乎无一例外地被配备大停车库的大型超级市场所占据。由于过大的交通负载和单一的功能建设，像这样的街道不仅谈不上逗留品质，超级市场等商业建筑的底层也无法进行商业利用，形成了"死区"。高层住宅区之间的开放区域主要设计成不具备行人休憩功能的停车、交通功能带。总体说来该区严重缺少社会活动的功能空间。

Gonghangro街道西南部分的区域对面是小街区的建设方式，来自于20世纪七八十年代保留下来的密集街道网。这些街道拥有大量沿街分布的小商店、小服务业店面和熙熙攘攘的行人。一些历史上被用作市场的小广场，对生动的、活跃的、城市化的居住区景象形成进一步的补充。

只有一些新建成的高层住宅区破坏了小尺度街道上生动的街区景观。如何将这些高层建筑构成的"岛"和谐地融入小尺度街区中，是本次规划的一个任务。

鉴于Gangseogu区位于市中心和金浦、仁川两个机场之间的便利区位因素，Gangseogu区的公共交通水平较高，整体形成了超出平均水平的密集的地铁线路网。此外，一年以前Gangseogu区同市中心在汉江上的水路连接也已经建立。

Traffic and spatial structure

The spatial structure in Gangseogu consists of two different systems. Northeast of Gonghangro Street running to the Gimpo Airport, are a series of high-rise residential areas built in 1990, constructed for the purpose of being a car-friendly city with wide traffic lanes. Located almost exclusively along these roads are large supermarkets with even more parking lots. These streets do not provide any living quality for pedestrians because of the heavy traffic loads and the building structure with mono-functional uses and unattractive ground-floor designs. The open spaces between the apartment buildings serve mainly as circulation and parking areas and are not suitable for spending any time. Overall, the area lacks functioning social spaces. In the area southwest of Gonghangro Street, however, the small-scale buildings and the small-scale street system from the 1970s and 80s have been preserved. Small shops, service units and many pedestrians are littered throughout the streets. Some historical squares that are exclusively used as market places complete a picture of lively and functioning urban quarters.

Only some of the newly formed high-rise residential buildings cut the network of small-scale daily-life streets. One of the tasks of the master plan lies in the question of how these islands could be integrated into the urban network.

In terms of public transport, the Gangseogu district has an outstanding and closely-knit network of underground metro lines due to its location between the city center and the two airports, Gimpo and Incheon.

从两种文化中学习 LEARNING FROM TWO CULTURES

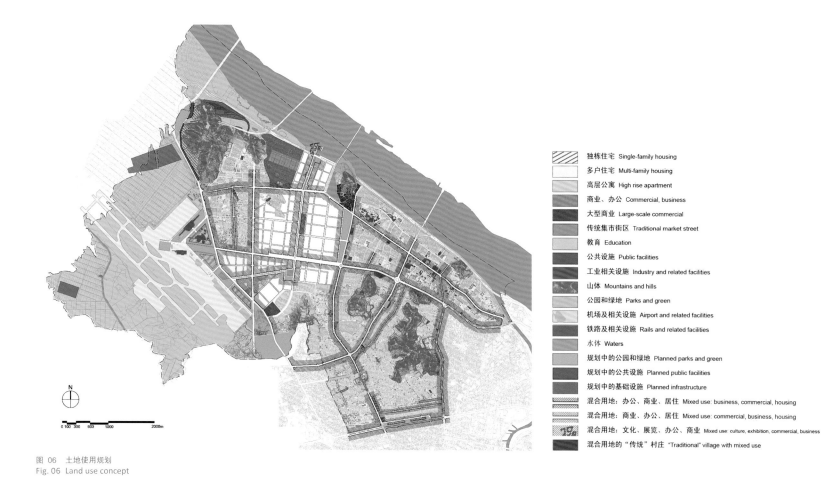

图 06 土地使用规划
Fig. 06 Land use concept

视角：机场城市与运河城市
Vision: Airport city and the canal city

A basic prerequisite and requirement of the concept is to end the total demolition of existing structures based on purely quantitative growth through residential high-rise settlements. Instead, existing structures are to be repaired, expanded and renewed. In addition, the environment and quality of the living environment should be improved.

The potentials of the two key image-makers of the district, the Gimpo Airport and the Hangang River, forms the starting point of mixed development for a city adapted to the needs of the population. The plan follows four superior goals. The first is to improve the quality of living in the environment of the existing neighborhoods. The second is to create an obvious connection between the Airport and the River in order to extend the district as an interface between different public transport systems: the airways, waterways, aboveground routes (roads), and underground (subway) routes. The third aim is to provide work places, which can benefit from their proximity to the airport and the well-equipped public transport system. The fourth and ultimate ambition is to develop the area Magok as a good place to live and work, on canals and on the water. Magok should be a role model regarding the living environment and thus give impetus to the renovation of existing residential areas.

这个方案的基本要求和前提是：终止以摧毁现状结构为代价、建设高层居住群的纯量化增长的城市增长策略，改为对现状结构进行修复、补充和更新，改善其自然、居住环境的品质。

该地区两个起决定性作用的潜在意象载体——金浦机场和汉江，将成为Gangseogu区建设综合、宜居城市这一目标的起点。

这个规划有四个上级目标。第一个目标是改善现状居住区的居住环境品质。第二个目标是建立金浦机场和汉江的有效连接，将Gangseogu建设成水、陆、空、地上（道路）和地下（地铁）等多种交通方式并行的、高可达性的地区。第三个目标是通过机场等大型基础设施的优势，给周边地区带来就业机会，并因此受益。第四个目标是在"运河及河畔的居住与工作"指导原则下发展麻谷新区。麻谷新区建设应当成为首尔居住方面的典范，带动对现状居住区的修复。

实现上述四个目标应遵循以下三个原则：第一是要以人性化的尺度，把Gangseogu区作为以步行为主导的地区来对待。因此在上一级的规划层面中根据下面提到的凯文·林奇的理论将Gangseogu分成了不同的"生活区域"。欧洲古城城市核心的平均最大直径约为1平方公里，首尔市历史城市核心平均最大直径约为1.5公里，这奠定了本区域"人居功能区域"的尺度标准。

第二个原则是在两个层面上实现保护与发展的均衡，其一是自然的开放空间与城市建设之间的平衡，其二是现状建筑保护与新建筑开发之间的平衡。这两种平衡都追求质量型的效益，而非数量型的效益。

自然与城市和城市建设空间之间的平衡意味着，通过分散式的布局和更佳的绿化、水体网络连接，让市民对城市与自然的感知更加明显。

新旧建筑维护的平衡被有意识地重新界定，如同韩国城市历史实践中普遍的发展轨迹。截至被日本人占领以前的李氏王朝，韩国传统的纪念建筑都得到小心翼翼地维护。日据时代开始后，所有建筑都被视作普通建筑来对待，被看做是无保留价值的群体。而Gangseogu区域自20世纪60年代末以来的发展，如同树木的年轮般，形成了一系列文化遗产，展示出不同发展时期的特征。这些"年轮"形成该区城市文化的集聚以及生动的集合。

以上述观点而言，年份较新的建筑也具有保留的价值。新建筑、新结构不应该像20世纪90年代的高层住宅岛漂浮在旧结构中，而应该作为现状结构的补充和完善部分来处理。新建筑群体应该形成一个整体，发挥新的"年轮"效应。新建筑要有新的独立品质，成为将来的文化遗产。

规划原则的第三点是混合集约型城市——即使韩国规划实践中目前仍然强调功能分区的原则（zoning concept）。混合的结构不仅允许住区对应控规条件的更改（例如经济条件）作出具有弹性的适应，这种灵活性也是可持续发展城市原则——最短路径、步行者友好型城市建设的一个重要前提。

In order to realize the four superior goals, the following three principles of planning were applied:

First, Gangseogu would be developed as a pedestrian-friendly district on a human scale. Therefore, at a higher planning level, the district would first be divided into diverse neighborhoods, following Kevin Lynch's theory. The size of each neighborhood would be based on the average size of European city centers with a maximum diameter of about one kilometer, and the size of the historic town center of Seoul of about one and a half kilometers in diameter.

A second design principle would be the balance between new development and preservation on two levels. It was a qualitative, not quantitative, term. First, a balance between natural open space and covered areas was sought. This meant that a more decentralized distribution and better network system of the green and water areas in the daily lives of residents needed to be appreciated. Secondly, a balance between the preservation of building structure and new constructions would be aimed for. This balance was deliberately defined differently than how it usually is for Korean practice. Usually the existing preservation law focused its attention exclusively on monuments dating back to the Lee Dynasty, which ended with the Japanese occupation. All buildings since the beginning of Japanese occupation, and all the anonymous buildings for every day life, are not considered to be worth maintaining. But the Gangseogu district has been evolving since the late 1960s and represents a cultural heritage of its own kind, in which the different stages of development can be read as growth rings on a tree. These rings are producing a dense and vibrant mixture of urban culture. Based on this view, the recent history of the buildings is considered worthy of preservation. New buildings and structures should not form isolated islands like the high-rise housing estates of the 1990s, but complete and improve the existing structures. They are supposed to be understood as new "growth rings" and should form a part of the whole. New buildings should in turn have a new, distinctive architectural quality and should be considered as inheritors of tomorrow's future.

The third of the design principles would be that of a mixed and compact city, even if the Korean planning practice still follows the ethos of land use separation: the Zoning Concept. But mixed structures not only allow a more flexible adaptation to the changing conditions of a neighborhood, like economic ones for instance, they are also an important prerequisite for a city with short distances in order to create a pedestrian friendly city.

> 取代以摧毁现状结构为代价、建设高层居住群的纯量化增长的城市增长策略，改为对现状结构进行修复、补充和更新，改善其自然、居住环境的品质。
>
> *A basic requirement and prerequisite of the concept is the end of total demolition of existing structures based on purely quantitative growth in the form of residential high-rise settlements. Instead, existing structures are to be repaired, expanded and renewed, as well as improving the environment and quality concerning the living environment.*

规划方案 Planning concepts

绿化与水体规划

以从平原中抬升出的丘陵地形为绿地的起点。为了更加明显有效地利用城区的绿色丘陵，将这些丘陵编入绿色长廊的网络中，构成沿街的绿化和林荫大道的小型空间轴线。我们放弃继续形成大型中央绿化的想法，而选择了这种小尺度、与居住区紧密结合的绿化构想。现状的小尺度水道作为设计元素融入城市空间的设计中。采用以上措施的目的是，从每一处房屋都能感受到绿化和水体。

Green and water concept

The topographical position of the hills rising from the plain was a starting point for dealing with green space. In order to make better use of the green hills and to perceive them in a better way, they were integrated into a network of green corridors, sequences of small green spaces located along streets and tree-lined avenues. By abandoning the idea of a large central green area, small-and-residential green areas became possible. Moreover, the existing small waterways were to be incorporated into the urban areas as design elements. The aim of these measures is to experience both green and water from each house.

Land use concept

The land use structure of a city is subject to continuous changes. It changes more frequently than the physical building structure, or even more than the land plot structure. Industrial buildings, for example, are to be converted into museums or lofts and houses are converted into offices. The urban planning practices thwart the modern architecture design principle "form follows function."

The land use structure can be influenced through spatial improvements of individual streets and squares by public authorities. This approach was a great opportunity for the Gangseogu district.

Instead of separated mono-functional zones according to the usual Zoning Concept, the concept of mixed use – multifunctional land use structure along a single street – has been applied. In this sense the land use structure should be managed by effective public interventions.

For example, internal streets within 'the high-rise residential island' would open up to the surrounding area so that the streets cut by these islands could be repaired. These car-oriented streets could be converted into a pedestrian-friendly design so that commercial and other non-residential functions would settle there.

Thoroughfares would be downgraded and turned into boulevards or tunneled to enhance the surrounding areas and to strengthen the residential function therein. The tunneling of

土地使用规划

用地结构应当作为一座城市的变化因素来看待。用地功能的转变明显比建筑物理结构或者土地权属结构的转变要更加频繁。比如工业建筑被改造成博物馆或者Loft，住宅楼被改成办公楼等。城市规划实践并不完全遵守建筑设计中公认的规则——"形式服从功能"。

通过公众力量对街道和广场的改善提高，能够对一个地区内的功能结构进行干预。这种操作方法对Gangseogu区是个很好的机会。

作为对韩国普遍功能区分离和区域性功能制定（zoning concept）的替代，该方案追求混合的街道空间功能结构。从这个意义出发，借助一些干预措施可以将功能结构引到公共空间中加以延伸阐发。例如对高层居住区内部的街道，通过强化其开放特性，借此实现对被分割的都市网络结构的修复。这些街道将从汽车主导型向步行者主导型转变，从而形成对商业与其他非居住型的都市设施的吸引。

多条穿越型街道的交通级别被降级，转化为林荫道甚至转为地下街道，以提高周边地区品质，强化居住功能。首尔市政府现已通过在连接市中心和仁川国际机场的Jemulpo快速路地下挖掘隧道的决议，下一年该项目就可付诸实践。

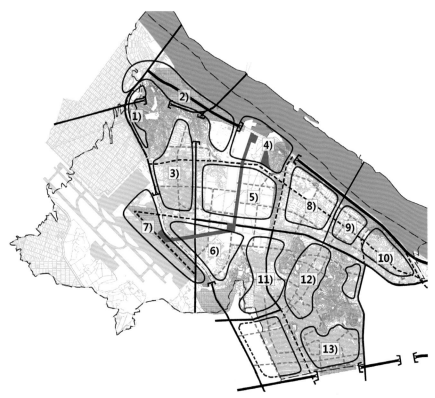

图 07　13个片区
Fig. 07　13 Neighbourhoods

区域功能规划

Gangseogu区被分成13个片区，每个片区都有自己的特点，这些特点又统一成一个整体的风格。各个地区彼此之间的界定首先是地形，特别是丘陵和河流以及穿越型街道。此外其标准还包括功能结构、街区结构特征以及尺度的限制——整体而言其最大直径为1.5公里。

每个分区都制定了单独的发展目标和相应的规划措施，为未来的项目引入形成框架与参考。

交通规划

交通方案的特别之处在于提出了一个动议：为金浦机场铺设一条运河，由此直接联系汉江，并与市中心的水上巴士网联系。同时在汉江上也建立类似的系统，区政府以运河的建设作为机场连接的第一步。

在Gangseogu区内，将来内部还应当将自行车和步行发展成最重要的出行方式。和首尔其他城区相比，本区平坦的地形是个良好的前提。通过好的功能混合和步行区域的良好界定，实现日常短途道路的安排。此外还应仿照欧洲的做法，在地铁站、码头和机场修建自行车交通站。这些站点同汉江边的生态公园相连，构成吸引周末度假者的驻足空间。

Gangseogu区一个大的问题在于旧有基础设施的超负荷，其原因相当部分来自于由于寻找停车位而形成的迤巡交通。这个方案计划，在13个片区的边缘尽可能地提供停车设施，在地区内部只设置短时停车位，以达到优化公共运输体系的目的。

分期规划

一座现状城市的结构改造需要用长远的眼光来看待。这明显与需要快速响应的市场规律和经济发展需求形成了矛盾。基于这些背景，一个长期的规划是非常必要的，同时应该在整体方案中植入短期内可进入的项目，并给这些短期项目制定出共同的发展方向。

整体方案贯彻的长期需求，要求对具体措施的优先性作排序。这里的问题在于，措施的广泛性及成本情况如何。为取得快速的成功并形成后续发展的推动力量，应该直接先开始实施便于推广的举措。其他标准来自于对问题紧迫性的评判以及后续发展的推动效应情况。

在分期规划中将对这些举措进行分类，并根据上述准则确定比较有利的可行性项目以及区内重要位置的建设战略。这样范本项目能够嵌入整体网络之中，对整体结构起到带动作用。

the expressway Jemulpo, connecting downtown with the Incheon International airport, has already been approved by the head office of Seoul.

Neighbourhood concept

The whole Gangseogu district would be divided into 13 neighborhoods, each with its own character and yet perceived as part of a whole. The boundary of the neighborhoods would be defined primarily by the topography, especially hills and rivers, as well as thoroughfares. Other defining criteria would include existing characteristics such as the land use structure, the building structure, and a restriction to a maximum size of one-and-a-half kilometers in diameter. Goals and measures for each neighborhood would be developed, which would serve as a master plan for future projects.

Traffic concept

The distinctiveness of the traffic concept lied in the suggestions to link the Airport in Gimpo to the Hangang River by means of a channel, and then to incorporate a continuous network of water taxis to the Seoul city center. Meanwhile, such a system has already been set up on the Hangang, so the realization of a channel as a premise for linking the airport could be a realistic goal.

Within the district, pedestrians and cyclists should be privileged. In comparison to other districts of Seoul, the Gangseogu's topography is very flat and provides good conditions for walking and cycling. Additionally, the mixed land use and walkable size of the neighborhoods would make the travel distance for daily life much shorter and encourage citizens to walk. In particular, cycling traffic would be encouraged through bicycle renting stations at metro stops, ship landing sites and the airport. This might, in association with the ecological park, even become an attraction for weekenders on the banks of the Hangang.

A major problem in the Gangseogu district are the amount of alleys and streets cramped by parked cars and the congested traffic caused by searching for parking lots. The traffic concept suggests parking facilities at the edge of the 13 neighborhoods and offers only short-term parking within. In this context, an optimization of public transport systems was sought.

Priority concept

The restructuring of an existing city requires a very long-term planning horizon. This seems to contradict the logic of the market and economic development, which requires a short-term response. But it is precisely despite this outlook, why a long-term oriented program is useful and necessary in order to embed the resulting short-term projects in an overall scheme and to give them a common direction.

But the long-term realization of the overall concept requires the prioritization of actions. A criterion for this is the question of how extensive and complicated an operation will be. Simple actions should be started immediately in order to achieve quick results and to provide motivation for those steps that require a long time. Other criteria include the urgency of the measure and its impact on incentives for further development.

Following this methodology, measures were categorized within the priority concept and defined in pilot projects, which are favorable according to the above criteria and also occupy strategically important locations so that the models reveal a well-networked and distributed structure throughout the district.

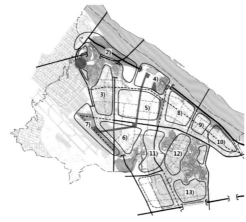

图 08　Gaehwadong 分区控制性规划
Fig. 08　Master plan for the area of Gaehwadong

先锋项目 Pilot projects

Gaehwa-residential area

This neighborhood was built in the 1970s as part of a subsidized housing program initiated by the South Korean government as a quiet detached housing area nestled in a green hillside. The existing housing typology can be an alternative to the residential skyscrapers and refined into a high-quality and unique residential area by enacting a few measures.

The most urgent measure would be the renovation of existing buildings, particularly in terms of heat and sound insulation. Incentives for homeowners should be offered in order to accomplish this measure, since the right of private properties cannot be intervened by public powers. First, the allowable ground floor and floor-area-ratio have to be increased if the owners renovate and improve their homes according to the design guidelines. This ensures that the smallness and homogeneity of the existing buildings be preserved during modernization. In addition, the public streets and plazas would be improved as to increase the value of real estate as a whole and to enhance the owner's readiness to invest.

Gaehwa 居住区

Gaehwa居住区归属于韩国政府推出的一项居住促进项目，始建于20世纪70年代。它是一片清静的独栋住宅区，被绿树所环绕。这种独栋建筑类型不失为普通高层住房类型之外的另一种选择，只需采取少量措施就可成为独一无二的高档居住区。

最急迫的措施是对建筑现状的修复，尤其是隔热和隔声功能。由于区政府无法直接干预私人财产，对房屋所有者提供改善修复的鼓励是最重要的。首先，如果私人业主根据设计导则修复和优化他们的房屋的话，法定的建筑密度和容积率将会被提高。由此也可以保证在现代化过程中，现状建设的小街区结构和统一性得以延续。同时政府将主导公共街道和广场品质的提升，直接形成房产价值的提高，增强物业持有者的投资热情。

图 09　通过对新建筑和对现状溪流功能与景观的提升，实现 Gaehwadong 区内部主要道路的价值的提升
Fig. 09　Improving the main street with a new building line, as well as structural and functional improvements of the existing but unused creek

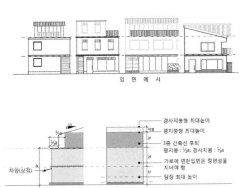
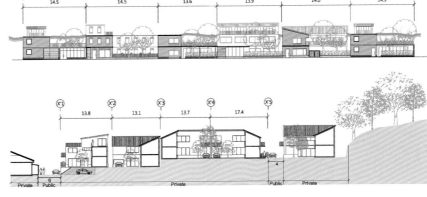

图 10　Gaehwadong 居住区的设计原则
Fig. 10　Design guidelines for residential buildings in Gaehwadong

Hwagokro 街周边地区

Hwagokro街周边地区规划建设于20世纪七八十年代，是Gangseogu地区发展的起点。地区中央是Hwagokro街，它是整个地区功能和地理意义上的中枢。这条已有40年历史的林荫道上，有着很好的功能混合布局——商业、公共设施、政府办公楼和服务业设施一应俱全。尽管此处交通流量大，它的步行功能还是很完善，展示出作为良好的社会空间的品质。

规划方案要强化原有的这些潜质，将Hwagokro街建设成地区公共生活的中心。通过引导过境交通绕行来降低这条街道的交通负荷，从而减少车行路面宽度，同时维持现状绿化。以这一方式形成大片广场、步行区和自行车道空间，街道品质由此大幅度提升。周围居住区的修复工程也将因此受益。下述其他的先锋项目在具体实施方面也使用类似的方法贯穿始终。

The area around the road Hwagokro

This area was built in the 1970s and 1980s and was the starting point for the development of the Gangseogu district. The street Hwagokro is centrally located within the district and forms the functional and geographical backbone. This now 40-year-old avenue has a vibrant multifunctional development with shops, public facilities, office buildings and services. Despite the heavy traffic, it is well attended by pedestrians and seen as a properly functioning social space.

The concept envisaged exploiting this potential and converting the road into the district center for public life. The road would be relieved of the diversion of transit traffic, and could be dismantled with the installation of trees. In this way, generous space for pedestrians and cyclists would be created, and the street would be improved into a walking and bicycle axis. This would provide an incentive to redevelop the surrounding neighborhood. Its renovation would be carried out in a similar manner as presented in the following pilot project.

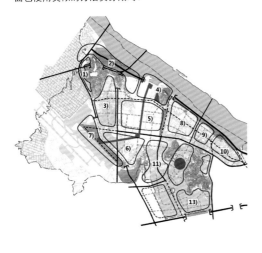

图 11　Hwagokro 分区控制性规划
Fig. 11　Master plan for the area of Hwagokro

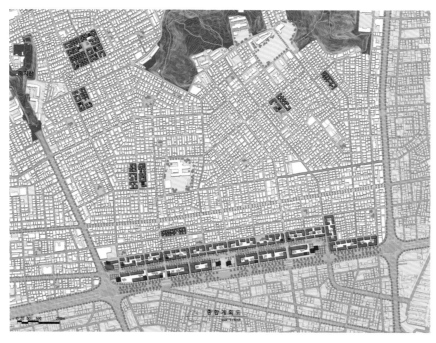

图 12　Hwagok南部地区分区控制性规划
Fig. 12　Master plan for the area south of Hwagok

Residential area of Hwagok

This area was originally built as a residential area for single-family houses in the 1970s and 80s, which was subsequently replaced by three- and four-story mansions. The density was increased significantly and the infrastructure became overburdened in many ways. For example, there is a significant shortage of parking lots and public open spaces.

The original small-scale urban fabric is largely preserved and not torn apart by high-rise estates. In this way, a relatively orderly image in human scale was maintained. The concept envisages preserving the existing infrastructure, land subdivision and urban character with three- and four-story buildings and making the different development phases of the area tangible.

Hwagok 居住区

这一区域建设于1970到1980年间，一开始是一个独栋房屋居住区，后来被多户型的住宅所取代。自此，地区的密度大大提升，造成基础设施超负荷运转（如出现停车位严重不足、公共开发区被挤占的现象）。

原始的小街区城市平面布局将被整体保留，并控制高层建筑的异化与干扰影响。区内保持一个有序的景观和适宜的人居尺度。规划整体保留了现状基础设施、土地权属分割和以3~4层的低层建筑为主体的城市特性，以便展现该区不同发展阶段的风貌。

图 13　改建前后Hwagokdong南部一条典型的住区街道
Fig. 13　Simulation, a typical residential street in the South of Hwagokdong before and after renovation

城市更新 URBAN RENEWAL

图 14 中国集市的现状情况
Fig. 14 Existing situation of the Chinese market

图 15 中国集市的重新设计
Fig. 15 Simulation, redesigned Chinese market

要达到上述的要求，有一些问题需要解决。地区南部有一个中国集市，类似于美国的唐人街，要对它作设计上和空间上的评估。在这个目前完全阻隔南部与附近地区联系的位置上，要建设上面提到的通往仁川机场的快速路隧道。街道沿线应该形成一个绿色空间，给中国集市和北边的集市街提供荫蔽，确保Gangseogu区和相邻地区的联系。

另一个主题是实现居住建设的升值。类似于第一个先锋项目Gaehwa居住区，通过提升公共空间品质和有限制的密度提升来鼓励个人业主修缮他们的房屋或者盖新房取代旧屋。规划导则中提倡一种新的建筑形式，它出自传统的韩式庭院房屋类型，又采用了现代多户住屋的形式。L形的房屋围绕内部绿化覆盖的庭院布局，这样即使在较大的建设密度下房屋仍有足够好的采光，远远优于三个方向都与周边建筑相邻的房屋。

一些其他的建筑类型的补充可以通过2～4个小地块的联合规划建设来实现，这样可以形成较大型的综合单元体。这些建筑允许有较高的密度和楼层，一部分用地可被用作公共空间，建设成小尺度的广场、绿化或者穿过居住区的横向连接的小型街道。通过这种方式可以增加该地段的空间多样性，优先突出紧密结合交通功能的密集公共空间，并对其品质加以提升。

To achieve this, it would be necessary to address certain regional issues. A Chinese market situated in the south of the area, roughly comparable to a "Chinatown," must be improved physically and functionally. Additionally, the highway to the Incheon International Airport, which currently completely cuts off Hwagok from the southern neighboring district, should be tunneled. This way, a green space could be created, which would ensure the linkage of the Gangseogu district with its neighbor to the south.

Another range of topics was the improvement of residential building structure. It is similar to the first presented pilot project in Gaehwa, which plans on improving the public space along with a slight increase in the authorized building density, dependent on conditions. This provides incentives for private owners who renovate their homes or replace them with better buildings. For this purpose, a new building typology was suggested, which took the traditional Korean courthouse typology and changed it into a multi-family house with a modern form. The L-shaped building opens up to a courtyard, thus ensuring significantly better light exposure even at a slightly higher density than existing residential buildings.

This typology would be supplemented by a few building types that are made possible by the combination of two or four lots. These buildings may be built in a higher density and with more stories, providing that a part of the land would be reverted to public space, either as a small square, green space or playground, creating an additional cross-connection through the block. This way, public space that currently serves as car traffic space and is not terribly varied would be improved.

图 16 中国集市的修复方案：以通过在街道中间设置步行区提升景观及功能品质为先导措施，以通过新建筑提升街道界面价值为长远措施
Fig. 16 Redevelopment concept for the Chinese market: functional and design improvement through the pedestrian zone in the middle of the road as an emergency measure, optimal use of road space by new edge as a long term measure

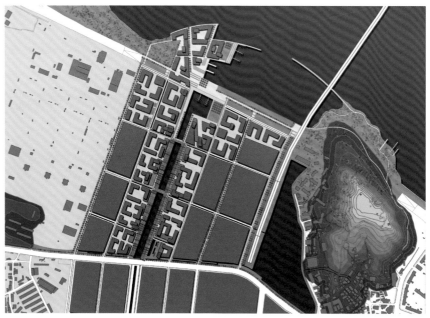
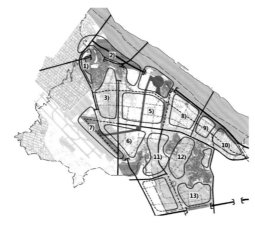

图 17　城市结构规划方案和 Hangang 区域鸟瞰效果
Fig. 17　Urban structure concept of the entrance at the Hangang'

Entrance to the city on the banks of the Hangang River

The current agricultural area of Magok would be developed as a new urban area, according to the plans of the Seoul head office. Notwithstanding that planning, the master plan for Gangseogu suggests establishing a connection axis from the Gimpo Airport to the Hangang River with a Grand Canal, which would form the backbone for the Gangseogu district and ends in an urban park on the riverbank. This urban park with its mixed use of services, housing, commercial facilities and restaurants is planned around a harbor. It would be a landmark with a few tall buildings. This is an artificial counterpart to the existing Gungsan Hill, and marks the new river side entrance to the district and to the city of Seoul, together with Gungsan Hill.

汉江堤岸入口

根据首尔市政府的规划，现在还是农业用地的麻谷将规划为住宅新区。和中央政府现有规划不同的是，Gangseogu总体规划建议：要在金浦机场和汉江之间修建一条大运河作为连接轴线，构成Gangseogu地区的脊柱。运河与汉江的交汇点形成轴线的终端，成为规划的重点。这是一个集服务、居住、商业和餐饮于一体的"头部"地区（混合功能用地）环绕港口排开，形成高楼地标群体，和历史性人文自然要素——Gungsan山丘一起标志了新区在河流的入口位置以及首尔市西边的城市入口。

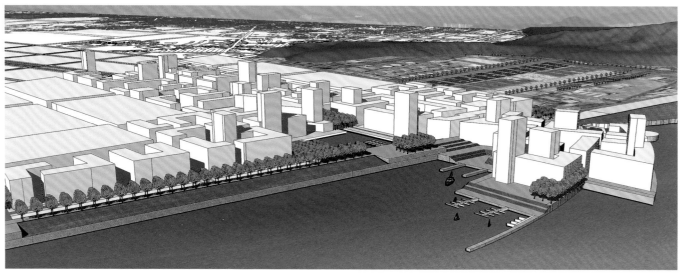

图 18　城市结构规划方案和 Hangang 区域鸟瞰效果
Fig. 18　Bird's-eye view of the entrance at the Hangang'

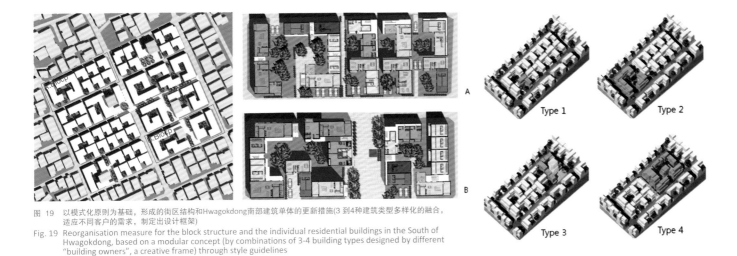

图 19 以模式化原则为基础，形成的街区结构和Hwagokdong南部建筑单体的更新措施(3 到4种建筑类型多样化的融合，适应不同客户的需求，制定出设计框架)

Fig. 19 Reorganisation measure for the block structure and the individual residential buildings in the South of Hwagokdong, based on a modular concept (by combinations of 3-4 building types designed by different "building owners", a creative frame) through style guidelines

总结与展望 Summary and future prospects

In this way, the master plan was completed in June 2008. The district government wants to implement the pilot projects as suggested, but this is dependent on support from the central government of Seoul. As already mentioned, the tunneling project from the Jemulpo expressway was absorbed into the long-term plan by the central government. The same applies for the dismantling of the street Hwagokro. First, the pilot project in the Gaehwa residential district will be carried out. It might prevail in a competitive process in the context of a newly launched program to support projects in other districts. Crucial to the central administration's decision regarding pilot projects in the Gangseogu district was that they are embedded in long-term and comprehensive planning.

The Building Design Code Plan for the area was created in April 2010, and is being conducted for the first time in Korea with citizen participation workshops. The ISA international urban planning studio supported this whole subsequent process as the author of the master plan.

在此介绍的总体规划于2008年6月完成。区政府计划实行上述先锋项目，并在许多领域得到了首尔市中央政府的直接支持。例如政府长期规划中纳入的Jemulpo快速路地下隧道项目，以及Hwagokro大街的降级改造。第一个将实施的先锋项目是Gaehwa居住区，它将代表Gangseogu区，针对国家新颁布的支持计划，与其他城区的新项目形成强有力的竞争。在中央管理部门决定将其纳入计划的过程中，至关重要的要素是，它作为Gangseogu地区先锋项目之一，植根于本区域长期和全面的总体规划工作中，具有超越个体项目的意义。

该项目后续的详细规划工作已经展开，也将是韩国民众以研讨会的形式参与其中的首个规划项目。ISA是总体规划和后续工作的执行者。预计第一期公共空间改善措施的预算为10亿韩元（50万欧元）。此外首尔中央政府业已专项拨出津贴赞助建筑物的修复工作。

城市更新
URBAN RENEWAL

01 伦茨堡控制性规划
Framework Development Planning of Rendsburg

02 "霍费尔天空"：
提升城市品质的范例项目
"Heaven of Hof":
an Initiative-project for Upgrading the City

03 埃尔旺根步行区
Pedestrian Zone of Ellwangen

04 埃斯林根总体规划和公共空间城市设计导则
Master Plan and Design Guidelines
for the Public Spaces of Esslingen am Neckar

05 "粉红岛"：韩国首尔方背洞街区广场
"Pink Islands":
a Street Square in Bangbae-dong, Seoul, South Korea

06 Gangseogu城市更新规划
Urban Renewal Gangseogu

07 北京市昌平区奥运自行车训练馆
及周边地区城市设计
Urban Design of the Area Surrounding
Changping Olympic Cycling Training Center, Beijing

07 北京市昌平区奥运自行车训练馆及周边地区城市设计
Urban Design of the Area Surrounding Changping Olympic Cycling Training Center, Beijing

规划面积 AREA		990 hm²
人口规模 POPULATION		614821
完成时间 PROJECT DATE		2005.12

从两种文化中学习 LEARNING FROM TWO CULTURES

项目背景和基础分析 Background and Analysis

On the occasion of the general planning revisions of the city of Beijing in 2004, a series of planning research inquiries were conducted under the guidance of the general regulations of the city, into Changping New Town, one of 11 new towns surrounding Beijing. Of particular importance was the urban design of the area surrounding the Changping Olympic Cycling Training Center.

The main planning area was Changping Old City, an area with a complex set of preexisting opportunities and challenges. As planning began, the group first conducted a comprehensive analysis of the urban development, land use, transportation, green space and landscape systems, tourism market, cultural character, etc. throughout Beijing as well as in the Changping District and the planning area. Then, the planning area was divided into six blocks and two roads in order to develop a detailed study, analyzing the existing characteristics and problems, and offer a preliminary development proposal.

以2004年北京城市总体规划修编为契机，作为北京周边11个新城之一的昌平新城也在总规指导下组织开展了一系列规划研究与制定工作，本次规划的奥运会自行车比赛昌平训练馆及周边地区城市设计项目就是其中的一个重要组成部分。

本次规划范围基本为昌平老城地区，现状情况复杂，优势与劣势突出，机遇与挑战并存。因而规划之初，我们首先从北京市层面、昌平区层面直至规划范围层面，对于城市发展定位、土地使用、交通、绿地及景观体系、旅游市场、文化特征等方面，进行了全面分析；其后将规划范围划分为6个街区及2条道路进行详细研究，分别分析其现状、特质、问题，并提出有针对性的初步发展建议。

以"活力运动之城，生态绿色之城，文化魅力之城"为本次城市设计指导意向。

A city of dynamic movement, ecologic green, and charming culture - set up the functional framework of planning areas
The development of three functional centers: a commercial center, tourism service center and sports leisure center.

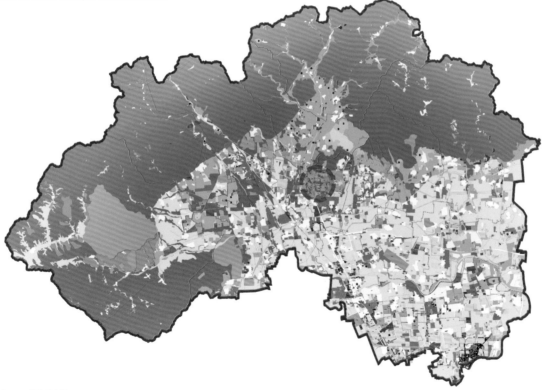

图 01 区位示意图
Fig. 01 Location of the planning area

图 02 航拍照片
Fig. 02 Aerial picture

图 03　用地现状
Fig. 03 Status quo

图 04　用地规划
Fig. 04 Planning

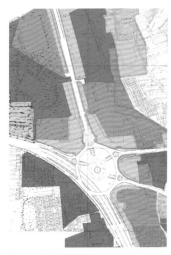
图 05　可考虑改造地区
Fig. 05 Considerable redevelopment area

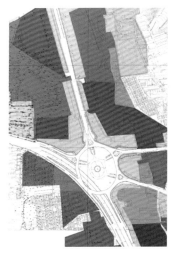
图 06　首期宜于改造地区
Fig. 06 First phase redevelopment area

规划理念 Planning concept

在此基础上，我们提出了城市设计指导意象——活力运动之城、生态绿色之城、文化魅力之城，并建立了规划区的功能结构骨架：

- 依托老城商业街区的基础发展商业中心
- 依托西关环岛的区位交通优势发展旅游服务中心
- 依托奥运自行车训练馆发展体育休闲中心

First, the group identified the design themes as: a city of dynamic movement, ecologic green and charming culture. These themes were then applied to the development of three functional centers defined as follows:

- A commercial center based on the Old Town commercial district
- A tourism service center making use of the traffic advantages of Xiguan around the island
- A sports and leisure center relying on the Olympic Cycling Training Center

图 07　可改造用地分析
Fig. 07 Redevelopment area analysis

图 08 鸟瞰图
Fig. 08 Bird's-eye view

并围绕此3个中心组织城市功能。同时通过生态景观廊道建立了3个中心之间的紧密联系，以此为基础组织城市空间形态。借鉴北京老城的城市空间特色，在规整的街区结构之内引入自由形态的绿化及开放空间体系，并结合生态措施的采用，使得城市在完成功能更新与发展的同时，建立起优美实用的绿色生态网络。

Individual space patterns were established according to these three functional centers, which would be connected by an ecological landscape corridor. Free form, open greenbelt systems were introduced into the regular district structure, using the urban space characteristics of Beijing Old City. Combining with the ecological measures, a beautiful and practical green ecological network was established with the completion of urban renewal and development.

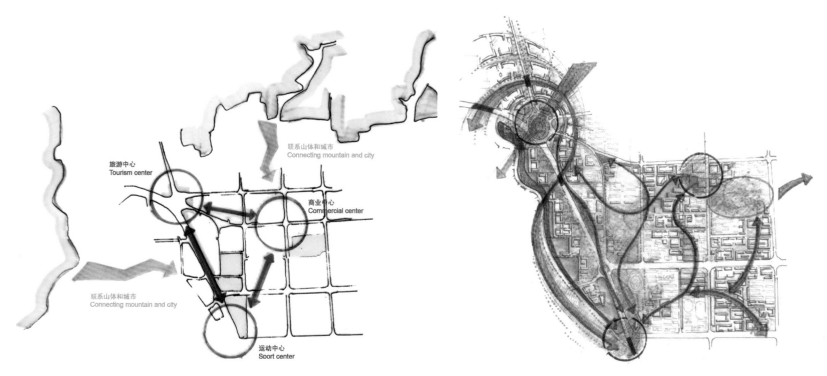

图 09 空间结构规划草图
Fig. 09 Sketch of the spatial structure

图 10 空间形态规划草图
Fig. 10 Sketch of the three centers

城市更新 URBAN RENEWAL

图 11 城市设计总平面图
Fig. 11 Site plan

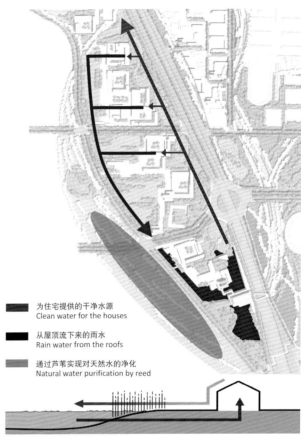

为住宅提供的干净水源
Clean water for the houses

从屋顶流下来的雨水
Rain water from the roofs

通过芦苇实现对天然水的净化
Natural water purification by reed

图 12 生态水体系统规划图
Fig. 12 Ecological water system plan

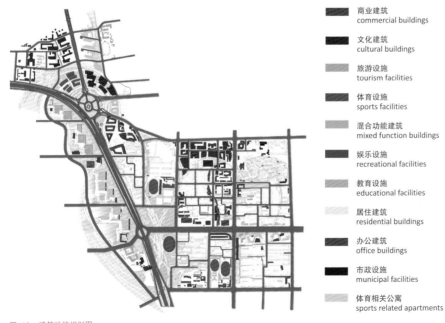

商业建筑 commercial buildings
文化建筑 cultural buildings
旅游设施 tourism facilities
体育设施 sports facilities
混合功能建筑 mixed function buildings
娱乐设施 recreational facilities
教育设施 educational facilities
居住建筑 residential buildings
办公建筑 office buildings
市政设施 municipal facilities
体育相关公寓 sports related apartments

图 13 建筑功能规划图
Fig. 13 Building function plan

重点强调的入口 important entrances
半公共空间 semi-public space
公共空间 public space

图 14 公共空间系统规划图
Fig. 14 Open space system plan

289

图 15 八达岭高速鸟瞰图
Fig. 15 Bird's-eye view of the Badaling Highway

借鉴北京老城的城市空间特色，在规整的街区结构之内引入自由形态的绿化及开放空间体系，并结合生态措施的采用，使得城市在完成功能更新与发展的同时，建立起优美实用的绿色生态网络。

Free form, open greenbelt systems were introduced into the regular district structure, using the urban space characteristics of Beijing Old City. Combining with the ecological measures, a beautiful and practical green ecological network was established with the completion of urban renewal and development.

Having completed the overall concept, a comprehensive design was then crafted with the representative tourist, commercial and residential areas in mind. A detailed research and redesign of the two key roadways was also conducted, from the Badaling expressway to the Great Wall as well as the South Loop.

Finally, a series of guiding principles for urban planning were created. Controlled factors included the location and partitioning of blocks, roof shape, building height, openness of elevation, color system, etc. These principals represented a summary and supplement to the overall urban design, and formed the basis for future developments in legal planning and urban administration.

The planning was approved by competition jury and received first prize in an international competition. The group was entrusted by the Changping District Government to complete the guidelines of urban design, which was recognized along with the overall planning of Changping New Town as a special research achievement.

整体城市设计之后，针对有代表意义的旅游片区、商业片区、居住片区进行了深入城市设计，并对八达岭高速、南环路两条重点街道进行了详细研究和设计。最后，以城市设计为基础，制定了一系列规划指导原则，包括对街区建筑立面水平及竖向分区、屋顶形态、高度控制、立面开放度、色彩体系等，作为对城市设计的总结与补充，以及日后制定法定规划和进行城市管理的基础。

本次规划受到评审专家的高度认可，取得了国际竞赛的第一名。我们继续接受昌平区政府的委托完成了城市设计导则的制定，该工作最终作为专项研究成果之一列入了昌平新城总体规划。

图 16 居住街区三维模型
Fig. 16 3D model of the residential

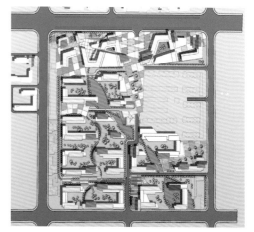

图 17 居住街区总平面图
Fig. 17 Site plan of the residential blocks

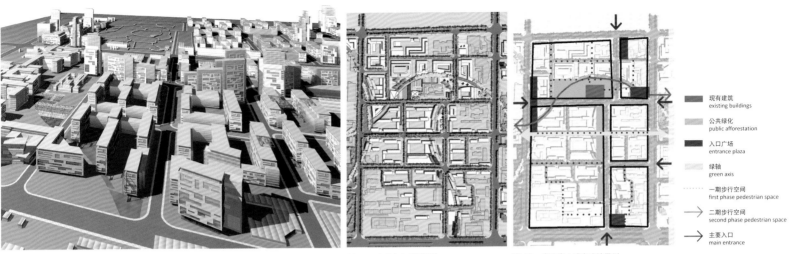

图 18 商业中心三维模型
Fig. 18 3D model of the commercial center

图 19 商业中心总平面图
Fig. 19 Site plan of the commercial center

图 20 商业中心城市设计导则
Fig. 20 Urban design guidelines for the commercial center

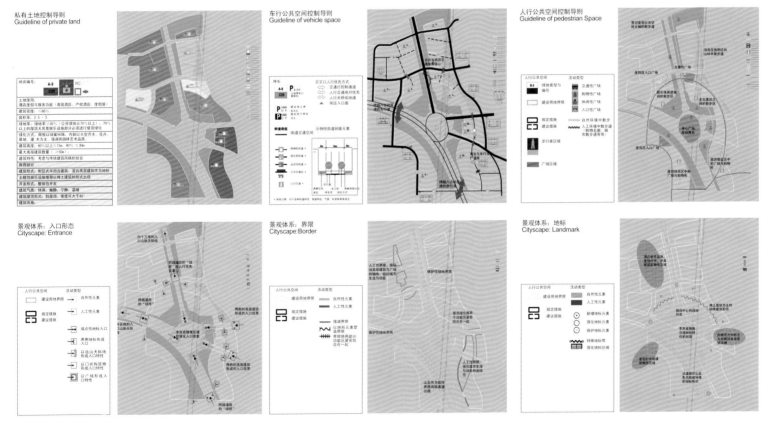

图 21 设计导则
Fig. 21 Design guidelines

城市保护及管理
URBAN PRESERVATION AND MANAGEMENT

01 世界文化遗产城市施特拉尔松：城市景观规划
World Cultural Heritage City of Stralsund: Townscape Planning

02 新勃兰登堡城市景观规划
Townscape Planning Neubrandenburg

03 世界文化遗产城市波茨坦瑙恩郊区
及耶格尔郊区的设计导则
Design Code for the Nauener Vorstadt and
the Jaegervorstadt in World Cultural Heritage City of Potsdam

04 泉州市法石街区保护
与整治规划及环境设计
Urban and Landscape Design for the
Fashi District in Quanzhou, Fujian

05 泉州市城市新区公共空间控制规划方法研究
Research of the Planning Methods for
Open Space in Quanzhou, Fujian

世界文化遗产城市施特拉尔松：城市景观规划

World Cultural Heritage City of Stralsund: Townscape Planning

规划面积 AREA	55 hm²
人口规模 POPULATION	60000
完成时间 PROJECT DATE	1994

问题 The problem

The 750 year old Hanseatic city of Stralsund was once a powerful trading city, a member of the League of Hanseatic cities of Luebeck, Stralsund, Wismar, Rostock and others during the Middle Ages. Today it is a UNESCO World Heritage city thanks to its rich architectural and urban heritage.

But after the reunification of Germany, the city was in decay. The structure of area buildings was deeply affected, leaving many either destroyed or uninhabitable and rendering the overall quality of housing unacceptable. A new building boom began, which brought with it the danger of affecting or losing the unique townscape of Stralsund to uncontrolled building projects. A historic building composition as valuable as the one found in Stralsund requires a carefully managed overarching plan that exerts control at even the level of individual building projects, and manifests itself in everything from small repairs to modernization. New buildings should be designed in such a way that they do not detract from the city as a whole, but compliment it, making it more valuable and beautiful. Simple ignorance or a lack of attentiveness can still lead to the destruction of the existing quality of the townscape.

汉萨同盟城市施特拉尔松拥有超过750年的历史，是一个强大的商贸城市。在中世纪，它和吕贝克、维斯玛、罗斯托克等一起组成汉萨同盟。今天，它因其丰富的建筑和城市建设遗产被联合国教科文组织列入世界文化遗产名录。

两德统一以后，该城的许多建筑物遭到破坏或处于无人居住的状态，落后的居住品质令人难以接受——城市处于衰败状态。新两德统一带来了新的建设浪潮，然而兴盛的建设浪潮曾一度带来建设尺度失控，而损害了施特拉尔松的城市景观。因此，无论是颁布完整的法令还是激进的城市现代化进程，施特拉尔松市要求每一条涉及有价值的历史建筑的建设措施都要谨慎处之。新建筑不仅要不妨碍城市原有的景观，更要设法为原有的景观锦上添花。然而由于知识的匮乏或是缺乏足够的关注，新的建设很可能会危及现状城市景观品质。

作为世界文化遗产整体保护城市，汉萨城市施特拉尔松的未来有着无与伦比的发展潜质。

Stralsund abounds with important development potential for the future indicated by international recognition as a world heritage city, exemplary pilot urban renewal projects in the Federal Republic of Germany, not to mention its ability to attract visitors from around the world.

The impressive downtown area and basic urban structure of Stralsund is still largely intact despite profound losses during World War II and the careless handling of historic buildings in the years since the war. Stralsund abounds with important development potential for the future indicated by its international recognition as a world heritage city, exemplary pilot urban renewal projects in the Federal Republic of Germany, and not to mention its apparent ability to attract visitors from around the world. But, Stralsund is not just an outward trademark city; it also holds special meaning for the people who call it home as an essential part of their self-esteem, quality of living and civic identity. The city architecture of Stralsund represents not only the priceless heritage of past generations, but is also an essential contributing factor in the overall quality of life in the city. It must therefore be considered an important economic factor as well.

汉萨城市施特拉尔松令人印象深刻的地方，在于它仍整体保留了原有的城市空间结构，虽然这些空间结构由于"二战"的毁坏和其后对建筑现状的不善维护而遭到了严重的破坏。作为国际著名的城市古迹、国家级的城市更新示范项目、富于吸引力的世界知名旅游目的地，施特拉尔松的未来有着无与伦比的发展潜质。这样看来，施特拉尔松的城市景观不仅仅对外具有标志性的意义，而且对当地居民来说也具有重要的精神意义——家园、自豪感和生活品质。该市的建筑不仅是上一辈人留下的文化遗产，也是城市生活品质的必要组成元素和重要的经济因子。

图 01 施特拉尔松：看得见的历史，感受得到的现在
Fig. 01 Stralsund: a visible history, a presence to be experienced

城市保护及管理   URBAN PRESERVATION AND MANAGEMENT

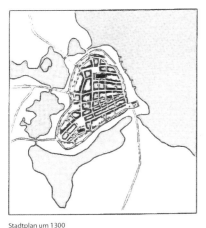
Stadtplan um 1300

Stadtplan um 1715

Stadtplan um 1880

Stadtplan um 1990

图 02　施特拉尔松：700年历史的传承与变迁
Fig. 02　Stralsund: Continuity and change over a 700 year history

现代与历史城市景观的融合

施特拉尔松的位置是独一无二的：位于波罗的海堤岸的一个岛屿上，仅仅通过一片狭长的陆地与大陆相连，如同大地景观中的皇冠。中世纪的城市平面，多样化的街道和广场，宏伟的教堂，巨大的古老仓库，精致的市政厅，汉萨式建筑类型，丰富的汉萨式立面以及材质、色彩的高度统一，共同构成了城市的景观，形成独一无二的城市剪影。

两德统一后，与施特拉尔松市乃至梅克伦堡—波莫瑞州政府的经济、社会、文化可持续发展目标背道而驰的是，历史遗产的价值面临出于短期经济利益而遭到轻视甚至牺牲的危险。在提供住房及工作岗位、满足投资需求及确定购物中心选址、建设停车位及交通设施等城市建设的博弈过程中，应该始终关注与城市景观相关的城市建设和建筑设计品质。

一方面要维护城市景观，另一方面要为新的城市建设提供发展空间，这正是州政府及市政府对意厦进行此次城市景观规划所提出的基本任务。规划要基于对城市建筑的历史和现状情况进行的清晰深入而脚踏实地的分析，提出施特拉尔松市未来城市景观发展的要求，既能保护城市及建筑的历史价值，又能兼顾新的城市建设，使之不仅不能破坏，还要传承和发展原有城市景观。因此施特拉尔松城市景观规划的任务在于：为将来的城市发展确定清晰易懂的指导原则，并作为城市的经济、法律和政治措施的重要基础。

Integrating modernity into the historic townscape

Even the shape and location of Stralsund is unique. Situated on an island on the banks of the Baltic Sea and connected to the mainland only through narrow peninsulas, the city area resembles a crown made of stone. The medieval city layout, with its streets and squares, magnificent churches, massive storehouses, sophisticated city hall, Hanseatic building typology, wealth of Hanseatic facades and powerful materials and color define the townscape, creating a unique city skyline.

After reunification, the city of Stralsund and the state government of Mecklenburg Pommern ran the risk of recklessly squandering the intrinsic value of the city, its cultural interests and long-term potential for stable economic growth, in the name of short-term progress. The struggle to supply the city with sufficient housing, employment, department stores, parking lots and roadways became a careful balancing act between concerns for the urban and architectural quality of the townscape and complying with the wishes of investors.

The city and state governments needed to find a way to secure the city, while at the same time leaving plenty of room to create new urban and architectural developments. To this end ISA-Stadtbauatelier was brought in to develop a townscape plan for Stralsund. The goal was to developing a vision for the future based on comprehensive analysis of the city's architectural past and present. The old urban and architectural values needed to be preserved and clear rules developed regarding how new architecture could be incorporated in a way that did not destroy the existing townscape, but instead developed it further.

The urban planning task for Stralsund was therefore to develop a set of guiding principles for future urban development that were easy for everyone to understand and could serve as a practical basis in future economic, legal and political matters.

一方面要维护城市景观，另一方面要为新的城市建设提供发展空间，这正是州政府及市政府对意厦进行此次城市景观规划所提出的基本任务。

The city and state governments needed to find a way to secure the city, while at the same time leaving plenty of room to create new urban and architectural developments.

295

 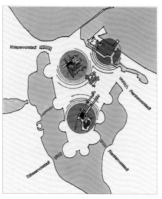 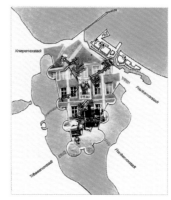 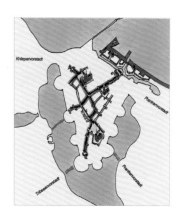

图 03 目标：历史、自然、生活品质和对城市的感知进一步加强
Fig. 03 Goals: history, nature and city life to be experienced

过程 The path

The history of urban architecture

The history of urban architecture in the city Stralsund stretches back more than 800 years, and represents the consistent development and growth of a basic architectural idea over many centuries. As a tree grows from a seed, so the most important parameters of urban architecture have been developed more or less as a harmonious whole: the urban layout and spatial structure, the urban fabric and townscape, the typology of streets and squares, building types and structure.

The synoptic study of the development of these elements of urban architecture reveal in an impressive way how they were modified, refined and expanded upon from century to century, while continuing to represent the same basic concepts of urban architectural development. In doing so, the clear rules, which were followed by urban development in all phases, allowed a variety of architectural styles to remain in harmonious co-existence. This created a beautiful, intriguing and increasingly diverse whole that grew from century to century. Yet with the dawn of industrialization, came the danger of neglecting this concept of a city, subjecting it to short-lived fashions, economic interests and traffic engineering dogmas.

如同种子长成大树，在城市景观各个重要层面上的要素共同体现出了一个高度的和谐整体。
As a tree grows from a seed, so the most important parameters of urban architecture have been developed more or less as a harmonious whole.

Vision of urban development

Still, arguably the largest and most sustainable development potential of Stralsund lies in its historical urban architectural concepts, but not in the economic terms that are generally imposed. The city was founded as a trading town and was later of great importance thanks to its shipyards. It was also a participant in the tide of industrialization as well, but that has now all but lost its significance. The city, however, has managed to retain its unique character forming the basis for the newly expanding industry of tourism.

A unified vision was essential to establish a set of criteria to assess the current state of Stralsund's townscape and debate a direction for its future. This was especially important in the planning of individual urban development parameters such as traffic engineering measures and economic strategies, as well as building height, mass, shape and location. The below goals were defined subject to the evaluation of the present townscape and the assessment of inquiries for construction permits, among other factors relating to future structural development.

城市建筑的历史

施特拉尔松城市建筑的发展已经连贯统一地延续了800多年，并充分展示了城市建筑发展的基本思想。如同种子长成大树，在城市景观各个重要层面上的各个要素——城市平面、城市空间结构、城市建筑体量和立面形态、街道和广场空间类型、建筑类型和建筑结构等——都体现出了高度的和谐。

对城市建筑发展历史这一具有"基石"意义的问题的研究，清晰地展现出城市建筑在数百年间的变化及发展，并且是以相同的城市建筑基本理念为基础而进行的发展。城市的发展在所有阶段都遵循这样的原则，即不同的建筑风格呈现了一种亲缘关系，多样化与和谐统一之间形成一个整体，随着时间的更迭而更加丰富和美丽。然而工业化时代开始以后，城市发展的历史原则和思想渐渐被忽视，而屈从于对时尚形式、经济利益和交通技术至上的追求。

城市设计理念

这种城市建筑塑造的基本理念在今天是城市最大的财富和可持续性的发展潜能，而不是人们通常所强调的那些经济技术因素。作为汉萨同盟城市而建立的施特拉尔松城，后来凭借其造船业而闻名，并参与到现代工业化进程中，这些在今天却已然失去了意义。不过它至今仍保留着独一无二的城市特色，并以此为基础形成了一个稳定的新的经济分支——旅游业。

一个简洁明了的城市设计指导原则，包括对于现状景观的评价——城市景观分析、城市景观的未来发展——城市景观规划，都是不可或缺的，同样在市政基础设施特别是交通技术措施的制定，经济决策，或建设项目决策等过程——涉及高度、规模、形式和区位等要素的决策。

这样就形成了一系列的目标，作为评价现有城市景观，以及未来建设发展中的所有决策——批准建设项目以及颁发建筑许可等——的基础。

- 继承并保护城市景观的独特性
- 强化自然空间的引入
- 增加历史的可感知度
- 实现可能的现代化
- 使对城市空间的感知多样化
- 提高生活品质

城市景观分析：现状景观的描述和评价

城市景观分析描述并评估了城市景观的现有状况。它研究了城市景观的基本构成因素，如岛屿型区位和拱形的龟背形状地形，以及由低层的一般建筑和较高的水塔、市政厅、教堂等特殊建筑构成的城市立面，反映了普通与特殊相结合的塑造原则；城市平面包括一系列非常具有贯通性的街区类型，但是在具体的建设重点与广场等要素方面又有种种特殊之处；

城市空间形态，包括街道和广场空间特征，由"龟背"主轴线向水体延伸而成的肋形形态，以及纵向的曲线形街道空间，形成了对于水体、绿化或者特殊建筑物的多样化视角，从而具有特殊的空间特性；

城市的建筑形态，以有限的建筑类型——山墙型、坡顶型和阁楼型——为基础，形成城市景观的基础形态，与教堂等功能上、体量上、形态上具有特殊性的建筑相结合。我们同时对建筑类型的各重要特性进行分析，如比

- Preserve and transform the uniqueness of the townscape
- Include an increasingly natural civic environment
- Bring history to life
- Make the current life style visible
- Experience urban space in many ways
- Improve the quality of life

Townscape analysis: description and evaluation of the existing townscape.

The townscape analysis described and evaluated the existing urban form. The city of Stralsund is an island with topography similar in form to a turtle's back. The city's silhouette consists of low buildings overall, with intermittent higher protrusions such as a storehouse, town hall or church, representing the design principle of synthesis of the universal and specific. The urban layout is also composed of relatively homogeneous block typologies with the exception of a handful of special features such as squares or densely built up block fields.

The direction of urban space, as well as of streets and squares starts from a main axis, and is oriented primarily toward the water with longitudinal curved streets, which are characterized as a special visual axis leading through different zones of water, greenery, or specific buildings.

The structures in the general cityscape also conform to a limited number of representative building types that includes those with pediment, eaves and attic. However, there are other more specialized buildings, such as churches, that don't conform to this paradigm because of their specific function, size and shape. Building types were analyzed in terms of their prominent attributes, including proportion, zoning and structure of facade, as well as the position and proportion of openings such as doors and windows.

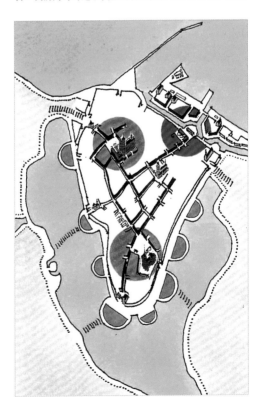
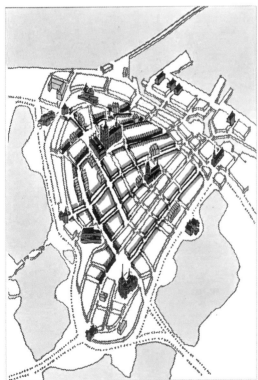

特殊建筑
Special building

街道形象代表建筑或是连续山墙建筑
Street image characterising building or gabled house series

山墙建筑出现的密度：一级
Frequency of the gabled houses: level 1

山墙建筑出现的密度：二级
Frequency of the gabled houses: level 2

山墙建筑出现的密度：三级
Frequency of the gabled houses: level 3

图 04 建筑类型分析
Fig. 04 Analysis of the building typology

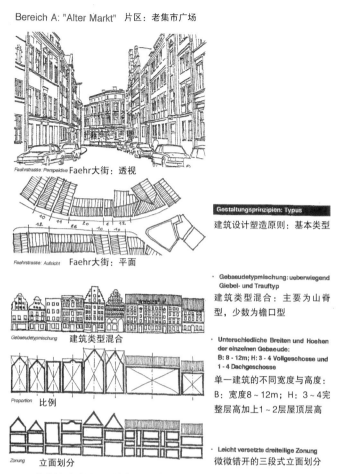

图 05 街道空间序列分析、建筑的划分和比例
Fig. 05 Analysis of urban space: building succession, structure and proportion

This process helped to highlight specific underlying design principles at work in the city's overall construction, including the use similar proportions throughout a wide variety of formats, building types and styles; be they Romanesque, Gothic, Baroque or Art Nouveau.

Finally, an investigation was done of the succession of buildings and facades that make up the quality of the streetscape. The inner city was also investigated in terms of urban design and subdivided into different areas. This analysis revealed a number of positive architectural components throughout the townscape, but assessment and presentation of many negative aspects also revealed many urban design failures, from empty lots and ugly facades to false proportions and density.

Townscape planning: from vision to concrete measures

The townscape planning of Stralsund developed the three-dimensional future form of the city based on an urban development and architectural history, a vision, and an inventory and assessment of the current situation. The aim was not simply to preserve the specific urban architecture, but to further developed it as well.

例、立面分区、立面划分、门窗等洞口的位置和比例，及其所依据的设计原则。举例来说，以竖向开窗为基本比例特征，在各种建筑风格——无论是罗马式、哥特式、巴洛克式还是青年风格派——的各种建筑类型中，形成了无数的变化形式。

最后还对街道建筑景观中的单体建筑及立面要素共同组成的序列进行了研究，中心城市按照不同的塑造类型和区域类型分别进行了上述分析。这些对城市建筑构成要素的分析，显示了城市景观的积极因素。此外还对城市景观的负面因素——包括与城市景观指导原则不相符的情况以及城市建设中的一些错误情况——进行了分析、评估和表述，如空置地块、丑陋的立面、不恰当的比例与过于密集的建筑等。

城市景观规划：从愿景到具体措施

施特拉尔松市的城市景观规划基于城市建筑的历史、城市形象提出指导原则与对现状的评估分析；其目标在于

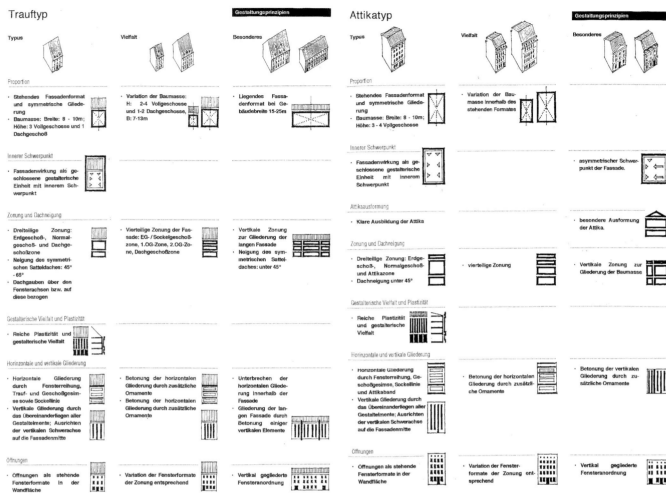

图 06 建筑类型分析：山墙型、坡顶型和阁楼型
Fig. 06 Analysis of the building typology, type of pediments, type of eave and attic

对未来的城市景观进行三维尺度的设计塑造，不仅仅要保护施特拉尔松市独一无二的城市建筑形态，同时还要有所发展。

城市形态设计导则

主要设计目标是增强岛屿特有地形特质：通过对建筑高度的严格限制来控制城市天际线效果，尊重城市不同区域的特色，保护和延续闭合的道路空间特性，以及路面铺装、照明、景观植被和街区内院的设计。

建筑形态设计导则

从建筑的层面来说，要为普通建筑和特殊建筑的形态提出相应的设计建议，以及为未来可以采用的建筑类型制定详细的设计导则。基本形态包括：山墙型、坡顶型、阁楼型。导则内容包括建筑比例、屋顶形态、分区、分段以及立面洞口比例、材料和色彩等各个方面。

对于普通的观者来说，街道中的建筑设计和城市空间设

Design guidelines for the townscape

The major urban design goals were to strengthen the topographical character of the island, to control the skyline by strictly limiting building height, to respect the unique nature of individual areas within the city center, and to preserve and develop the closed character of the urban space; including the design of street paving, lighting and vegetation and the design of block interiors.

Design guidelines for buildings

On an architectural level, design guidelines were elaborated for general building typologies such as pediment, dormer, eave and attic, and for other special types; then, more detailed design principles were developed. Guidelines range in scope dictating various elements from building proportion, molding of pediments, zoning and structure, to the proportions of openings, material and color of facade.

The typical observer on the street will see the architecture and urban form as one unified element of the city. This is why the design guidelines must work within these categories to unify the townscape and its individual buildings within each individual street space.

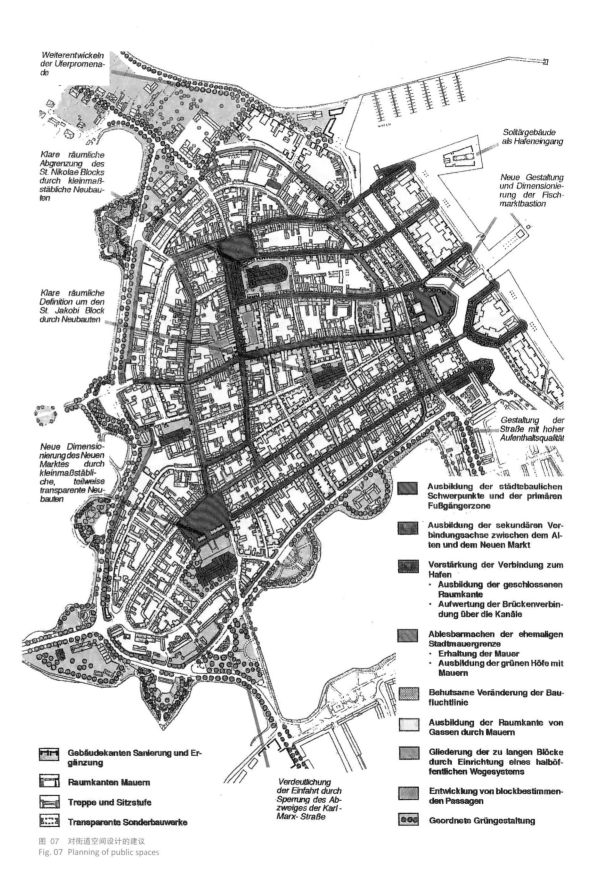

图 07 对街道空间设计的建议
Fig. 07 Planning of public spaces

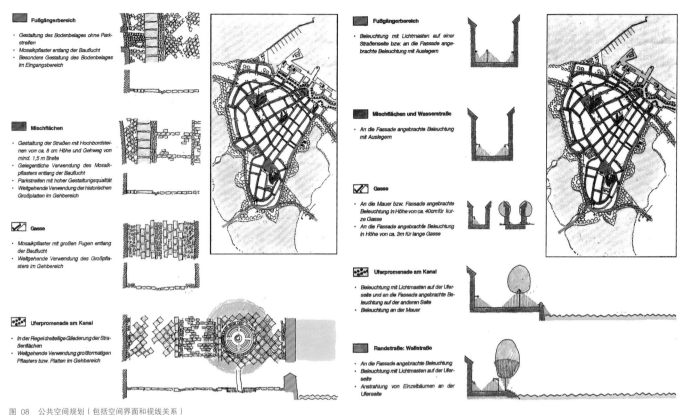

图 08 公共空间规划（包括空间界面和视线关系）
Fig. 08 Design guidelines for street pavement, lighting and planting

计是一个整体，因此必须为每条街道分别提出城市设计和建筑设计两方面的导则。

建筑类型、立面拼接序列的确定及示范设计

设计导则的基础不仅包括清晰的城市空间设计指导方针，也包括对建筑类型进行详细定义的图纸和文本，例如准确说明什么是坡顶型、什么是山墙型等。改建和新建建筑不应该模仿历史风格来建造，而应该是与时代相符的现代建筑。为了验证导则与现代建筑匹配程度，我们系统化地进行了多处示范性建筑设计。整体而言导则既针对建筑单体也针对其共同构成的立面序列，但建筑从来不是单独出现的，而永远是某个建筑群体的组成部分，作为立面序列中的一员，因而也需要制定对建筑群体的控制导则，避免出现单调的或混乱的群体形象。

城市设计和建筑设计的导则只针对普通建筑，而不适用于教堂、船坞、博物馆或其他重要公共建筑，这些建筑的设计要求更为特殊，需具体情况具体分析，无法提出笼统的设计要求。

汉萨城市施特拉尔松的城市景观规划以上报参议院的议案为最终版本，包括城市控制性规划、文字导则以及相关的高度和体量控制等规划图纸。在三十年后我们的回访中，施特拉尔松中心城区得到了良好的保护，直到今天仍然没有高层建筑。

Definition of building types, building successions and test design

The basic principle of these design guidelines was therefore meant not only to provide a clear instruction on the urban space itself, but also to establish a detailed definition of building types. The inquiry needed to answer the questions of what is meant by eave type, pediment type, etc. Any renovated or new buildings were not to be built in a historical style, but should instead utilize more a contemporary architectural mode. In order to verify the span and width, any new architecture concepts were systematically subjected to the established design guidelines. These guidelines applied to individual buildings as well as successions, since buildings are never perceived alone but as a sequence of individual facades. The recommendations included clear design guidelines developed for building successions that were intended to prevent both monotony and disorder.

It was crucial however, that the planning and architectural design guidelines be applied only to functional everyday buildings like homes, and not to more specialized buildings such as churches, hangars, museums or other facilities of public interest. Updates to these more specialized types of buildings should not be generalized, but dealt with individually due to their special position in the larger urban context.

The townscape planning of the Hanseatic city of Stralsund ended with the submission of an urban development master plan to the city senate. This master plan included the previously discussed building height control concept; which is the reason no skyscrapers exist in Stralsund to this day.

> 为了验证导则与现代建筑匹配程度，我们系统化地进行了多处示范性建筑设计。
>
> *In order to verify the span and width, any new architecture concepts were systematically subjected to the established design guidelines.*

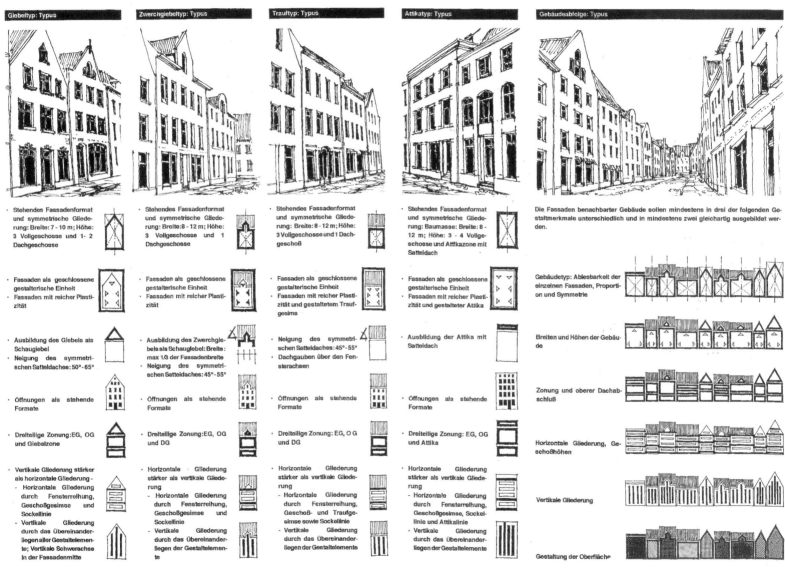

图 09 建筑单体的设计建议
Fig. 09 Design guidelines for individual building types

城市保护及管理　URBAN PRESERVATION AND MANAGEMENT

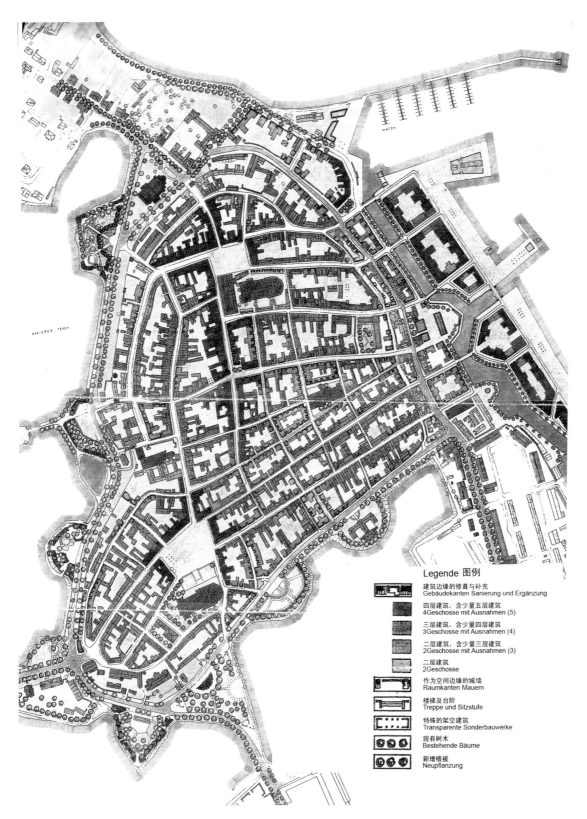

图 10　城市景观规划：街道和广场空间、绿化、高度控制和体量控制
Fig. 10　Townscape planning: master plan

城市保护及管理
URBAN PRESERVATION AND MANAGEMENT

01 世界文化遗产城市施特拉尔松：城市景观规划
World Cultural Heritage City of Stralsund: Townscape Planning

02 新勃兰登堡城市景观规划
Townscape Planning Neubrandenburg

03 世界文化遗产城市波茨坦瑙恩郊区及耶格尔郊区的设计导则
Design Code for the Nauener Vorstadt and the Jaegervorstadt in World Cultural Heritage City of Potsdam

04 泉州市法石街区保护与整治规划及环境设计
Urban and Landscape Design for the Fashi District in Quanzhou, Fujian

05 泉州市城市新区公共空间控制规划方法研究
Research of the Planning Methods for Open Space in Quanzhou, Fujian

02 新勃兰登堡城市景观规划
Townscape Planning Neubrandenburg

规划面积 AREA		60 hm²
人口规模 POPULATION		66000
完成时间 PROJECT DATE		2000.06

问题及任务 The task of townscape planning

图 01　新勃兰登堡：一个位于内陆的海港城市
Fig. 01 Neubrandenburg: a seaport city in the midland

The townscape planning of Neubrandenburg was developed for the historic inner city of that municipality. It is a "new town" that was founded during the middle of the 13th century, as early as 1248. The townscape of the city is unique, distinct and has an almost elegant, cool and clear atmosphere. But these things could have been jeopardized by rampant modernization and construction measures if not regulated or coordinated throughout the process of a long-term townscape design.

新勃兰登堡城市景观规划是为新勃兰登堡的市中心量身定做的。虽然新勃兰登堡在12世纪中期（1248年）就建立了，她是作为一个"新城"被整体规划和建成的。由此形成其城市景观风格独立明晰，具有良好的可识别性，不会被混淆，而且她拥有一个近乎完美的、凉爽、清新的自然环境。如果个别措施不按照一个长期的城市景观构想来实施和配合，这些城市与环境的改建和重建会由于将来各种各样的现代化措施而受到威胁。所以这个城市景观规划面临一项和施特拉尔松市相似的任务——展现城市面貌的个性与品质。

图 02　过去的城市景观
Fig. 02 Townscape in the past

The aim of the townscape planning was to coordinate, in combination with an urban framework plan regulated mainly by a land use and traffic system strategy, various public and private construction measures according to an overall design concept for the city of Neubrandenburg. The task was therefore to develop a long-term urban design concept based on a comprehensive analysis of the townscape that consists of elements such as city skyline, city layout, public spaces and building typology. The individual characteristics of those elements served as the starting point for future development. The essential contents of this townscape plan are design guidelines that include statements not only about the city layout, city skyline and public spaces, but also about the design of private and public buildings regarding remodeling and new development. Building types, height and mass of structures, facade design, openings, roof types, as well as materials and colors would have to be regulated.

城市景观规划的目标是，与都市发展空间框架规划——尤其是在土地使用规划和交通规划方面——共同发挥控制作用，以总体城市景观框架为基础，对新勃兰登堡内城的众多官方和私人措施进行协调。因此本次规划任务是在全面的景观分析基础上形成一个长期的城市景观体系，将新勃兰登堡的城市景观特征——城市天际线、城市肌理、城市空间和建筑类型——作为未来发展的出发点。这个景观规划的重要内容是制定设计方针，不仅有对城市肌理、城市天际线和城市空间的控制，也特别包括了对改建及重建的私人和公众建筑的塑造要求，具体包括建筑类型、高度和体量划分、立面形式、开口、屋顶形态、材料和色彩。

图 03　典型城市景观及典型现代化城市景观对其影响
Fig. 03　A typical townscape with atypical modern architecture

城市景观规划必须和都市发展规划一起，在城市景观维护及未来发展方面作为地方行政、经济和法律层面的调控基础。换句话说，就是服务于建设计划评估，协调公共空间的塑造措施，以及对更多的相关个别措施进行控制。所以景观规划的结果就是要将这些设计建议应用于内城的实践建设，可以直接作为市政府日常决策（例如对建设意向和咨询项目的评估）的基础。除此之外从城市景观规划延伸进行进一步的深入研究，如街道和广场空间设计、街区研究、色彩规划或者景观和广告法规制定等。

Together with the urban framework plan, the townscape planning would serve as a legal basis for the preservation and development of the existing urban landscape and guide the decision-making process of local government. Individual building projects and permissions would be controlled, diverse implementation measures in public spaces coordinated, and large-scale individual projects regulated. The results of the townscape planning would be conceived as a series of practical design guidelines for the whole city. They could then be applied to daily planning and municipal decision-making, such as in the assessment of an application for planning permission or in general guidelines for diverse projects. Further detailed studies could be derived from this townscape plan if appropriate, such as street and open space designs, block studies, color design as a part of urban design, and design codes for buildings and signboards.

研究方法 The Method

为形成一个可行的、有依据的以及能让每个人理解的对现状和对新勃兰登堡未来城市景观的完整表述，本次规划还借助计算机模拟技术应用了三维尺度上的城市景观分析和规划方法，这是意厦1970年以来，将研究工作与实践工作相结合发展而成的新实验。

In order to create a comprehensible and well-founded presentation of the current situation as well as the vision for Neubrandenburg's future townscape, specific methods of city analysis and townscape planning were developed. In fact, they had been developing through applied research on specific tasks and practical experience from the ISA since 1970.

城市景观历史：存在了800年的城市思想的力量

Townscape history: The power of a city idea, alive for 800 years

工作伊始首先进行了新勃兰登堡市城市景观的规划建设发展研究。对于城市建设历史研究的任务，一方面要注重其连续性，另一方面要展现数百年间城市景观的变迁。基本问题是：哪些城市景观的特征历久弥新，并且同所有其他城市相比是独特和不可替代的？哪些城市景观的特色已经随时间而消逝，哪些是新增的特点，这种变化在今天有着怎样的意义？

例如，在城市天际线中，教堂和城门这些元素保留了显著的中世纪特征；竖向形体的高层建筑可以作为积极元素，而超尺度的板式高层住宅则是不和谐元素。对城市街道和广场空间的发展历史的研究显示，空间序列在街道和广场不断相互转化的发展过程中原则上保持不变，例如在互交网络体系，内部街道微微弯曲、对建筑角部进行强调设计等特点，基本上被保留下来；但是随着城市的大规模扩张，街道与建筑的尺度提升，许多城市空

At the beginning of the project, the urban and architectural development of the historic inner city of Neubrandenburg was analyzed. The goal was to determine both the continuity of the townscape on the one hand and what changes have occurred over the centuries on the other. The essential questions were: what features of the townscape have remained the same throughout history and make the city individual and unique compared to other cities? What kinds of characteristics have been lost over time, which were added, and what is the impact of these changes today?

For example, the preserved medieval structures such as the churches and gates that rise prominently in the city silhouette have been evaluated as positive vertical landmarks. But the negative elements are the high-rise residential buildings in the form of "slabs" which cut into that silhouette. The history of the roads and public spaces showed that the spatial sequence of streets and squares remained basically the same. Characteristics such as

基本问题是：哪些城市景观的特征历久弥新，并且同所有其他城市相比是独特和不可替代的？哪些城市景观的特色已经随时间而消逝，哪些是新增的特点，这种变化在今天有着怎样的意义？

The essential questions were: what features of the townscape have remained the same throughout history and make the city individual and unique compared to other cities? What kinds of characteristics have been lost over time, which were added, and what is the impact of these changes today?

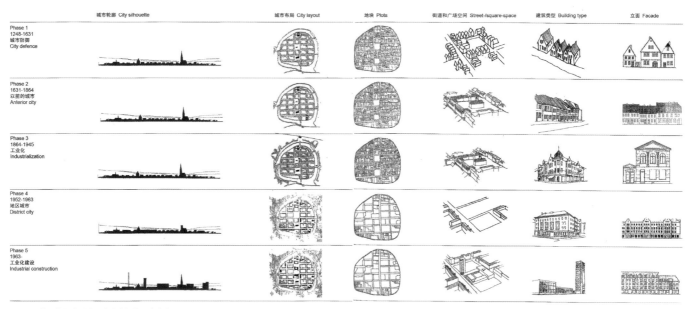

图 04 从13世纪直到今天城市建筑的历史演变
Fig. 04 History of urban architecture from the 13th century until today

the slightly curved building lines based on the rectangular grid system of the urban layout, and the accent of the building corner at block intersections are all still there. But due to substantial widening of streets at the same building height, many streets and squares appear much larger and have lost their spatial qualities.

Nevertheless, Neubrandenburg demonstrates a relatively clear picture of the urban-architectural idea and its positive and negative developments to date. Negative cases being where a measure did not contribute to the evolution of the townscape, but caused a disturbance instead.

Vision of townscape: to increase the quality of life in a city like champagne

Neubrandenburg has its own unique character that is reflected in the townscape. This is characterized by the natural situation of the city – determined by its social, economic and cultural history – which has been shaped in part by its architecture. The crucial point is that the shape of the city is created by the interplay of landscape, urban layout, city skyline, streets and squares, building types and characteristics of the facade.

For decades, not enough attention was being paid to the townscape, so tasks such as the reconstruction and maintenance of structures or the general improvement of infrastructure were of top priority. Today the

城市景观塑造是决定性的因素，它由自然景观、城市平面、城市天际线、街道和广场空间、建筑类型和立面特征共同作用形成。生活品质与城市景观的品质密切相关。

The crucial point is that the shape of the city is created by the interplay of landscape, urban layout, city skyline, streets and squares, building types and characteristics of the facade.

间的尺度被扩大了，失去了其空间品质。

整体而言研究的结果形成了一幅城市规划和建筑设计发展的清晰图画，以及至今的积极和消极的城市景观的状况。消极方面尤其表现在一些城建措施不仅不利于城市景观的发展，反倒对其造成了困扰。

城市景观目标：提高城市生活质量

新勃兰登堡具有独一无二的特色，并被反映在城市景观上。其以城市的自然景观为基础，由城市的社会、经济和文化历史所确定，并以建筑形态为体现。城市景观塑造则是其中决定性的因素，它由自然要素、城市平面、城市天际线、街道和广场空间、建筑类型和立面特征共同作用形成。

这一城市景观体系在过去的几十年发展中没有获得足够的重视。城市建设优先考虑的是重建、旧建筑维修或基础设施的改善等。今天，城市景观正变得越来越重要。市民再次将城市景观作为其家乡的明显象征，城市面貌具有重要价值。人们越来越清楚，一个地区的吸

图 05 城市单体与城市天际线
Fig. 05 Urban body and city skyline

城市保护及管理　URBAN PRESERVATION AND MANAGEMENT

Straßenraum 街道空间

Das Querprofil stellt sich weitgehend geschlossen dar.
Im östlichen Bereich jedoch sind offene Raumkanten prägend, im Norden durch fehlende Bebauung und im Süden durch den Marienkirchplatz. An der Marienkirche wird der Straßenraum durch Großgrün gefasst. Zur Stadtmauer hin nimmt die Gebäudehöhe ab. Auf der südlichen Straßenseite ist gleiches auch für die Gebäudedichte zu verzeichnen.

Der Raumabschluss ist an der Westseite durch Mauer und Großgrün gegeben. Nach Osten erfolgt der Übergang in die Neutorstraße.

Gebäude 建筑

Als Baukörper treten überwiegend 3-geschossige Gebäude in Erscheinung. In der Länge ist zwischen den Blockrändern der 50er Jahre und der Kleinteiligkeit auf historischen Parzellen am westlichen Ende zu unterscheiden.
Dort sind Satteldächer typisch, ansonsten ist die Dachlandschaft durch Walmdächer geprägt.

Östlicher Abschnitt　Westlicher Abschnitt

Bauflucht 建筑基线
■ weitgehend geschlossene, regelmäßige und geradlinige Bauflucht
■ leichte Verjüngung an den Straßenquerungen
■ Hochhaus Waagestraße als störender Höhensprung

Horizontale Gliederung 水平划分
■ durch ausgeprägte durchlaufende Kanten, Fensterreihen, vereinzelt Fensterbänder

Vertikale Gliederung 垂直划分
■ durch Gebäudekanten in großen Abständen
■ vereinzelt vertikale Fensterbänder

Grünbereiche 绿化
■ Großgrün im Bereich der Marienkirche (Linden), gegenüber Kastanien
■ Vorgärten im nördlichen Bereich des Westendes und auf der südlichen Straßenseite auf Höhe des mittleren Quartiers jeweils mit Baumreihe
■ z.T. begrünte Fassade
■ Sichtbeziehung zum Großgrün des Walls

Stadtmöblierung 街道小品
■ Straßenbeleuchtung beidseitig
■ Verkehrsschilder und Müllbehälter

建筑基线 Building line

水平划分 Horizontal formation

垂直划分 Vertical formation

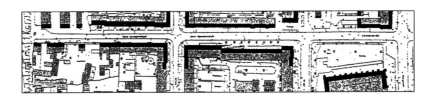

图 06　城市景观分析范例：基准线、外观街接、绿化及街道小品
Fig. 06　Example of a townscape analysis: building line, facade structure, greenery and furnishing of a street

Öffnung 开口
■ Lochfassade
■ überwiegend gleiche Fensterformate

Proportion 比例关系
■ liegende Rechtecke, im Verhältnis bis max. 1:5

Zonierung 分区
■ ausgeprägte Traufkante
■ durch Gesims getrenntes Ober- und Untergeschoss

Farbe und Material 色彩与材料
■ helle Putze in Pastellfarben
■ dunkle Betondachsteine
■ weiße Kunststoff- und Holzfenster
■ Holztüren

开口 Opening

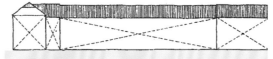
比例关系 Proportion

分区 Zoning

图 07　城市景观分析范例：立面开口、比例、分区、色彩、材质
Fig. 07　Example of a townscape analysis: openings in facade, proportions, zonings, colors and materials

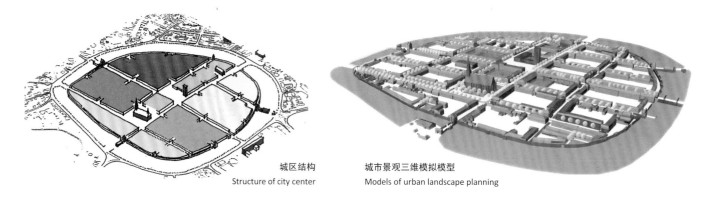

城区结构
Structure of city center

城市景观三维模拟模型
Models of urban landscape planning

图 08 城市景观规划：用地结构与功能
Fig. 08 Urban planning: the field structure

townscape has become increasingly important. The citizens once again recognize the townscape as a visible symbol of their hometown. It has become increasingly clear that the attractiveness of a community – even in economic terms – is not only dependent on the number of employment opportunities, shops, building sites or recreational facilities, but also on the quality of life in general.

The quality of life is closely linked with the design quality of the city. It is the result not only of the economic, social, and technical provisions of a city, but also its cultural and aesthetic quality. In terms of design, the historic inner city of Neubrandenburg was still unsatisfactory in some parts and aspects. To improve the quality of life, the city needed to realize new design ideas as an essential part of urban revitalization, with increasing the attractiveness for downtown residents, workers and visitors as top priority. The primary aims of the townscape in regard to developing design quality was to strengthen the unique identity of the city, increase the visual attractiveness, improve the quality of stay and the feeling of orientation, and to keep the urban history alive.

There was a lack of identity in some of the architecture and among parts of the city center. Street spaces were unattractive, facades were faceless and interchangeable, and vacant plots led to unattractive streets without any unique identity. Therefore it was necessary to add new identifiers, such as attractively designed buildings, interestingly designed streets, redesigned block interiors, new facades, attractive lighting and updated materials and colors. Yet, everything always had to be part of an overall design concept. It was therefore necessary not only to preserve the unique character of Neubrandenburg, but also to give each city district and their components an increasingly new and modern identity.

Also, the inclusion of visual motivation proved to be necessary. People need some inspiration, which implies diversity and ambiguity of place. Such inspiration must be experienced in a multi-faceted way, both in order to continuously discover something new at the same place; but also in order to change over the course of time, through the use of familiar and timeless elements or contemporary additions. Diverse overlapping functions, activities and land uses support the design of facades and public spaces with multi-layered structures giving a rich visual ambiguity. Meanwhile the street furniture – such as bollards, street lighting and tree arrangements – provides a second spatial structure and increases the sense of visual excitement. The visual excitement of certain neighborhoods, streets and buildings in the inner city was not sufficient and produced

引力——包括经济方面的吸引力——不仅取决于工作位置、商业、用地或娱乐设施，而且还取决于普遍的生活品质。一个城市的文化遗产、城市现有的或未来规划的城市景观都属于上述软性吸引力的范畴。

生活品质与城市景观的品质密切相关。它不仅是一个城市经济、社会、技术方面的体现，也是其文化和审美品质的体现。新勃兰登堡内城的城市景观塑造还有一些方面不尽如人意。作为城市品质的重要组成部分，提高人居品质是内城设计的原动力。增加新勃兰登堡内城对城市居民、劳动者和游客的吸引力，乃是当务之急。为维护和提高城市景观品质的城市景观设计的重要目标包括：加强城市的独特性，增加视觉吸引点，改善逗留品质，缓解内城压力，复兴城市历史，使其焕发出勃勃生机。

就现状而言，内城的部分地区或元素并未体现出城市的独特性。街道空间没有吸引力，部分建筑立面缺乏个性，部分用地空置，导致道路空间缺乏特色。因此，有必要增加新的地标，如美观的建筑、有趣的街道、重新设计的街区内部空间、新的立面、有吸引力的照明、新材料和色彩——作为整体城市景观设计理念的一部分。通过这些手段，不仅能保留新勃兰登堡的独特性，而且也能给每个城区及其建设项目引入新的现代影响。

此外，增加视觉吸引点的想法被证明是必要的。人们需要新的感知刺激，地区的多样性和多义性是以多元化的感知为前提的。视觉的刺激不仅在多种层面都可被感受，在同一个地方也总能带来新的发现，并且可以随着时间改变——通过使用所留下的痕迹以及随时间推移进行的完善。功能的重叠，多样性的活动和功能分区，服务于多层次的立面塑造及道路、广场空间设计，交通设施、路灯、树木通过透明的空间界定提高了视觉印象的多样性。内城的一些居住区、街道和建筑，都存在相对类似缺乏视觉感知品质的问题，因而给人以空虚、单调、死气沉沉的印象。富有魅力的城市景观最重要的

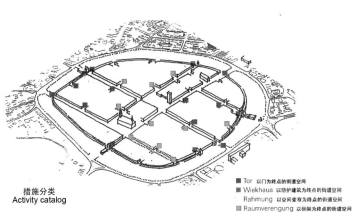
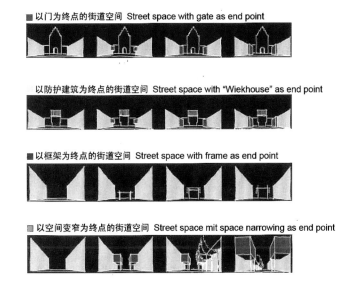

措施分类
Activity catalog

- Tor 以门为终点的街道空间
- Wiekhaus 以防护建筑为终点的街道空间
- Rahmung 以空间变窄为终点的街道空间
- Raumverengung 以框架为终点的街道空间

图 09 城市景观规划：焦点
Fig. 09 Urban planning: focal points

"机密"，在于多层次多种含义的视觉感受。因而重要的城市景观目标之一，就是通过增加视觉体验的可能，来形成生动的、令人兴奋的甚至是富于神秘感的城市景观效果。对此，精心设计的街道空间与富于吸引力的、细节精巧的立面能够发挥同样的作用。

因此，城市景观规划的指导原则，是通过对城市景观进行客观、清晰的提炼，发展出一种令人激动的理性。这意味着将新勃兰登堡的城市景观提升为一种独特的美丽，如同葡萄酒发酵成香槟时产生的飞跃，途径是：将新勃兰登堡景观塑造特质中的对立面进行有意识的叠合——一面是浪漫的理性，另一面是现代的功能性；流淌于历史中的现代城市文化，被天然壮丽的风景所包围的人造艺术世界。目标是：呈现简单中的多样，平静中的生动和开放中的封闭。

城市景观分析：物质和非物质的城市解读

我们同样为城市景观进行了"SWOT分析"。在本次城市景观规划中，对新勃兰登堡的每条街道都进行了详细的分析，包括对现状情况的中立性的分析，城市居民及游客对街道印象（气质、氛围和意象）的描述，通过与城市景观目标的比较而对其积极和消极特性作出评价，并对改善城市景观的机会进行表述。

现状情况中立性的分析要素包括城市空间剖面、建筑基准线、建筑序列的水平和纵向分区、绿化区、城市小品特殊的空间形态和建筑形态。建筑分析要素包括建筑类型、开口、分区、色彩和材质。此外还有从物质和非物质的角度对城市印象的描述，例如功能，以及街道或广场空间的个性、气质、氛围的塑造。以上述各项因素作为积极和消极特性的评价基础，从工作内容上涵盖了从功能、交通技术、前庭设计以及建筑单体的形式和体量等各方面。

feelings of emptiness, monotony or sterility. One of the most important "secrets" of attractive townscapes is the ambiguous complexity of visual experience. A major goal of urban landscape revitalization is thus to increase the visual possibilities and make the townscape vivid and inspiring, even mysterious. Carefully designed street spaces, as well as affectionate and visually attractive details of facade design, can contribute to that goal.

The guiding idea of townscape planning and design therefore is to take the sober clarity of an urban landscape and develop it into "sparkling" sobriety, by providing the existing townscape with a specific beauty, like the beads of bubbles that arise when wine turns into champagne. The contrasting properties of Neubrandenburg have been overlapped consciously – for example, romantic sobriety on the one hand, modern functionality on the other, history intertwined within history, and magnificent landscapes surrounded by an artificially constructed world – to create diversity in the simplicity, liveliness in peace and seclusion in openness.

Townscape analysis: how to understand the material and immaterial city

A townscape analysis is not that much different from a primarily visual "SWOT" analysis of the current state of a city. Within the townscape planning of Neubrandenburg, it proceeded street by street. Initially it consisted of a neutral inventory, then a description of the impressions – i.e. mood, atmosphere and memorable images – from residents and visitors, followed by an assessment of their positive and negative properties which are measured in terms of the goals of the townscape and how many opportunities are given for improvement.

The parameters of a neutral inventory of public space include such elements as cross-sections of streets, building lines, horizontal and vertical zoning of building facades, along with roads, greens, street furniture and special figures such as a building mass or form that differs from the average. The parameters of a building inventory, on the other hand, primarily cover building type, facade openings, facade zoning, color and material. The description of impressions involves both material and immaterial aspects, functions, design, character, mood and atmosphere of the street and square spaces. The assessment of positive and negative aspects is based on the same parameters as above. The atmosphere of a street, its functional and technical aspects, the design of a front garden and the form and mass of a building are what should be evaluated.

一个中心城市景观分析的结构与对城市整体景观结构的"SWOT分析"是同一源泉与结构。

A townscape analysis is not that much different from a primarily visual "SWOT" analysis of the current state of a city.

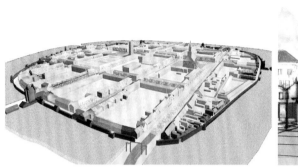  

图 10 整体布局：局部街道的实验性设计
Fig. 10 Townscape planning: facade design

Townscape planning: casting a vision of the future townscape in words and pictures

城市景观规划：以文字和图片展示未来城市景观

The townscape is always the overall effect of a three-dimensional urban structure, of an "urban body," the spatial structure of streets and squares as well as the architecture of the buildings. Therefore, taking the following steps enhanced the Neubrandenburg townscape plan.

一个城市的城市景观，是三维尺度的都市结构整体，一个城市的有机体，是以三维的街道、广场、空间结构为整体系统而形成的城市建设结构。因此，新勃兰登堡城市景观规划也是分步骤、循序渐进铺开的。

规划的深度达到了可以清楚地表示出每一个地块内部建筑布局指导方式。

The strictness of the site plan is chosen in such a way that allows a very precise containment of the planning statements per plot.

At the beginning there was the "urban body as a whole," then its division into individual separate areas that differ from each other in a creative way – that is to say each area has a unique character within the whole all-encompassing townscape. This was followed by open space planning for the whole: the creation of interesting spatial sequences, design principles for the entire urban space system, from urban structure to street furnishings, to the design recommendations for particular buildings as well as general design principles for individual buildings.

In doing so, all building types in the city were divided into four categories. For each category there are different suggestions regarding preservation and design principles. These measures are not compulsory law, but serve as inspiration in order to easily identify the nature and scope of design possibilities. In each specific case, a separate draft must then be developed. The townscape planning provides a framework for the position, nature and scope of design possibilities.

首先是城市总体层面的景观设计，而后是各区域的分别塑造，也就是说在总体城市景观的基础上为各个区域形成各自的特性，以及进行整体公共空间设计，寻求塑造富于趣味的空间序列的措施，建立总体城市空间系统乃至城市小品的设计原则，单体建筑设计导则和特殊建筑设计建议。其中内城建筑被划分成四个类别，每个类别相应有专属的维护、设计举措，这并不是强制性的规定，而是作为一种设计指导，让人能够更加清晰地理解各种可能的设计方式和范围，在此基础上，每个具体方案的建筑师再进行单独的设计。

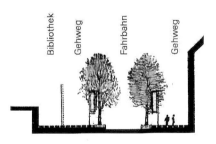

a-a
Stargarder 大街，HKB片区图书馆向北看
Stargarder Street, HKB-area library to the north

b-b
Stargarder 大街，集市广场向北看
Stargarder Street, area market square to the north

c-c
Stargarder 大街，圣玛利亚大教堂向北看
Stargarder Street, area Marien Church to the north

图 11 城市规划：街道横截面
Fig. 11 Townscape planning: cross sections of the street

城市保护及管理　URBAN PRESERVATION AND MANAGEMENT

图 12　不同区域立面设计的实验性设计
Fig. 12　Test designs for facade design in various areas

总而言之城市景观规划的成果，对建筑及街道空间设计建议进行了总结。它是为新勃兰登堡市中心的每一条街道制定的，包括对街道未来力求形成的特色的表述，街道空间的设计，以及建筑的设计建议，特别是建筑立面及其材料和色彩设计。

导则的目标是：以一种具有新勃兰登堡特色的独特的形式提升内城的感知度和逗留品质。它们是如此的详细和具体，因此它们既可以作为所有城市、城市设施以及公共建设方面的指导原则，又可以作为私人建设咨询和决策的基础指导原则。其中文字和平面图一起形成各街道的设计建议，照片和草图作为辅助指导。规划平面清楚地表示出每一个地块的深度。

The results of the townscape planning were then summarized in guidelines for building and street space design. They were developed for each street in downtown Neubrandenburg and included statements about the requested future character for that area, the design of the public space, along with its peculiarities. In addition, guidelines were made on the design of the individual buildings, including the facades, with regard to materials and colors.

The aim of these design guidelines was to enhance the experience of Neubrandenburg in a characteristic and unique way, and to improve the quality of stay. They were so detailed and particularly developed that they could be used both as a guideline for city building, municipal facilities and public issues, as well as a basis for advice and decision-making for all actions by private sectors.

Thereby, the text and site plan map form a legitimate set of guidelines for each street. Some photos and drawings are only illustrative in nature. The strictness of the site plan is chosen in such a way that allows a very precise containment of the planning statements per plot.

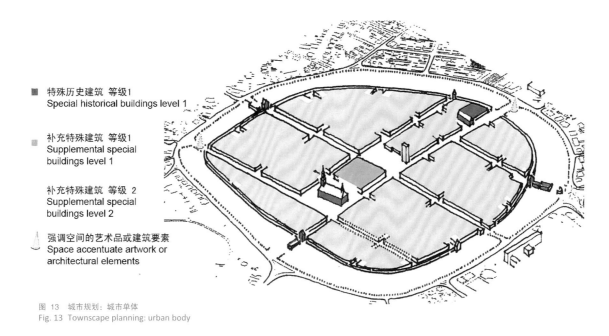

■　特殊历史建筑　等级1
　　Special historical buildings level 1

■　补充特殊建筑　等级1
　　Supplemental special buildings level 1

　　补充特殊建筑　等级 2
　　Supplemental special buildings level 2

▲　强调空间的艺术品或建筑要素
　　Space accentuate artwork or architectural elements

图 13　城市规划：城市单体
Fig. 13　Townscape planning: urban body

城市保护及管理
URBAN PRESERVATION AND MANAGEMENT

01 世界文化遗产城市施特拉尔松：城市景观规划
World Cultural Heritage City of Stralsund: Townscape Planning

02 新勃兰登堡城市景观规划
Townscape Planning Neubrandenburg

03 世界文化遗产城市波茨坦瑙恩郊区
及耶格尔郊区的设计导则
Design Code for the Nauener Vorstadt and
the Jaegervorstadt in World Cultural Heritage City of Potsdam

04 泉州市法石街区保护
与整治规划及环境设计
Urban and Landscape Design for the
Fashi District in Quanzhou, Fujian

05 泉州市城市新区公共空间控制规划方法研究
Research of the Planning Methods for
Open Space in Quanzhou, Fujian

03 世界文化遗产城市波茨坦瑙恩郊区及耶格尔郊区的设计导则
Design Code for the Nauener Vorstadt and the Jaegervorstadt in World Cultural Heritage City of Potsdam

规划面积 AREA		360 hm²
人口规模 POPULATION		64000
完成时间 PROJECT DATE		2000

从两种文化中学习　LEARNING FROM TWO CULTURES

设计导则的必要性 The necessity for a design code

The city of Potsdam as artwork

Potsdam, the former home of Prussian kings and later the German Emperor, located a few miles southwest of Berlin, is today the state capital of Brandenburg. The city covers an area of approximately 190 square kilometers with a population of roughly 157,000 people. But above all, the city is famous for its numerous monuments and gardens, which collectively, comprise a unique "artwork" made up of area castles, landscapes and the city itself. In 1990, this entire ensemble was included on the UNESCO list of World Cultural and Natural Heritage sites.

The history of preservation areas

Immediately after reunification, the city of Potsdam attempted to create a future-orientated development that was in accordance with the culture of the city's existing historical buildings. In order to carry this out necessary plans were created, e.g. a monument and an urban development plan which served as regulation tools for urban development. In this context, the preservation statutes for the Nauener Vorstadt and Jaegervorstadt were also developed.

The area concerned is characterized by its special location in the immediate vicinity of Sanssouci Park and the baroque town center. This area offers a distinctive and recognizable municipality determined by its unique cultural and spatial development, individual history, distinctive multifunctional facilies and also - not insignificantly - its broad mix of architectural styles from post-classicism through historicism and eclecticism up to art nouveau. The majority of the buildings in the preservation area were built over a period beginning in the mid-19th century and ending with the close of that century.

在两德统一并列入人类世界文化遗产这一转折点之后，波茨坦就开始寻求一种与历史的建筑文化相协调的面向未来的发展模式。
Immediately after reunification, the city of Potsdam attempted to create a future-orientated development that was in accordance with the culture of the city's existing historical buildings.

A great number of these historic buildings have survived largely untouched, imbuing the area with a strong character. Most buildings are residential, having been originally constructed mainly for officers and senior officials. They represented an alternative to the contemporary upper-class mansion of the day, ranging in form from detached villas to elaborately decorated apartment buildings. However, despite these various building types and styles, a homogenous townscape of high quality has evolved over time, thanks in large part to relatively uniform building mass, similar subdivision practices and the arrangement of buildings.

Development aims for the preservation areas

The pressure to invest in two preservation areas rose sharply due to their prominent location in the city after reunification. The owners' understandable desire to modernize the buildings, rebuild roofs and to make use of undeveloped land had to be complied with, and was necessary to maintain a vibrant city that adjusts to the needs of the present day. On the other hand, the history, it was important that the architectural culture and artistic value of the structures not be sacrificed in the name of short-term economic goals. In the long run, the townscape of Potsdam still represents an important world capital.

For these reasons, the city of Potsdam decided to protect the image of the two areas by drawing up preservation statutes. The concept of the ISA urban planning studio is based on the following three principles:

作为整体艺术作品的波茨坦

波茨坦市曾经是普鲁士王国和德意志帝国的皇宫所在地，位于柏林西南方向几公里，如今是勃兰登堡州首府。城区占地近190平方公里，居民达到15万。城市因拥有众多的文物古迹和公园而著名，这些古迹和公园塑造了一个由城堡、山水和城市组成的无与伦比的整体艺术作品。1990年，整个城市建筑群被联合国教科文组织列入人类世界文化遗产和自然遗产名录。

规约范围的发展历史

在两德统一并列入人类世界文化遗产这一转折点之后，波茨坦就开始寻求一种与历史建筑文化相协调的面向未来的发展模式。因此就必须进行一些必要的规划来控制未来的发展方向，包括城市发展规划和文物保护规划——瑙恩郊区和耶格尔郊区的设计导则同样也是在这个框架内进行的工作之一。

规约区域以其紧邻无忧宫（Sanssouci）公园和巴洛克式城中心的特殊地理位置而显得尤其重要。区域通过其独特的文化区域发展、特殊的历史、独特的功能混合，特别是通过其独特的从后古典主义到历史主义、折中主义直至青年风格派的多种建筑形式，形成了一种具体唯一性的、令人印象深刻的城市意象——混合性风格——虽然其建设的主体部分产生于19世纪中后期。

这些历史建筑大部分保留了下来，并对区域产生了深刻影响。这些建筑主要是住宅，它们原本主要是为军官和高官准备的，是处于独立别墅和公寓之间的一种上层阶级别墅，并发展形成了直至奢华公寓的多种过渡形式。即使有这些不同的建筑类型和风格流派，因其相对统一的尺度，划分的形式和建筑次序，还是形成了一种高品质的整体的城市意象。

规约区域的发展目标

两德统一之后，两个规划区域内的投资压力都因其在城市中的突出位置而大大加强。为了创造一个有活力的城市，以不断适应新条件，所有者希望建筑物更现代化、重修屋顶、利用空置场地，这些合理的愿望应该得到支持。而另一方面，也不能草率地忽视区内建筑的历史、文化和艺术价值，使之成为谋求短期经济利益的牺牲品。因为长期来看，城市景观意象正是波茨坦的一项重要资产。

出于此考虑，波茨坦政府决定通过制定设计导则来保护这两个区域的意象。对此，德国ISA国际设计集团提出以下三个原则：

Fig. 01 Typical building types

- 让历史重现：通过逐步保护性的更新，重现相应区域的建造历史。不仅仅是因为我们这个时代的快速变革需要与历史的稳定联系，以找到定位和识别的依据；只有与过去对比，进行研讨，才能以一种面向未来的方式应对现在。

- 保留和进一步发展其城市意象的独特性：城市更新不仅仅要考虑功能性和实际可用性，也要考虑以合理的形式和塑造来突出其识别性。这就是说，应该保留、强调、进一步发展现有建筑和城市建设的特征，并允许形成新的特征以形成原有特征的新生。

- 建立与现在的联系：一个城市的发展介于延续和改变之间。城市——作为有机体——不应该保持在被建成时的状态，而是应该不断地发生一些小变化，城市不是博物馆，而是现代社会有活力的一部分。除了过去，现在的情况也应该体现在历史片区。

在相应区域不应该过于鼓励某种特定建筑风格或历史性的建筑。旧城的品质和魅力在于几百年对应不同时代所形成的相应的新建筑。

现有的历史建筑应该被保留和更新，必要时也可以通过补充与恢复重建进行功能的完善。而缺失的建筑应该根据现有的街区结构谨慎建造。

- Make history come alive: The constructed history of the preservation areas is to be kept intact by piecemeal renewal of the town. But amid today's barrage of rapid changes, will need to stay connected to the past in order to have a reference point for orientation and identification. Only by confronting our past will the basis for navigating our future developments become clear.

- Preserve and develop the townscape's uniqueness: In the renewal of the town, not only the applied functionality and usability should be reflected, but the form must also be considered in relation to civic identity. This means preserving the existing architectural and urban structural characteristics, and developing them further in ways that allow them reemerge.

- Establish a relationship with the present: The development of a city is a process that takes place somewhere between continuity and change. Cities are much like organisms in that they do not remain static once built, but are always in a constant state of change. Therefore, cities must not be thought of as museums but as living extensions of the present. Separate from the past, the present day of preservation areas must also be reflected and experienced.

Along with these design guidelines, no historicised architecture is to be encouraged. The quality and charm of old cities is precisely due to the fact that for centuries, older construction has been allowed to comingle with the relevant contemporary architecture wherever possible.

The existing basic historical structure of these buildings should be preserved and renewed. If necessary, they may also be improved upon or restored with additions that support the intended use of the building. Missing buildings should be carefully constructed according to the established architectural norms of the area.

在规约区域不应该推崇某种建筑风格或历史性的建筑。旧城的品质和魅力正是在于几百年来在各个时代所形成的相应的新建筑。

Along with these design guidelines, no historicised architecture is to be encouraged.

Interplay of the monument conservation law, preservation law and design code

The design code fits into a bundle of several relevant pieces of legislation. Firstly, primary buildings, private gardens and public parks in the preservation areas are subject to special provision of the monument conservation law, which states that the demolition of buildings or any alterations deviating from the buildings' original design, are prohibited. In addition, the design code in connection with the preservation law under §172 of the federal building code also prohibits any demolition or structural alterations of buildings that 'merit preservation'.

Factors that must be taken into account when assessing if a structure 'merits preservation' include the position of the buildings on the estate, the succession of buildings, and the construction methods used. Individual buildings with little intrinsic value as monuments should be protected and maintained if they help contribute to the overall character of the area by themselves or in conjunction with surrounding structures.

However, a city can never be completely preserved as it is always changing, even if this only means a small extension or a repainted facade. The important question is how these individual changes are designed to preserve the original characteristics that made the structures worth preserving in the first place.

The design code follows contemporary building design through the preservation of a building's historical structure, while at the same time adhering to area regulations. In this way, it does not replace existing rules, but rather complements them.

文物保护法、维护法律和设计导则的相互协调

设计导则要与很多法律规章相协调。一方面，规约区域内的很多建筑要受到文物保护特殊规定的限制。这其中就包括适用于总体相关区域的以建筑法第172条为依据的维护规约。据此，禁止对有保存价值的建筑进行拆除或者改变建筑本质——所谓保存价值包括用地上建筑物的位置、建筑次序和建筑方式等。特别是对于本身并没有文物价值，但是独自或者与其他建筑一起形成地区特色的建筑，应该加以保留和保护。

无论如何，一个城市不能被收藏起来，它总是在改变自己，例如一次小的扩建或是房屋门面重漆。问题是，怎样进行这些个人性的改造，同时保护这些有保存价值的独特性，从而整体首先保证都市结构的价值。

设计导则力图在避免结构性改变的基础上，结合具体措施的制订，形成对于历史片区的富于活力的现代建筑设计塑造要求。从这一意义而言，规约不是要替代规约区域内现有的规定，而是要对其进行补充。

导则的制定 Development of the design code

Putting the building design code to work through analysis and test design

通过分析和示范设计来制定导则

设计导则的规定以对城市意象的分析和一系列示范设计为基础。包括哪些城市景观的塑造要素，哪些建筑设计，哪些形式、标准、建筑规模和建筑组成，哪些材料和颜色等等构成了规约区域的城市景观意向，而应作为具有区域特色的艺术要素被保留。

The provisions of the design code are based on an analysis of the existing townscape and a range of test designs.

The provisions of the design code are based on an analysis of the existing townscape and a range of test designs. Urban design elements of the townscape including, structure, type, scale, condition, building mass, facade composition, material and color were all taken into account when setting the guidelines for matching buildings to each other and establishing which structures should be preserved as typical design elements in the preservation areas.

This analysis, along with the test designs, verified ways in which the existing characteristics of area buildings can be transferred into a contemporary design language when it comes to renovations or new construction.

设计导则的规定以对城市意象的分析和一系列示范设计为基础。城市景观的塑造要素，包括结构、类型、尺度、标准、建筑规模和建筑立面，材料和颜色等被综合考虑构成了规约区域的城市景观导则，以使建筑个体和谐相处，而结构本身也被作为典型设计要素，应用于规划区域，应作为具有区域特色的艺术要素被保留。

这一分析之后就借助示范设计进行检测，以哪种方式能够在改建或重建的过程中，通过合乎时代的形式语言来传递这个区域内建筑的现有特征。

图 02 现状情况和两种示范设计
Fig. 02 The existing situation and two test designs

设计导则中的重要性与效率性控制

由分析和示范设计制定形成的设计规则被分为三类,并将对建筑总体的影响、在城市意象中的重要性与导则的效率性控制考虑在内,包括:

• 适用于整个规约范围的一般原则。
• 适用于主立面从公共区域可见的临街建筑或建筑区域的设计规则。
• 适用于侧立面从公共区域可见的临街建筑或建筑区域的设计规则。

在不损害城市意象目标、不触及公众利益的前提下,这个分类是很有必要的,可以避免对个体设计自由的不合理影响。

公共区域首先是指所有的大街、道路、广场和公共绿地。此外,规约区域内开放式的建筑方式,使得从公共街道空间就可以看到很多街区内部的院落,地形因素也提供了观赏屋顶景观的可能。因此,设计规约也应该对屋顶景观塑造作出规定。

导则内容之建筑组合设计规则

一个城市不是由单个的建筑组成的,而是各个建筑的整体组合。规约区域的独特性和识别性不仅仅取决于各个建筑类型的品质,更多的是由建筑组合和城市空间的品质决定的。规约区域内的建筑应该根据现有建筑组合的规律和设计原则进行发展,以形成与区域内各个时代历史建筑的联系。

所以整体设计导则来自于对现有建筑组合的规律和设计原则归纳而成。它们成为适用于所有方面的一般要求,其中包括对建筑类型进行划分,对建筑组合和序列中的各个建筑的塑造要求,特别是建筑形体和建筑立面的塑造要求,以及门前花圃、门廊建筑间隔和围栅的设计要求。

The importance and efficiency of control in the design code

Provisions of the design code derived from analysis and test designs are differentiated in terms of their level of importance and efficiency of control following guidelines below:

• General design guidelines intended to apply throughout the entirety of preservation areas.
• Design guidelines that apply only to building elements that are visible from public spaces.
• Rules for main facade facing street, street-side roof zone and side elevation.

This distinction is necessary to avoid undue interference with individuals' freedom of design, as long as the aims of the townscape and the interests of the general public are not jeopardized.

Public spaces are considered all roads, streets, squares and public parks. It should also be noted that since many area buildings have wide alleyways between them, backyards are often visible from public streets. Additionally, there are many locations providing clear views of rooftops, making it therefore necessary to regulate the design of the roofscape as well.

The principals of building succession in the design code

A city is no t simply the sum of disparate architectural parts, but an overall composition made up of individual buildings. The uniqueness and identification of the registered area is not based solely on the quality of each building type, but rather on the quality of building groupings in the overall urban space. For example, architecture should b e developed according to the rules of the preservation area and the provisions of the design principles for existing building groups to achieve the highest possible degree of integration of contemporary architecture into historical areas.

This is why the design code has been drafted with existing building groups in mind. The code is considered to be a guideline for all aspects of the architectural design process, including the definition of building types, the design of individual buildings as it applies to their effect on the succession of surrounding buildings, as well as the design of individual building shapes and facades, front yards, open alleyways / spaces between buildings, and the design of fences.

图 03 建筑立面序列导则
Fig. 03 Guidelines for the sequential arrangement of buildings

For example, one requirement for building succession stipulates that each building must be designed as an individual structure within a common design framework, thus assuring a cohesive ensemble and succession. Adhering to this requirement is a matter of adjusting material, proportion, scale and plasticity to match the existing structures in the area. Resultant de sign guidelines pertaining to facade design in a succession of buildings are as follows:

(1) Adjacent facades of the same building type must differ in at least three of the following ten design features : Wide-sections of the facade and eave height, the horizontal and vertical design structure, quantitative proportion of the wall to the openings, shape of the openings, height of balustrade, type and dimension of plasticity of the facade and material color.

(2) Adjacent facades must meet at least two of the design features indicated in paragraph (1). The same level of eaves and attic and similar horizontal structural elements may not exceed two consecutive buildings. All other common design features may occur over three consecutive buildings at most.

Further statements are made on the design of elements including building type, roof and window shape, building equipment such as roof mounted solar panels, building mass elements such as bays, balconies and loggias, structural elements of the facade, openings in the facade, material and color of the facade, canvas blinds and shutters, front gardens, building lines, fencing and signs.

规约区域的独特性和识别性不仅仅是取决于各个建筑类型的品质，更多的是由建筑组合中整体城市空间的品质决定的。

The uhicaeness and identification of the registered area is not based solely on the quality of each building type, but rather on the quality of building groupings in the overall urban space.

An excerpt of the statutes is listed below:

Definition of building types: It is crucial for the building design code, that the existing building types be modified within a common design framework and using features which a re typical for the area. They are intended to serve as an inspiration for individual contemporary architectural solutions. For this reason, it is necessary to define some key building types in terms of their form and function, irrespective of style epochs.

例如建筑立面组合序列的要求之一是建筑形体和立面一方面应当形成共同的设计框架，同时又可以很清楚地识别每个建筑：通过与现状建筑的材质、比例、尺度和体量的协调，实现建筑群体的整体设计目标。由此产生的建筑立面序列设计规则如下：

(1)相同建筑类型的相邻立面必须在以下10点设计特征中的3点以上有所区别：立面区段的宽度、檐口高度、立面的水平分区、立面的垂直分区、实墙面与门窗的比例、门窗的形式、底层基础的高度、立面形体凸出凹进的形式和规模、材质、颜色。

(2)相邻立面必须在条款(1)中所列设计特征中的两点以上保持一致。檐口、拦墙和其他水平分区元素的相同高度最多允许出现在两个相邻建筑中。其他的一般设计特征最多允许出现在三个相邻建筑中。

其他的陈述阐释了以下要素的设计：建筑类型，屋顶形式和屋顶天窗，屋顶附属结构如太阳能集热器，建筑主体部件如凸窗、阳台和敞廊，立面分区，立面开口，材质和颜色，立面上的建筑附件如遮阳篷和百叶窗，门前花圃和建筑间隔，围栅，广告设施。

部分导则法令的摘录

建筑类型的定义：对于设计导则来说是至关重要，现有的建筑类型结合一个相对标准的框架体系被修订，赋予相关标准下的典型性特征。在此背景下允许一定的变化，以适应时代及建筑个性的要求。因此，确定特定的建筑类型，而不是风格时期，对建筑形式与功能进行定义，是很必要的。

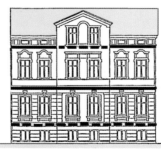 

图 04　檐口型典型特征
Fig. 04 Characteristics of the dormer type

三种主要的建筑类型是：

(1)檐口型建筑以临街的水平檐口和水平线脚为主要特征，形成上部立面的结束，例如贯穿整个立面宽度的女儿墙。

(2)山墙型建筑通过垂直于沿街檐口的山墙面打断了连续的临街水平檐口线条，其宽度不可超过建筑立面宽度的40%。

(3)塔楼型以塔楼为标志，它明显高于屋脊。塔楼的基本立面与平面形式均是正方形的。它的最大宽度和深度分别是5米。塔楼的上边缘高度不超过临街檐口高度的150%。

立面设计：除了作为"主体"的建筑类型的设计，从公共空间可见的外墙面的设计房屋立面设计也决定了一个建筑物的外部意象，进而决定了街道和城市意象。因此需要对立面设计制定相应规定。

立面设计

(1)立面应被设计为洞口式立面。对于临街立面，实墙面积应占立面总面积的60%～80%。
(2)从视觉上，立面至少要被清晰地分为三个水平区域基座，底层和楼层。
(3)装饰性要素如水平勒脚线、竖线条或者凹缝可向外突出或向内凹进3～25厘米。

建筑物的形体划分：另外一种明显的特征是通过门廊、阳台和敞廊实现建筑物的形体划分。这些突出的建筑部件在原则上应该出现在临街立面，并把建筑物分为不同的部分，形成一种有活力的建筑和立面意象。

门廊、凸窗、阳台和敞廊

(1)外侧立面只允许有门廊、凸窗、阳台和敞廊。
(2)临街立面的门廊或者凸窗最多只能有2米进深，宽度占所处立面总宽度的比例不超过40%。上边缘至少要在临街檐口的1.5米以下。

The three main building types are:

(1) The eave type is characterized by a gutter along the street or by the shape of the moldings, which are plastically and continuously shaped as a distinct upper front end of the facade, such as parapet on the entire width of the facade.
(2) The dormer type is characterized by an interruption of the gutter along the road by a transverse gable or ornamental gable, the width of which does not exceed 40% of the building's total width.
(3) The tower house is dominated by a tower that rises well above the ridge height of the roof. Towers have a square shape, as well as their floor plans, with a maximum width and depth of 5 meters. The height of a tower shall not exceed 150 % of the height of the gutter facing the street.

Design of facades: Apart from designing the types of buildings as a 'body', the design of the exterior wall areas, or facades, visible from public spaces forms each building's overall appearance, and in turn dictates the appearance of the streetscape and cityscape as a whole. The design of facades is regulated in the following ways:

Design of facades

(1) Facades are to be formed as perforated facades. Windows and doors in the main facade facing the street must constitute at least 20-40% of the total facade's surface area.
(2) Facades must be optically and clearly divided into at least three horizontal zones – base, ground floor and first floor.
(3) Relief - structured elements such a s cornices, pilaster strips and cuts must jut out or rebound with a minimum depth of 3cm and maximum depth up to 25cm.

Design of building mass elements: Another notable feature is the structural layout of building mass through colonnades, balconies and loggias. Such components are usually built into the facade facing the street, subdividing the building volume into sophisticated components, and make up a vivid building and facade.

Porticos, bays, balconies and loggias

(1) Only porticos, bays, balconies and loggias are allowed in the exterior facade.
(2) For porticoes or bays on the facade along a street, a maximum depth of 2m and width of up to 40% of the facade is allowed. The top edge must be at least 1.5m below the eaves facing the road.

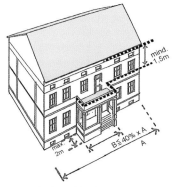

图 05　门廊示例和设计导则
Fig. 05　Examples and Guidelines for porches

(3) Up to two street-side balconies and loggias are allowed per story, but the depth is not to exceed 1.6m and width not to exceed 2.5m.

Design of openings in the facade: The shape of the perforated facade is derived from the prevailing pattern of the solid structure. The bigger picture is lost if, for example, a wall is dissolved into a column grid or replaced by a band of windows. This is why the quantitative proportion of wall surface to window area and the design of openings such as windows and doorways is important to the overall design of the facade. Size, format, shape, number and arrangement of openings are all regulated.

Design of roofs and skylights: The design of the roof - shape and pitch, material and color, gable board, eave construction and the depth of the overlapping roof - defines the character of a building. The overall effect of different roofs together in turn shape the city skyline, which in the city of Potsdam is easily visible due to topography and its Belvedere lookout towers. But roofs remain an important factor even if only partially visible from the street level.

Roofs and skylights

The design of roofs is regulated in the following ways:

(1) The main structure must have a pitched roof with a continuous ridge. The allowable roof pitch range for a main roof is anywhere from 25° to 45°. Eave roof types with a parapet are an exception with a minimum roof pitch of 5°.
(2) Eaves and gable boards may have a minimum overlap of 20cm and a maximum of 50cm.
(3) Skylights and roof dormers are permitted. Dormers are only allowed on roofs with a pitch of at least 35°. The entire elevation width of the dormers may have a maximum of up to 50% of the affected eave length, etc.

Design of front gardens and fencing: front gardens in the preservation area are designed as ornamental gardens. Their surfaces are primarily horticultural and not paved. Even the open spaces between buildings are harmoniously designed to be a part of the front gardens. Only the necessary access spaces to the backyard or main access paths to doorways are paved with cobblestones or mosaic paving.

Almost as important as the gardens are the wrought-iron fencing and garden entrances. Along with the gardens, they are an additional expression of each individual building or estate, and dominate the streetscape in the preservation area.

This is why the design of front gardens are as important as the design of the building itself in terms of preserving the extant streetscape.

（3）每层最多能有两个临街阳台和敞廊，不得超过1.6米深、2.5米宽。

立面开口的设计：对一个洞口式立面的设计主要是由墙体承重的建筑形式所决定的。当实墙面因采用了柱网结构而被取代，或使用条形窗，就不会形成这种形式的立面。实墙面与开口的比例关系，以及门窗等开口的设计对于立面设计来说也很重要。因此规约中对于立面开口的大小、形式、数量、布置及其详细设计，也进行了规定。

屋顶和天窗的设计：屋顶的设计——形式和斜度、材料和颜色、水平和斜檐口设计以及屋檐的深度——也决定着一个建筑的特征。由于城市的地势和现有观望塔和望景楼，在波茨坦可以很清楚地看到各个屋顶集合在一起形成的屋顶风景。同时屋顶对街道意象来说同样是很重要的一个方面，即便从街道上只能直接看到很小的一部分屋顶。因此，应该对屋顶和天窗的设计也作出规定。

屋顶和天窗

（1）主要建筑的屋顶应该建成有贯通屋脊的斜屋顶。主要屋顶斜度应在25°～45°之间。对于女儿墙檐口类型可放宽至最少5°的屋顶斜度。
（2）水平和斜檐口的进深应在20厘米到50厘米之间。
（3）天窗的形式可以是凸出天窗和平天窗。采用凸出式天窗的屋顶其斜度应大于35°。凸出式天窗的总宽度最多可至檐口长度的50%等等。

门前花圃和围栅的设计：规约区域内的门前区域被设计成观赏花园。其表面尽量用来进行园艺种植。建筑间隔也被和谐地设计成了门前花圃的一部分。只有必要的通往区域内部的通道或者通往建筑入口的主要通道用小石块或彩石铺成。

铸铁的围栅和花园大门与花园一样重要。它们与花园一起，是街道建筑和用地的另外一种独特的表现方式，是规约区域街道意象的重要组成部分。

因此，为了维护现有街道意象，门前花圃的设计像建筑自身的设计一样重要。

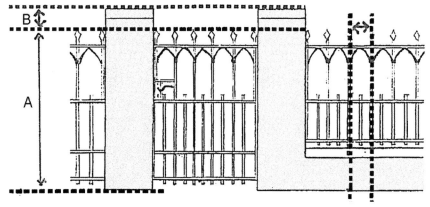

图 06 围栅示例和设计导则
Fig. 06 Examples and guidelines for fencing

城市保护及管理　URBAN PRESERVATION AND MANAGEMENT

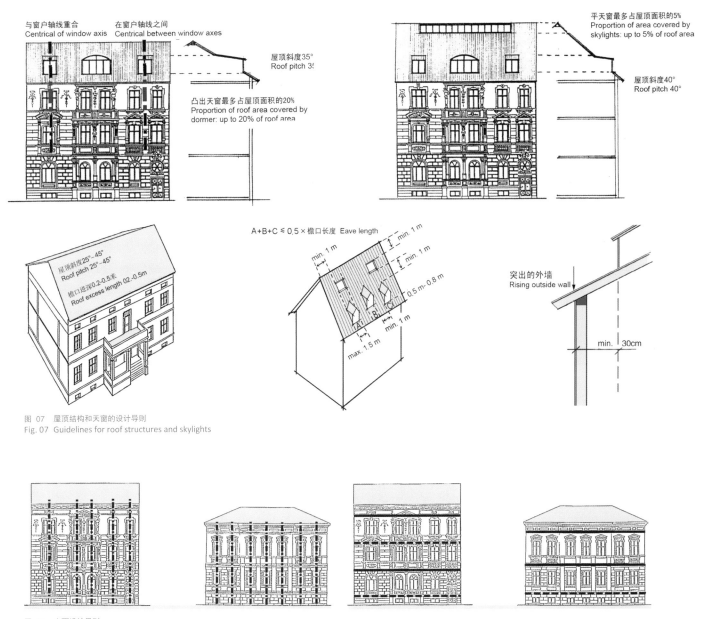

图 07　屋顶结构和天窗的设计导则
Fig. 07　Guidelines for roof structures and skylights

图 08　立面设计导则
Fig. 08　Guidelines for design of facades

城市保护及管理
URBAN PRESERVATION AND MANAGEMENT

01 世界文化遗产城市施特拉尔松：城市景观规划
World Cultural Heritage City of Stralsund: Townscape Planning

02 新勃兰登堡城市景观规划
Townscape Planning Neubrandenburg

03 世界文化遗产城市波茨坦瑙恩郊区
及耶格尔郊区的设计导则
Design Code for the Nauener Vorstadt and
the Jaegervorstadt in World Cultural Heritage City of Potsdam

04 泉州市法石街区保护
与整治规划及环境设计
Urban and Landscape Design for the
Fashi District in Quanzhou, Fujian

05 泉州市城市新区公共空间控制规划方法研究
Research of the Planning Methods for
Open Space in Quanzhou, Fujian

04 泉州市法石街区保护与整治规划及环境设计

Urban and Landscape Design for the Fashi District in Quanzhou, Fujian

规划面积 AREA		486.31 hm²
人口规模 POPULATION		4000
完成时间 PROJECT DATE		2009.03

项目背景 Background

The Fashi District is located at the joint of the East Sea Group and the current downtown of Quanzhou. The district is located in one of the most protected historical sites of the Jiangkou area in Quanzhou, and represents a section of the historic Maritime Silk Road. Throughout history there have been frequent maritime business exchanges between Quanzhou and foreign countries that left behind a great deal of important cultural heritage. There are many historic relics of these maritime business exchanges in Quanzhou Jiangkou, which are extremely important reminders of the area's unique history and merit protection. The Fashi District is the most striking and important area within the Marine Silk Road in Jiangkou because it holds the Jinjiang seaport, a land-water transportation hub in the ancient city of Quanzhou. This is one of three gulfs and twelve harbors in the city where the shipping and shipbuilding industries flourished during the Song and Yuan Dynasties. Yet, with historical development during the Ming and Qing dynasties demolishing many traditional buildings and sites, few remain today, making what is left just that much more significant.

The scope of planning in the district takes into account a total area of approximately 44 hectares, with the Donggan Canal to the west, the Sanjing road to the east, the Sugang Road to the north, and the sea to the south. There are three important cultural relic sites under state landmark protection, three sites under city landmark protection, and 136 traditional buildings protected at both the state and city levels. The most important historical sites and buildings are distributed along Stone Street, yet cultural relic sites and historical buildings represent only 6% of the total planning area due to demolition and new construction in recent years. For this reason, the Fashi District has fallen short of being named a historical district meriting general protection.

法石街区位于泉州市东海组团与现状中心城区衔接处，是海上丝绸之路泉州史迹保护规划中江口片区的最集中地区，历史上是著名的商贸街。泉州自古以来就与国外有着频繁的海上商贸交往，孕育了以海上商贸为基础的特色文化，留下了众多文化遗产。现存的"海上丝绸之路"史迹中，较突出且集中的地区就是法石街区。法石街区扼晋江海口关要，是古代泉州城区与港区水陆转运的枢纽，也是泉州三湾十二港之一。宋元时期其航运、造船业就相当发达，至今仍然保留有较多历史遗迹。由于村落受到明清两代的历史变革影响，现存传统建筑保存较少，保留情况较差，因此其历史场所意义远大于实体建筑保护意义。

本次规划范围南到沿海大通道北侧，总面积约44公顷。法石街区有国家级重点文物保护单位3处，市级文保单位3处，本次规划经调研确定传统保留建筑（包括国家级、市级文保单位）136座，主要历史场所和历史建筑沿石头街分布，由于历史原因及近年来加建拆迁严重，导致现状文保单位和历史保留建筑仅占现状总建筑面积的6%。鉴于此种情况，法石街区已不足以作为国家历史街区进行全面保护发展。

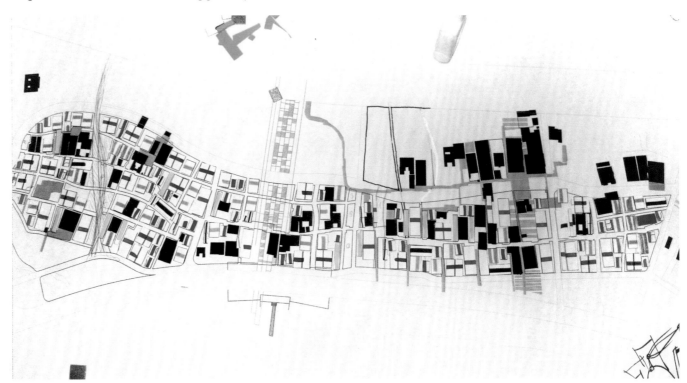

图 01　现状调研情况
Fig. 01 Situation of status quo research

城市保护及管理　URBAN PRESERVATION AND MANAGEMENT

保护类型 Type Of Protection	保护信息 Information	保护类型 Type Of Protection	保护措施 Measure	保护强度 Intensity
自然地理信息 Physical Geography	① 周边山体 Mountain	宝觉山和石头街的关系，由于采石而形成的石头街。	通过重塑山体和法石村的关系达到恢复原有风貌	合理利用
	② 周边水系 Water	海洋、晋江与法石的航海历史关系，海洋文化与船舶文化	通过重塑水体和法石村的关系达到恢复原有风貌	合理利用
	③ 区域内地形 Topography	区域内因地势高度差异自然形成由南向北高度上升的台地	将新老城区以台地结构区分	合理利用
城市建设层面 Urban construction	④ 文保单位及石头街全部 Focus & Stone street	建筑本身及周边环境	拆除搭建，严格根据建筑原貌修缮破坏部分，控制周围一定比例建筑高度	严格保护
	⑤ 传统古厝石屋西洋厝 Old building	建筑本身及周边环境	根据传统建筑的典型面貌修缮（修旧如旧）重建，同时注意设计风格的多元性	合理利用
	⑥ 公共空间 Public space	绿地、洗衣空间、广场及大型植被形成的公共空间	修缮损毁部分，复原传统样式，继续沿用传统功能	适度调整
	⑦ 历史街巷 Old street	道路铺装、道路边界肌理、建筑沿街立面	修缮道路铺装，保持原有石材铺装，将铺上水泥的地面恢复为石材地面	适度调整
	⑧ 内部水系 Well	渠道、井口	复原水网结构，形成体系，可保持作为浇灌用水的风俗。整治古井周边环境	适度调整
	⑨ 村镇肌理 Texture	原有村镇建筑基底形成的肌理结构	新建建筑形成的建筑肌理与原有建筑肌理形式相同，仿照原有肌理形式进行区域设计和新建筑布局	适度调整
文化层面 （非物质文化） Non-material cultural heritage	⑩ 承载非物质文化的物质元素载体 Carrier Intangible heritage	戏台、绿地、洗衣空间、祠堂、码头空间、各类社区交往空间各种公共建筑和服务业、商业设施	保护或恢复文化空间功能，鼓励恢复传统艺术活动和传统节日鼓励村民保存原有村落民俗，同时为活动提供各项服务和场地	严格保护

图 02　历史片区整体保护信息、保护类型与保护措施目录
Fig. 02　Preservation elements, type of protection and protection measure in the historical area

规划原则 Planning principles

本次规划坚持"保护为前提的城市发展"的方针，本着整体控制、重点保护、合理利用、适度发展的原则，坚持可持续发展原则，在深入调研和综合分析之后，将空间保护信息作如下界定：除文保单位和传统保护建筑外，还将自然地理环境（原山水结构），以石头街为核心的公共空间结构、水渠结构以及村庄肌理等列入保护范围中；根据其现状品质以及对城市生活和建设的影响进行分类，进行保护信息的提取与保护级别的确定，共分为严格保护、合理利用、适度调整、科学规划四个层次，以不同保护方法及强度进行保护和开发。

The planning adhered to a general sustainable development policy of "protection first, development second" that was in line with the principles of overall control, special protection, rational utilization of resources and proper development. After a period of comprehensive research and analysis, the method of approach toward protecting the district was defined as follows:

Along with the cultural relic sites and historical buildings, the following categories will be included in the scope of protection: structure of the natural landscape, structure of public spaces in the Fashi District's core, structure of drainage systems, structure of villages, etc. Each preservation element will be classified and its level of protection determined based on its existing status, the quality and impact on urban life and building structure. Each district is intended to be protected and developed using varying methods and intensities divided into the four categories of: strict protection, rational utilization, appropriate adjustment and critical development.

由于村落受到明清两代的历史变革影响，现存传统建筑保存较少，保留情况较差，因此其历史场所意义远大于实体建筑保护意义。

With the historical development during the Ming and Qing dynasties demolishing many traditional buildings and sites, few remain today, making what is left just that much more significant.

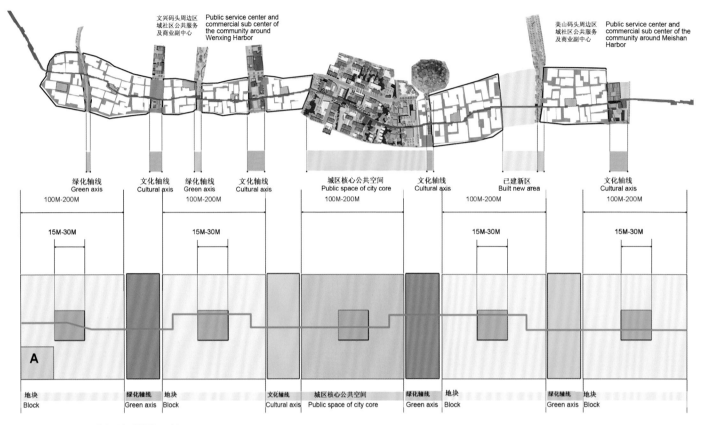

图 03 新社区空间结构与社会结构的相互融合
Fig. 03 Integration of new community spatial structure and social structure

规划设计 Urban design

Regarding the preservation of the natural landscape, historical relics, traditional buildings and future development areas in the Fashi District, the spatial concept of "three centers, three horizontal axes and eight vertical axes" was developed.

Three centers : In the middle of the District, the state-owned paper mill would be converted into a central tourist and entertainment area as a functional center. Wenxing harbor and traditional blocks north of Meishan harbor would become two sub-centers with commercial, touristic and public facilities.

Three horizontal axes : The three east-west axes running parallel to each other are the waterfront greenbelt north of the avenue along the southern coast, Stone Street at the center, and Hou Street and its greenbelt to the north. Hou Street would serve as the main social space, Stone Street would represent the traditional district, and the beautiful environment along the coast line to the south would offer diverse entertainment and leisure in a waterfront greenbelt north of the avenue.

Eight vertical axes : The north-south cultural axes and the axes of the greenbelt would connect every group of different quality to form a network, that balances the preservation and development of Stone Street, and makes full use of the natural, cultural and

在区域功能结构与城市设计结合方面，依据保护范围，以自然地貌形态、文物古迹分布特点、传统保留建筑位置以及可建设区域分析为基础，形成法石街区"三心三横八纵"的空间布局。

三心：以三处集中性改造节点作为核心型发展引擎

区域中部以市民政造纸厂厂房为主体现状，规划改造为中央旅游休闲娱乐区，同时作为本规划区域的功能中心，以文兴码头和美山码头北侧传统街区为核心规划两处副中心，功能结合商业旅游与社区服务。

三轴：以传统主街脉络作为公共空间核心职能（交通、文化承载、游憩）载体

以南侧沿海大通道以北滨水绿化带、中部石头街、北侧后街及两侧绿带三段东西向平行结构为主轴。其中后街为主体交通通道，石头街沿线为历史传统街区面貌，沿海大通道北侧辅道借助带状生态水面形成优美的娱乐休憩环境。

城市保护及管理　URBAN PRESERVATION AND MANAGEMENT

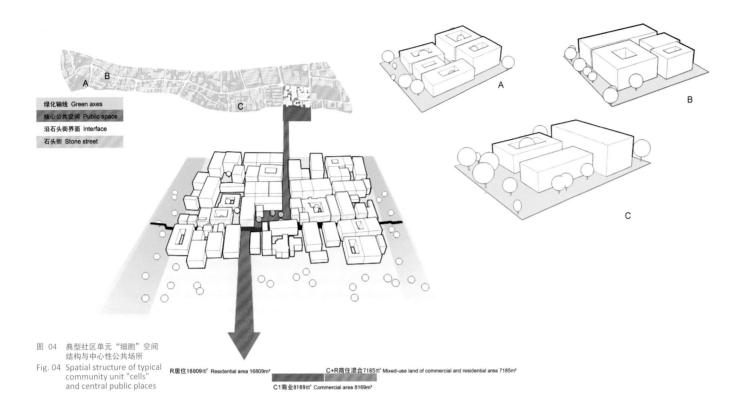

图 04　典型社区单元"细胞"空间结构与中心性公共场所
Fig. 04　Spatial structure of typical community unit "cells" and central public places

八纵：以传统公共建筑与相关公共空间作为社区服务与社区文化传承的载体

由南北向文化功能轴线和绿化轴线串联各个不同性质的组团，点、线、面共同形成网状结构。由区域主体结构中"八纵"将法石街区所划分出的7个地块作为子结构，每个子结构分别作为社区单元的"细胞"体，每个结构有50～100户居民，形成一个较好的邻里单位。每一邻里单位设置一处核心公共空间，同时使街区产生节奏和韵律，也为邻里活动与行人提供休憩空间。公共空间以广场为主要形式，设在主要公共建筑或者戏台等特殊位置，面积在100平方米左右，能起到承载公共交流与非物质文化的作用。

石头街和后街两侧区域隔30～50米设置南北向次小巷，构筑街坊形式，间隔设立相对私密的半开放性次要公共空间，主要为街坊居民服务，以此形成由石头街—主要轴线—核心公共空间—次要公共空间构成的公共空间系统，并通过林荫道联系成网。我们的意图是将石头街的文物保护与开发有机地结合在一起，充分发挥其自然、人文、社会等三大旅游资源的整体优势和功能作用，并保证石头街珍贵的旅游资源利用的延续性，确保其经济效益、社会效益和环境效益的协调发展。

区域空间景观规划以法石片区传统建筑文化为基础，体现其历史保护街区的特色和氛围，在突出重点历史文化遗迹的同时，在缓冲区内局部结合建筑设计，使街区呈现新的活力。

social advantages and functions of tourism resources. This would guarantee the constant utilization of tourism resources and the coordinated growth of economic, social and environmental benefits.

After the renovation of traditional districts, the regional structure and green axes would be defined as a back born. The downtown planned at the core would be supplemented by semi-open public spaces in the neighborhoods. The main structure of the region consists of seven blocks divided by the sub-structure of eight vertical axes. As a cell of the community unit, each sub-structure contains between 50 and 100 households, which would form a strong neighborhood. Each neighborhood would have a public square at the core, which generates the tempo and rhythm of life style in the district, and provides pedestrians with public space available for diverse activities. The square, the main pattern of public space, with an area of about 100 square meters, would be established at a special site, such as a main public building or stage, which helps cultivate social and cultural activities.

On both sides of Stone and Hou Streets, a north-south alley would be created, where every 30-50 meters a semi-open public space is constructed for the use of neighborhood residents. Linked to the network of avenues and the public space system of Stone Street, another sequence of interconnected public spaces came into being: a main axis, main public spaces and secondary public spaces. Landscape planning and urban design based on the traditional architectural culture of the Fashi District, attempted to maintain the character and atmosphere as a historic district, stressing the historical and cultural relics as well as new architectural design in a buffer zone showcasing vitality.

保护区当中严格复原原有街区形态。
In the conservation area, original block forms should be strictly recovered.

图 05　现状调研图片
Fig. 05　Status quo research photos

图 06　核心保护区街道场景
Fig. 06　Core preservation zone

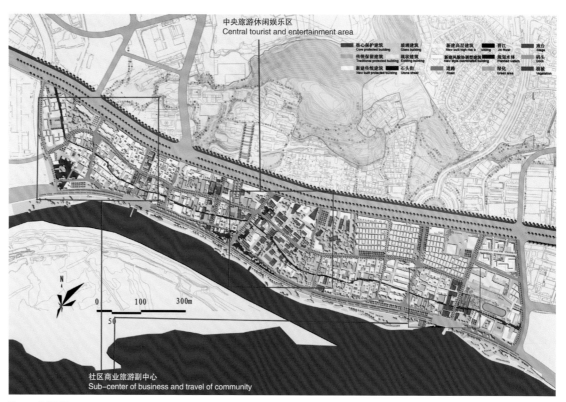

图 07　总平面图
Fig. 07　Site plan: Sub-center of business and tourism

建筑设计 Architectural design

As for architectural design, an emphasis was placed on the key historical and cultural relics by transforming the historical architectural elements into modern design elements, refreshing and invigorating old buildings, like a butterfly breaking free of its cocoon. Traditional buildings with the most characteristics represent the diverse qualities of the cultural genesis of the Fashi district. The planning of functional areas for tourism, business and high-grade residences should not interfere with each other, but instead each must enhance the active quality of daily life and encourage ecological living. With the help of multiform tourist facilities, the district would not only become the new waterfront tourist attraction in Quanzhou, but also reposition the region as a well-known cultural center for recreation. In addition, based on the study of the existing buildings and traditional residential houses in the southern part of Fujian, traditional architectural elements were refined and re-designed to meet modern architectural functions and patterns, thus achieving an integration of traditional and modern.

建筑设计方面，在协调片区中突显重点历史文化遗迹的同时，通过提取原有建筑元素后加入现代元素的设计方法，使新建筑如破茧成蝶般焕发新的生命力。法石街区最具代表特性的传统古厝建筑，文化底蕴浓厚，同时具有较好的多样性品质，通过功能区的划定，达到旅游功能区域与商业功能区域、高档居住功能区域互不干扰的目的，既具有活跃的城市生活品质，也具有高品质的生活环境与生态环境。借助多类型多档次的旅游基础设施，使这里不仅仅是泉州新型的重要海丝旅游景区，也将成为泉州地区的著名休闲娱乐文化中心。

泉州法石街区保护与整治规划及城市设计项目历时2年有余，前后参与设计人员多达30人以上，本项目在旧城更新、"传统"街区复兴方面作出了有益的尝试，试图从城市规划、社会经济推动力及文化魅力等多方面，阐述如何振兴那些我们记忆中的充满活力及历史文化认同的传统街区。

协调区当中，有意识结合旅游服务，现代生活的需求，融合现代元素。

Modern elements are in combination with the tourism service and modern life demands consciously integrated.

The project of historical preservation, renovation and urban design of the Quanzhou Fashi District took more than two years, with upwards of thirty people involved in the planning process. It was a useful attempt at renewing the old city and revitalizing the traditional district, in that it offered an example of how to revitalize the city's vigorous traditional districts while positioning cultural identity as a driving force of socio-economy and cultural charm.

城市保护及管理　URBAN PRESERVATION AND MANAGEMENT

图 08　中央旅游休闲娱乐区鸟瞰图
Fig. 08　Bird's-eye view of central tourist and entertainment area

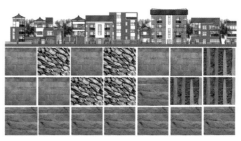

图 09　城市设计导则对地方材料的指导性应用
Fig. 09　Design guidelines for local materials

图 10　中央旅游休闲娱乐区场景
Fig. 10　Simulation scene of central tourist and entertainment area

图 11　社区商业旅游副中心
Fig. 11　Simulation scene of business and tourism sub-center

331

城市保护及管理
URBAN PRESERVATION AND MANAGEMENT

01 世界文化遗产城市施特拉尔松：城市景观规划
World Cultural Heritage City of Stralsund: Townscape Planning

02 新勃兰登堡城市景观规划
Townscape Planning Neubrandenburg

03 世界文化遗产城市波茨坦瑙恩郊区
及耶格尔郊区的设计导则
Design Code for the Nauener Vorstadt and
the Jaegervorstadt in World Cultural Heritage City of Potsdam

04 泉州市法石街区保护
与整治规划及环境设计
Urban and Landscape Design for the
Fashi District in Quanzhou, Fujian

05 泉州市城市新区公共空间控制规划方法研究
Research of the Planning Methods for
Open Space in Quanzhou, Fujian

05 泉州市城市新区公共空间控制规划方法研究
Research of the Planning Methods for Open Space in Quanzhou, Fujian

规划面积 AREA	184 km²
人口规模 POPULATION	450000
完成时间 PROJECT DATE	2009.04

规划设计理念 Concept of planning and design

"The study on public space planning for the new district in Quanzhou" was a systematic attempt to abstract and combine Chinese and Western urban theories, based on three decades of practical urban space design experience.

We as a European firm assumed that there was a great difference between space systems in Europe and China, especially in regards to public space. The expression of "public space" and many existing theories on public space originated in Europe, so Chinese theories on the subject are few and can often be ambiguous in nature. The aim of this study, however, was to analyze and define the typical characteristics of Chinese traditional space and contrast them with European counterparts.

In the Chinese traditional urban space, the relationship between city and nature is emphasized. Not only should urban space be adapted to nature, but the two elements should be engaged in a mutual dialogue of exchange. In the creation of urban landscapes with a central public space, our study emphasized aspects of ecological planning and the organic integration of nature, culture and urbanity. It was not simply the design of physical urban space, but a comprehensive design taking into consideration ecological succession, the development of an urban traditional culture and of social communication systems.

In addition, the study analyzed the reasons and manner in which public spaces are generated vis-a-vis the three aspects of social foundation, cultural origin, and public life; taking the relationship of material space and urban life as a main reference point. Because contemporary public urban life in China is in a state of flux, this study sought out adaptable public urban forms that are amenable to modern development, but are still able to support historical and cultural traditions.

在对比本土公共空间与欧洲的主要差异的基础上，对中国的公共空间典型特质加以分析和提炼，这是本次工作的兴趣点与创新点。

The aim of this study, however, was to analyze and define the typical characteristics of Chinese traditional space and contrast them with European counterpart.

《泉州市城市新区公共空间控制规划方法研究》这一研究课题，是本集团受泉州市规划局委托进行的城市空间研究项目，与此同时，也是意厦结合三十年来城市公共空间设计研究经验和对国内外诸多公共空间理论的梳理研究，对中西方公共空间理论和经验进行的一次系统性总结和融合。

意厦作为欧洲事务所，在研究过程中发现欧洲和中国的空间体系存在着巨大的差异，尤其在公共空间上体现极其明显，而"公共空间"一词及其相关理论本身源于欧洲，因此现有众多公共空间相关理论都带有过多的外域色彩，而中国本土化的公共空间理论相对含糊和欠缺。在研究工作当中，试图在对比本土公共空间与欧洲的主要差异的基础上，对中国的公共空间典型特质加以分析和提炼，这是本次工作的兴趣点与创新点。

中国城市空间传统强调城市与自然的关系。不仅仅是城市空间适应自然，更需要空间与自然之间的对话与交流，在对公共空间为主体的城市景观塑造中，本研究课题一再强调"城市空间设计生态化、城市传统文化继承和发展生态化、城市社会交往体系生态化"这一自然、文化、城市有机结合的生态规划理念。

此外，在公共空间非物质基础层面，即社会基础、文化溯源、公共生活三个角度深入分析和挖掘了城市公共空间形成的软质动因，并以物质空间与城市生活之间的相互影响关系线索作为研究的主体脉络，试图在中国发生巨大改变的城市公共生活当中，寻找真正与现代相适应同时被赋予历史文化传统生活意象的城市公共空间物质形式。

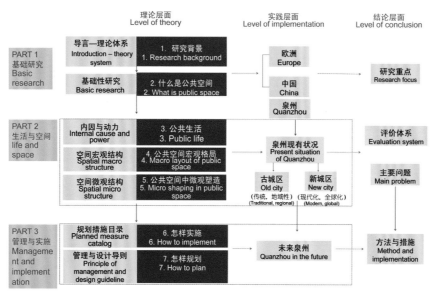

图 01 课题研究整体工作框架
Fig. 01 The overall structure of the study

图 02 泉州城市公共空间与自然格局
Fig. 02 Quanzhou urban public space and natural pattern

研究逻辑体系

为使本次课题研究重点明确化、结论清晰化，整篇采取了理论层面、实践层面、结论层面三轨并行的研究框架。在横向体系上，对于理论层面的各个重要理论点，均在实践层面以欧洲、中国的典型案例加以阐释与佐证，同时总结对于泉州而言的重要结论点，并直接指导泉州城市公共空间规划的具体实施工作。在竖向体系上，本研究采用缜密的逻辑推衍方法，以空间与生活的相互影响关系为主线，从上至下将城市公共空间的影响要素、城市社会生活、城市物质空间等诸多理论点有机地组织串联起来，在三个层面上分别形成了独立而完整的研究系统。

Logic system of the study

To help make the focus and conclusions clearer, the study utilized a three-track framework focusing on theory, practice and conclusion. In the horizontal system, the various important theoretical points could be explained and proven using typical case studies in Europe, China. In the conclusion, the important factors affecting the city of Quanzhou specifically, could be separated out in order to directly guide the planning of public spaces in Quanzhou. In the vertical system, major clues regarding the relationship between space and life could be defined through meticulous logical deduction, from top to bottom. These organically linked elements affecting public space, urban social life and many other theoretical points, form an independent and integrated research system on three levels.

本次课题采取了理论层面、实践层面、结论层面三轨并行的研究框架。

The study utilized a three-track framework focusing on theory, practice and conclusion.

图 03　泉州城市空间肌理对比分析
Fig. 03 Comparative analysis of texture of Quanzhou urban space

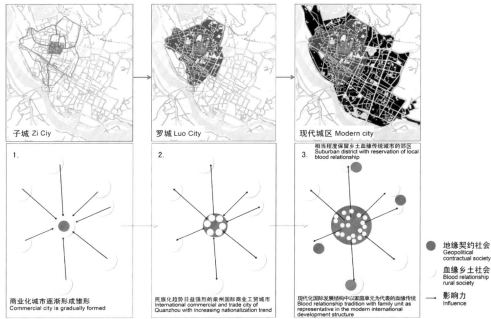

图 04　泉州城市社会形态与公共空间演变分析
Fig. 04 Development trends in public space

研究主体内容与成果

在上述规划理念和逻辑体系的指导下，本次规划研究首先对比了中西方自然、社会和文化特征，总结了中西方城市公共生活和空间的差异性，并为中国传统城市公共空间体系的深入研究抛砖引玉；其次，在对城市公共空间诸多影响要素广泛讨论、重点研究的基础上，以泉州为实例进行了城市公共空间宏观和微观层次上的物质空间研究，并确定了在广泛意义上，城市公共空间宏观和微观层次上的评价体系和评价方法；此外针对泉州城市公共空间的现状特征和问题，以典型案例和经验介绍的形式，为泉州介绍了国际上典型的城市公共空间相关问题的解决方法，并以措施目录的形式予以呈现；最终本次研究以城市设计导则的形式，总结提出了城市新区公共空间塑造方法和规划管理手段，并以此为泉州城市新区公共空间的规划和塑造提供一个具有可实施性意义的蓝本参照。

The main contents and achievements of the study

Under the guidance of the above-mentioned planning concepts and logic system, we first compared natural, social and cultural characteristics and summed up the differences between both Chinese and Western public life and public space. The hope was to attract more attention to in-depth study of Chinese traditional urban public space systems. Next, the public spaces of Quanzhou were analyzed and evaluated at the micro and macro levels with particular attention paid to which factors would make them more attractive and which would not. Then, the existing problems facing Quanzhou's public spaces were identified, solutions introduced, and a catalog of measures was created. Finally, in consideration of urban design guidelines the planning and management methods of urban public space were developed in order to serve as a blueprint of practical significance in the public space planning of the new district in Quanzhou.

图 05 泉州城市公共空间界面分析
Fig. 05 Analysis of urban spatial development of Quanzhou

开元寺 Kaiyuan Temple

传统城区公共空间界面风格分析
Interface style analysis of traditional urban public space

丰泽广场 Fengze Square

现代城市区域界面风格分析
Interface style analysis of modern urban area

府文庙 Fuwen Temple

传统城区公共空间界面风格分析
Interface style analysis of traditional urban public space

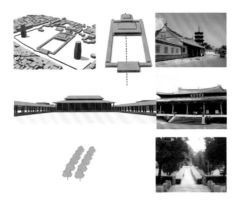

界面风格：传统寺庙院落空间的建筑边界风格对称而正式，给人一种严肃宁静的庄重感，同时之间又有丰富的自然型空间边界，在空间当中又增加了一些自然性和亲切感

界面风格：泉州现代城区广场截面风格与实体建筑的立面效果有很大关系，丰泽广场建筑边界较为自由活泼，具有强烈的商业感，同时一些传统元素的运用也能体现一定的地方特色。但自然边界明显不足，空间显得过于粗犷

界面风格：府文庙是重要的传统空间，其建筑和布局都有严谨的对称关系，甚至广场内的植物种植和绿地形态都是对称的，充分体现其空间性质的正式性。广场的序列性也较强，这些都受主轴线的控制

图 06 泉州城市公共空间构成要素分析
Fig. 06 Analysis of spatial elements in Quanzhou

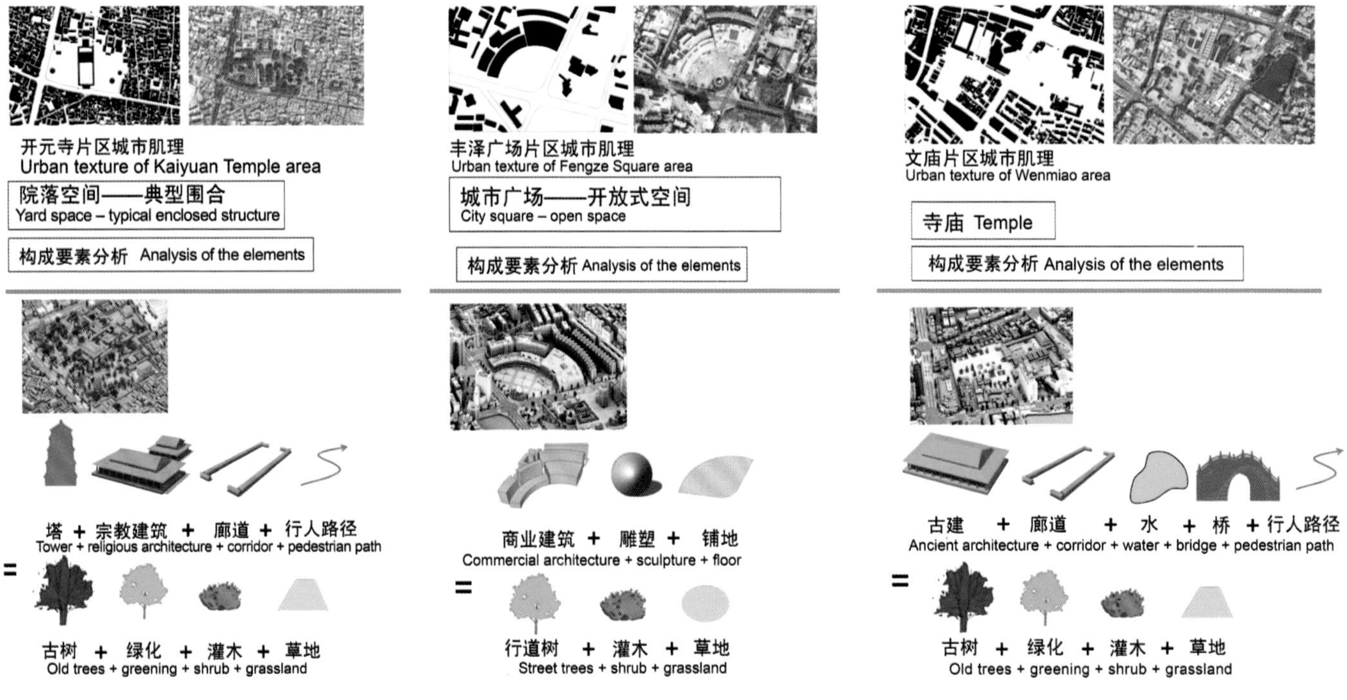

开元寺片区城市肌理
Urban texture of Kaiyuan Temple area

院落空间——典型围合
Yard space – typical enclosed structure

构成要素分析 Analysis of the elements

丰泽广场片区城市肌理
Urban texture of Fengze Square area

城市广场——开放式空间
City square – open space

构成要素分析 Analysis of the elements

文庙片区城市肌理
Urban texture of Wenmiao area

寺庙 Temple

构成要素分析 Analysis of the elements

塔 + 宗教建筑 + 廊道 + 行人路径
Tower + religious architecture + corridor + pedestrian path
=
古树 + 绿化 + 灌木 + 草地
Old trees + greening + shrub + grassland

商业建筑 + 雕塑 + 铺地
Commercial architecture + sculpture + floor
=
行道树 + 灌木 + 草地
Street trees + shrub + grassland

古建 + 廊道 + 水 + 桥 + 行人路径
Ancient architecture + corridor + water + bridge + pedestrian path
=
古树 + 绿化 + 灌木 + 草地
Old trees + greening + shrub + grassland

开元寺的空间要素是由塔、宗教建筑、廊道、行人路径、古树、绿化植物等构成。构成要素丰富多样，空间层次丰富

空间构成要素相对简单，除雕塑和铺地外，只有局部有景观性草坪花卉等植被

文庙的空间要素主要由广场、水、桥、宗教建筑、树木、绿化植物等构成，要素种类较多，空间层次丰富多样，富于变化

城市保护及管理 URBAN PRESERVATION AND MANAGEMENT

图 07　泉州城市用地图
Fig. 07　Urban land use of Quanzhou

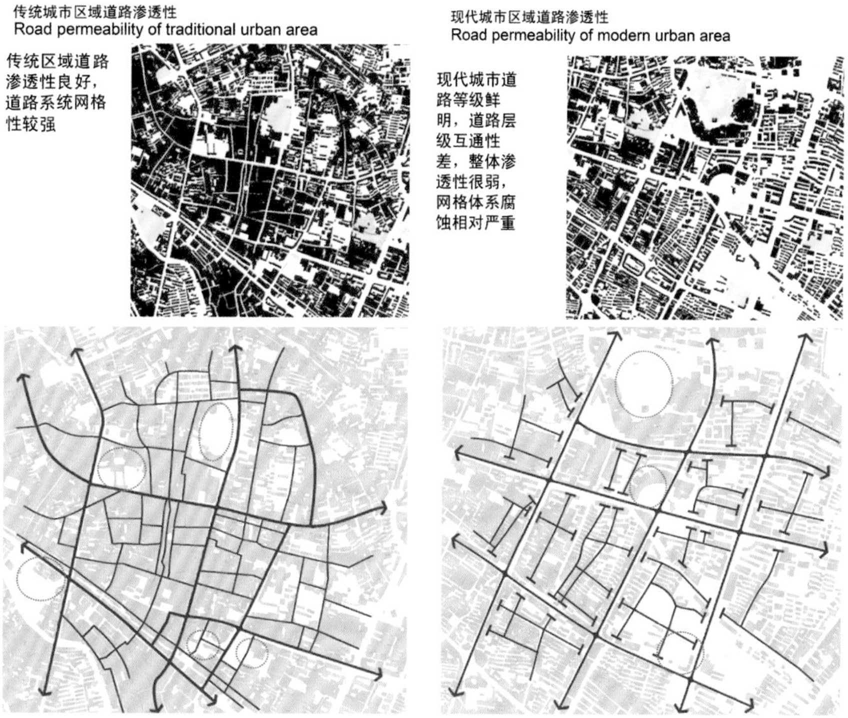

图 08　泉州传统与现代道路腐蚀程度分析
Fig. 08　Analysis of traditional and modern street patterns in Quanzhou

与研究并行的实践设计

此外值得一提的是在理论化的城市空间研究工作当中，加入了颇为有趣的工作实践内容，即对泉州本土城市公共空间改造和设计具有参照价值的概念设计工作。在此过程中以中国最具特色的道路公共空间改造作为主体，进行系列实验性设计，并为泉州道路空间的设计管理工作提供新的思路。

以泉州打锡街城市公共空间改造为例，打锡街是泉州古城保留了传统空间尺度的古街，但是其临街道面的风貌已经不复存在，同时街道南侧紧邻文庙公共空间密集区，而北侧空间状况却低劣而混乱。针对打锡街特有的空间品质和历史意义，在我们总结的研究模式下对其进行了相应的要素分析，以及道路交通等级和断面改造、周边区域公共空间体系改造发展建议两个空间方面的改造设计与建议。

当城市由最初的经济集中体开始回归宜居本质的时候，城市公共生活所散发的魅力就逐渐掩盖了"淘金"的诱惑。一座城市的生命力所在是城市生活，是公共交往，作为规划者当前所做的研究与设计工作就是在为人们营造一个生活的家园，为公共交往构建一片乐土。公共生活与公共空间的探讨在这里只是一个"节点"，在这条丰富的"轴线"上，相信还有更多的闪光点有待去发现和探索。

The practical design with Study

It is worth mentioning that learning from this fascinating practice enriched the theoretical study of urban space in general. Thanks to this process, a valuable concept was developed to rejuvenate Quanzhou's public spaces and redesign the city. A series of experimental designs were applied to help improve life in public spaces, and meant to help cultivate new ideas for the design and management of street spaces in the future.

One example was the reshaping of Daxi Street in Quanzhou.
Daxi Street is an old street retaining some of the traditional spatial elements of Quanzhou. However, with a densely populated area including a Confucian Temple to the south and chaotic sprawl to the north, a link to the landscape no longer exists. As for the unique quality and historical value of Daxi Street, relevant spatial elements were analyzed and a design concept for reshaping the street was suggested. First, the basic functions and traffic flow in the street space itself were to be reshaped; then, the structure of street patterns in surrounding areas would be reorganized as well.

When the city began to return from an initial economic centralized structure to easy living, the charms of urban public life began to overshadow the temptation of the "Gold Rush". The vitality of a city lies in its urban life and public exchanges. The current research and design goal is to create a living homeland for the people and build an inviting space for happy public exchanges. The study of public life and public space here is only the start. More suitable spaces must also be discovered and explored.

从两种文化中学习 LEARNING FROM TWO CULTURES

西街
街道东窄西宽，取平均宽度10米，临街建筑多为一层，取3米层高

次级巷
建筑层数多为一层，少量二层

台魁巷
巷宽约为3.5米，临界建筑多为一层或者两层，取4.5米平均高

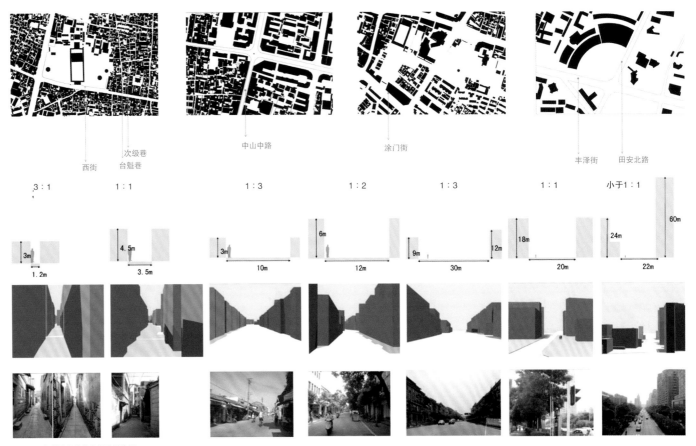

Fig. 09 泉州城市公共空间比例分析
Fig. 09 Analysis of shaded areas in public space

打锡街道路现状道路断面形式
打打锡街现有宽度12米，双向两车道，人行空间狭窄。两侧临街建筑为商业建筑，是泉州主要的商业街之一

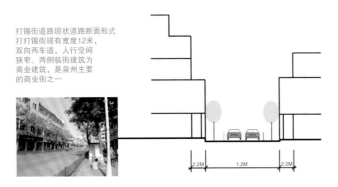

道路断面改造设计
a.中间设置单行车道，车道两侧布置人行空间，同时人行与临街商业建筑紧密结合，形成良好的商业步行看哦关键氛围

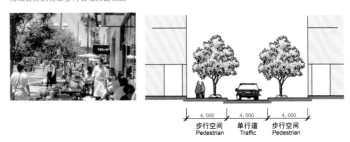

道路断面改造设计
b.在具有较好阴影效果的道路南侧设置较宽的人行空间与遮阳树木，单行车道设置于道路北侧，同时北侧保留具有步行通行能力的较窄的人行空间

图 10　泉州打锡街道路改造分析
Fig. 10 Analysis of Daxi Street in Quanzhou

图 11　泉州打锡街道路改造设计
Fig. 11 Reshaping of Daxi Street in Quanzhou

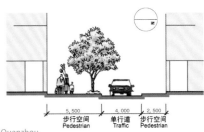

城市保护及管理 URBAN PRESERVATION AND MANAGEMENT

图 12 泉州打锡街区域公共空间体系改造设计
Fig. 12 Spatial reorganization of the Daxi Street area in Quanzhou

从两种文化中学习 LEARNING FROM TWO CULTURES

图片来源及参考文献
IMAGE SOURCE & REFERENCE

关于ISA ABOUT ISA

人即是城市 继承过去，规划未来 (所有图片来源：ISA)
The Human Being Is the City Learning from the past, for the future *(All Source: ISA)*

理论文章 ARTICLES

01 设计塑造生态：不使用高科技的生态新城
Ecology by Design：Ecological New Towns without High-Technology

图 01　图宾根洛雷托广场的生态住宅区 "法兰西社区"
Fig. 01　Ecological settlement "French Quarter" in Tuebingen, Loretto-square
(Images: City of Tuebingen)

图 02　马拉喀什（左图）与柏林（右图）的航拍图比较
Fig. 02　Bird's view of Marrakech (left) in comparison to Berlin (right)
(Images: left:"architecture without architects", Bernard Rudofsky, right: Google earth)

图 03　希巴姆（也门）的黏土高层建筑，格雷梅（土耳其）的穴居和迪拜的风塔
Fig. 03　High-rise buildings built of clay, Schibam/Jemen, cave-houses in Gorem/Turkey, and wind tower in Dubai *(Images: Jialiang Gao, Karsten Doerre, Dita Leyh)*

图 04　捷克共和国的泰尔奇地拱廊、德国的黑森林民居
Fig. 04　Arcades in Telc/Czech Republic, Black Forest House/Germany
(images: Wikipedia commons)

02 从经济起飞到生活质量：实践与研究中城市规划的经验和任务
From Ecolonomic Miracle to Quality of Life：Urban Development Experiences & Tasks of Practice and Research

图 02　弗里德里希港口城市意象：旅游与高科技城市
Fig. 02　Friedrichshafen: tourist and high tech city *(Created by Vaikunda Raja (talk), licensed under the terms of the GNU Free Documentation License, Version 1.2)*

图 04　设计无处不在：食物、材料、发型、浴室、汽车
Fig. 04　Everything is design: food, fabric, haircut, bathroom, car *(Ferrari Paris 2010 1.JPG Created by Cobra bubbles, licensed under the Creative Commons Attribution 3.0 Unported license.)*

图 05　维尔茨堡的城市景观：法制的和经济的城市建设下生活质量的提高
Fig. 05　City Image Wuerzburg – living quality by judicial and economic urban development stimulation *(Pictures from Congress-Tourismus-Wirtschaft Wuerzburg)*

图 08　城市生态：从自然风光到建筑
Fig. 08　City Ecology: From landscape to building *(Created by Christoph Muench, licensed under the Creative Commons Attribution-Share Alike 3.0 Unported license)*

图 10　城市团体：个人和集体间的合作关系
Fig. 10　City community – partnership between individual and community
*(Picture taken by user J van der Kasteele, licensed under the Creative Commons Attribution-Share Alike 2.0 Generic license) (Source: ISA) (Picture taken by user Maryam Laura Moazedi, licensed under the terms of the GNU Free Documentation License, Version 1.2) (Picture taken by user Adam Jones Adam63, licensed under the Creative Commons Attribution-Share Alike 3.0 Unported license).(Picture taken by user William Murphy, licensed under the Creative Commons Attribution-Share Alike 2.0 Generic license.)
(City experience as a reflection of physical, mental, intellectual and spiritual needs Picture taken by User Bruce Tuten, licensed under the Creative Commons Attribution 2.0 Generic license.)(taken from the film the kiss, public domain)*

03 我们应以何种风格建造？——21世纪初建筑学定位的探讨
In which Style should We Build? Architecture at the Beginning of the 21st Century -A Positioning Approach

图 01　Schinkel 的四个设计中，出现了四种不同的风格，这明确体现了 19 世纪建筑风格的不定性
Fig. 01　Four drafts by Schinkel, in four different styles, showing clearly the exchangeability of architectural styles in the 19th century. *(Image from Bergdoll 1994:91)*

图 08　在洛杉矶的城区 Playa Vista，许多住宅区都是历史化的欧洲风格。城市空间的质量由此得到改善。但是这些实例能不能成为未来的模式，还是有争议的
Fig. 08　In the "Playa Vista" quarter of Los Angeles, many residential blocks are constructed in historicizing European styles. While the quality of urban spaces seems to improve, the question remains, if these examples can be models for the future *(Photography: Seog-Jeong Lee)*

参考文献
Ref.　1) Bergdoll 1994_Bergdoll, Barry: Karl Friedrich Schinkel – Preussens beruehmtester Baumeister, Muenchen 1994 (Originalausgabe New York 1994)

2) Berlage 1905_Berlage, Hendrik Petrus: Gedanken ueber Stil in der Baukunst, Leipzig 1905 aus: http://www.tu-cottbus.de/theoriederarchitektur/Archiv/ Autoren/Berlage/Berlage1905.htm, download 16.11.2010
3) Collotti 2001_Collotti, Francesco: Architekturtheoretische Notizen, Luzern 2001
4) Frank 1931_Frank, Josef: Architektur als Symbol, in: Collotti 2001: 43
5) Gleiter 2002_Gleiter, Joerg: Urban Bodies, in: Wolkenkuckuksheim 1/2002, http://www.tu-cottbus.de/theoriederarchitektur/Wolke/deu/Themen/021/Gleiter/Gleiter.htm, download 27.07.2011
6) Haubrich 2008_Haubrich, Rainer: Viele deutsche Städte sind potthässlich, ein Gespräch mit dem Architekten Christoph Mäckler, WELT Online 01.06.2008, aus: http://www.welt.de/kultur/article2051644/Viele_deutsche_Staedte_sind_potthaesslich.html?, download 02.06.2008s
7) Huebsch 1828_Huebsch, Heinrich: In welchem Style sollen wir bauen?, Karlsruhe 1828
8) Hollenstein 2009_Hollenstein, Roman: Strategien fuer eine schoenere Stadt, NZZ Online 14.01.2009, http://www.nzz.ch/nachrichten/kultur/aktuell/strategien_fuer_eine_schoenere_stadt_1.1695233.html, download 19.08.2010
9) Kaufmann 1933_Kaufmann, Emil: Von Ledoux bis Le Corbusier: Ursprung und Entwicklung der Autonomen Architektur, Wien, Leipzig 1933
10) Le Corbusier 1926_Le Corbusier: Kommende Baukunst, Stuttgart / Berlin / Leipzig 1926 (Uebersetzung der zweiten Auflage von 1925)
11) Rauterberg 2005_Rauterberg, Hanno: Denkt endlich groesser!, ein Gespräch mit den Architekten Meinhard von Gerkan und Volkwin Marg, DIE ZEIT 34/2005, aus: http://pdf.zeit.de/2005/34/GMP-Interview.pdf, download 04.07.2011
12) Rauterberg 2009_Rauterberg, Hanno: Baut auf den Prinzen!, DIE ZEIT 21/2009, aus: http://www.zeit.de/2009/21/Architekturdebatte, download 04.07.2011
13) Schirrmacher 2008_Schirrmacher, Joachim: Luxurioese Weite, NZZ am Sonntag, 5.10.2008: 109
14) Schreiber, Moenninger 1995_Schreiber, Mathias; Moenninger, Michael: Die steinerne Mitte, ein Gespräch mit dem Architekten Hans Kollhoff, DER SPIEGEL 45/1995: 246-255
15) Schwarz 2008_Schwarz, Christoph: Blick zurueck nach vorn, WirtschaftsWoche 07.09.2008, aus: http://www.wiwo.de/lifestyle/blick-zurueck-nach-vorn-306143/print/, download 19.11.2010
16) Tiedemann, Schmoock 2007_Tiedemann, Axel; Schmoock, Matthias: Voscherau: Architekten versagen bei HafenCity, Hamburger Abendblatt 13. April 2007, aus: http://www.abendblatt.de/hamburg/article854084/Voscherau-Architekten-versagen-bei-HafenCity.html, download 26.06.2011
17) Thomas 2010_Thomas, Christian: Die Kunst des Bauens, ein Gespräch mit dem Architekten Christoph Mäckler, Frankfurter Rundschau 04.08.2010, aus: http://www.fr-online.de/kultur/architektur/die-kunst-des-bauens/-/1473352/4531982/-/view/printVersion/-/index.html, download 18.11.2010

04 从数量到质量：韩国城市规划控制的经验 (所有图片来源：ISA)
From Quantitiy to Quality: Experiences from city supervision in South Korea *(All Source: ISA)*

05 新城：有关城市的梦想 New Town: Dream City

图 01　描绘工人居住状况的版画
Fig. 01　The workers' living conditions *(source: Dennis Hardy: Tomorrow - tomorrow 1899-1999, TCPA 1999 P7)*

图 02　新和谐城
Fig. 02　New Harmonious City *(source: Chris Light GNU Free Documentation License & Wikimedia Commons*

图 03　Port Sunlight 村
Fig. 03　Village of Port Sunlight *(source: Wikimedia Commons)*

图 04　霍华德诞生地的纪念碑
Fig. 04　The monument of Howard's birthplace
 (Source: Dennis Hardy: Tomorrow - tomorrow 1899-1999, TCPA 1999 P9)

图 05　城市、乡村共同构成的第三极磁铁
Fig. 05　City and countryside together constitute the magnets of the third
(Source: Ebenezer Howard: Garden City of Tomorrow, Commercial Press, 2006 Beijing P7)

图 07　列契活斯
Fig. 07　Letchworth *(source: Creative Commons Attribution Share Alike 2)*

图 08　低密度建设的哈罗新城
Fig. 08　Low-density Harlow New City
(source: Rolf Nemitz GNU Free Documentation License & www.deutsche-wochenschau.de)

图 10　柯布西耶光辉城市
Fig. 10　Corbusier Glorious City *(source: http://www.tu-harburg.de/B/Kuehn/lecorb.html)*

图 11　汉莎小区
Fig. 11　Lufthansa area *(source: Rolf Nemitz GNU Free Documentation License & www.deutsche-wochenschau.de)*

图 12　法兰克福西部新城
Fig. 12　Western New City of Frankfurt *(source: http://www.schader-stiftung.de/wohn_wandel/285.php)*

图 13　德国海德堡 Emmertsgrund 新城
Fig. 13　Heidelberg New City in Germany Emmertsgrund *(source: http://www.heidelberg-ballon.de/Bildarchiv/Heidelberg/emmertsgrund/Thum1.html)*

图 14　城市逃亡：Cottbus 德国联邦环境协会以 107500 欧元研究资金发展的有效拆除大型板式居住区的体系与机器
Fig. 14　Fleeing from the city: Cottbus German Federal Environment Association effectively removed the large plate residential system and machine with the research fund of 107,500 euros *(source: http://www.innovations-report.de/HTML/BERICHTE/preise_foerderungen/special-2083.html)*

图 16　汉堡新城
Fig. 16　Hamburg New Town *(Source: Tsoe rendering Office)*

图片来源及参考文献 IMAGE SOURCE & REFERENCE

城市发展 URBAN DEVELOPMENT

01 世界文化遗产城市波茨坦城市总体规划
The Urban Development Planning of World Cultural Heritage City of Potsdam

图 01　在过去与未来之间的波茨坦
Fig. 01　Potsdam, between yesterday and tomorrow
(www.aip.de/image_archive/images/einsteinturm_7443_xl.jpg)
(www.flickr.com/photos/ordfabriken/1016271791/sizes/o/)
(www.flickr.com/photos/zustand/81742464/)
(www.bbs-denkmalschutz.de/fileadmin/templates/objekte/Potsdam/Au_Rivage/potsdam2.jpg)
(www.booking.com/images/hotel/org/624/624311.jpg)
(www.potsdam-magazin.de/picture/upload/NH-Voltaire-Hotel-Potsdam.jpg)

图 04　城市整体艺术的组成部分
Fig. 04　Components of an urban all-embracing art form
(www.biosim.agnld.uni-potsdam.de/src/Potsdam1.jpg)
(www.flickr.com/photos/wolfgangstaudt/640522931/sizes/o/)
(www.flickr.com/photos/checco/417471238/)
(www.maerkischeallgemeine.de/cms/bilder/133445/80/800/521/2f3101c5/statue.jpg)
(www.flickr.com/photos/96dpi/3633422134/sizes/o/
farm3.static.flickr.com/2290/2398153025_d41ef013b2.jpg?v=0)

02 丽江玉龙新城规划　Master Plan of Yulong New Town, Yunnan

图 03　原有用地规划
Fig. 03　Predetermined structural concept *(Source: Yunnan Institute of Design)*
图 14　一期典型建筑竖向功能混合布局
Fig. 14　Vertical mixed-use development *(Source: BAUM / ISA)*
图 17　北侧立面序列
Fig. 17　Facade sequences *(Source: BAUM / ISA)*

03 泉州市城市新区建筑天际轮廓控制规划方法研究 (所有图片来源：ISA)
Planning the City Skyline of Quanzhou, Fujian *(All Source: ISA)*

04 泉州市区和环湾地区空间发展战略研究 (所有图片来源：ISA)
Strategic Study on Development of the Urban Districts and General Area around Quanzhou Bay, Fujian *(All Source: ISA)*

05 广州市南沙光谷地区城市发展规划 (所有图片来源：ISA)
Urban Development of Nansha Guanggu in Guangzhou *(All Source: ISA)*

06 阳宗海旅游度假区西北部概念规划设计 (所有图片来源：ISA)
Conceptual Planning of the Northwestern Part of Yangzonghai Resort, Yunnan *(All Source: ISA)*

07 东山蝶岛发展战略规划 (所有图片来源：ISA)
Dongshan Butterfly Island in Fujian: the Strategic Planning of Urban Development *(All Source: ISA)*

08 十堰市东部新城概念规划及城市设计 (所有图片来源：ISA)
Conceptual Planning of the East New Town in Shiyan, Hubei *(All Source: ISA)*

09 福州市东部新城中心城市设计 (所有图片来源：ISA)
City Planning for the Center of the Eastern New Town of Fuzhou, Fujian *(All Source: ISA)*

10 杭州天堂鱼生活：运河新城概念规划 (所有图片来源：ISA)
A Fishing Life Paradise in Hangzhou: Concept Planning for the Canal New Town in Hangzhou *(All Source: ISA)*

11 唐山市南湖生态城概念性总体规划设计及起步区城市设计 (所有图片来源：ISA)
Concept Planning and Urban Design for an Ecological Zone in Tangshan, Hebei *(All Source: ISA)*

12 大同市御东新区概念性总体规划及核心区概念性城市设计 (所有图片来源：ISA)
Conceptual Planning of Yudong District and Urban Design of the Core Area, Datong *(All Source: ISA)*

13 文昌市抱虎角概念规划 (所有图片来源：ISA)
Conceptual Planning for Baohujia Area in Wenchang, Hainan *(All Source: ISA)*

14 昆明市绿地系统概念性规划 (所有图片来源：ISA)
Concept Planning for Kunming Green Space System *(All Source: ISA)*

城市更新 URBAN RENEWAL

01 伦茨堡控制性规划 (所有图片来源：ISA)
Framework Development Planning of Rendsburg *(All Source: ISA)*

02 "霍费尔天空"：提升城市品质的范例项目 (所有图片来源：ISA)
"Heaven of Hof": an Initiative-project for Upgrading the City *(All Source: ISA)*

03 埃尔旺根步行区　Peadestrian Zone of Ellwangen

图 01　修道院与集市广场，2008 年
Fig. 01　Monastery with canon's houses and market place in 2008 *(Source: City of Ellwangen)*
图 03　埃尔旺根城以及老城中心区
Fig. 03　Ellwangen city's historic old town *(Source: city of Ellwangen)*
图 04　从南部看埃尔旺根老城区
Fig. 04　Historical view of Ellwangen from the south *(Louis Zadig,1818; Source: http://www.natur-ostwuerttemberg.de/froelich.html)*
图 05　改建后的 Marien 大街
Fig. 05　Marienstrasse after implementation *(Source: City of Ellwangen)*
图 06　交通方案：红色为步行区，绿色为受交通管制的商业街，蓝色为受交通管制的住宅区巷道
Fig. 06　Traffic concept: pedestrian zones in red, commercial streets as restricted traffic zones in green, residential alleys as restricted traffic zones in blue *(Source: city of Ellwangen)*
图 07　改建后的 Marien 大街
Fig. 07　Marientrasse after reorganization *(Source: city of Ellwangen)*

04 埃斯林根总体规划和公共空间城市规划导则 (所有图片来源：ISA)
Master Plan and Design Guidelines for the Public Spaces of Esslingen am Neckar *(All Source: ISA)*

05 "粉红岛"：韩国首尔方背洞街区广场 (所有图片来源：ISA)
"Pink Islands": a Street square in Bangbae-dong, Seoul, South Korea *(All Source: ISA)*

06 Gangseogu 城市更新规划　Urban Renewal Gangseogu

图 02　两种对立的建筑结构：Hwagokdong 的小尺度街区结构、高层居住区的超级街区结构
Fig. 02　Two contrasting building structures in Hwagokdong with small-scale structures and high-rise housing estates *(Source: City of Seoul District Gangseogu)*
图 03　农业用水道
Fig. 03　Agriculturally used streams *(Photo Source: City of Seoul district Gangseogu)*
图 05　Hwagokdong 的建设结构
Fig. 05　Building structures of Hwagokdong *(Source: City of Seoul District Gangseogu)*

07 北京市昌平区奥运自行车训练馆及周边地区城市设计 (所有图片来源：ISA)
Urban Design of the Area Surrounding Changping Olympic Cycling Training Center, Beijing *(All Source: ISA)*

城市保护及管理
URBAN PRESERVATION AND MANAGEMENT

01 世界文化遗产城市施特拉尔松：城市景观规划 (所有图片来源：ISA)
World Cultural Heritage City of Stralsund: Townscape Planning *(All Source: ISA)*

02 新勃兰登堡城市景观规划　Townscape planning Neubrandenburg

图 01　新勃兰登堡：一个位于内陆的海港城市
Fig. 01　Neubrandenburg: a seaport city in the midland
(www.flickr.com/photos/nb-fotos/3022072674/sizes/o/in/set-72157608873066356)
(http://www.flickr.com/photos/7205019@N04/2793797136/sizes/l/)
(http://www.flickr.com/photos/kd85/2858923965/sizes/o/)
图 03　典型城市景观及典型现代化城市景观对其影响
Fig. 03　A typical townscape with atypical modern architecture
(http://www.flickr.com/photos/87422695@N00/3438256259/sizes/o/)
(http://www.hochzeitinmv.de/Standesamt-Neubrandenburg/files/blocks_image_3_1.png)
(http://www.feierabend.de/images/channel/web/3/7/g.223077.jpg)

03 世界文化遗产城市波茨坦瑙恩郊区及耶格尔郊区的设计导则 (所有图片来源：ISA)
Design Code for the Nauener Vorstadt and the Jaegervorstadt in World Cultural Heritage City of Potsdam *(All Source: ISA)*

04 泉州市法石街区保护与整治规划及环境设计 (所有图片来源：ISA)
Urban and Landscape Design for the Fashi District in Quanzhou, Fujian *(All Source: ISA)*

05 泉州市城市新区公共空间控制规划方法研究 (所有图片来源：ISA)
Research of the Planning Methods for Open Space in Quanzhou, Fujian *(All Source: ISA)*

从两种文化中学习　LEARNING FROM TWO CULTURES

项目概览

区域和城市发展规划

2011年　泉州, 中国
环泉州湾同城化研究
重庆, 中国
北温泉拓展区旅游发展规划
广州, 中国
南沙新区总体规划
西安, 中国
汉长安城遗址区概念规划设计
太原, 中国
南部区域概念规划及重点地段城市设计, 竞赛二等奖
宁波, 中国
象山港发展规划
龙游, 中国
"石窟"旅游功能区概念规划
哈尔滨, 中国
海航中国集项目修建性详细规划

2010年　泉州, 中国
北翼新城发展战略规划
福州, 中国
综合实验区概念性总体规划
北京, 中国
潭柘寺周边地区历史文化片区保护专题研究
广州, 中国
广州市黄埔中心区城市设计
首尔, 韩国
城市空间发展规划

2009年　丽江, 中国
泸沽湖三家村地块概念性设计
台州, 中国
临港新城核心区概念设计及启动区城市设计
唐山, 中国
东南片区震后重建区控制性详细规划与城市设计
唐山, 中国
小商品城与风情小镇修建性详细规划
唐山, 中国
杨柳一面街区域修建性详细规划
泸州, 中国
城市总体发展概念规划
成都, 中国
新都北部新区总体发展概念规划及核心区城市设计
北京, 中国
丰台科技园新东区整体定位及发展战略
太原, 中国
汾河两岸城市设计, 竞赛第一名
魏尔, 德国
历史城市中心发展规划
杭州, 中国
钱江经济开发区概念(战略)规划
杭州, 中国
运河新城概念规划
福州, 中国
马尾组团与快安组团中心区域项目前期研究
文昌, 中国
抱虎角片区总体规划

2008年　文昌, 中国
抱虎角片区概念规划
天津, 中国
蓟县月亮湾旅游规划
东山, 中国
东山岛发展战略规划
柏林, 德国
滕珀尔霍夫区Columbiaquartier与进程相关的城市发展规划
阿纳帕, 俄罗斯
黑海海滨旅游片区的城市规划前期研究
重庆, 中国
主城两江四岸滨江地带溉澜溪片区城市设计
株洲, 中国
湘江风光带滨水公共空间与环境景观设计

2008年　长沙, 中国
"两型社会"综合配套改革试验区大河西先导区规划
北京, 中国
平谷北部区域生态旅游发展策划

2007年　约勒松得, 丹麦
"约勒松得发展愿景2040"
阳宗海, 中国
帆船度假小镇修建性详细规划

2006年　阳宗海, 中国
旅游度假区西北部概念规划设计
阳宗海, 中国
西北片区控制性详细规划
太原, 中国
北部地区概念规划, 竞赛一等奖
福州, 中国
东部新城市发展规划及政府中心区建筑设计
巴黎, 法国
Plateau de Saclay地区城市发展研究及竞赛准备

2005年　昆明, 中国
绿地总体规划, 竞赛一等奖
九江, 中国
八里湖新区概念规划, 竞赛一等奖

2004年　滨州, 中国
区域规划

2003年　上海, 中国
崇明陈家镇东滩侯鸟保护区区域规划及城镇总体规划
上海, 中国
游艇业发展总体规划

1996年　黑肯高, 德国
土地使用规划与景观规划

1995年　伊林根, 德国
城市发展概念研究、生态规划及城市设计

1995-
1996年　伊林根, 德国
城市发展概念研究、生态规划及城市设计

1994年　阿穆博格, 德国
土地使用规划、生态规划及城市设计

1994 -
1995年　耶拿, 德国
洛伯达地区城市发展规划

1993年　赫林斯多夫, 德国
乌泽多姆班泽土地使用规划
塔卡哈那, 智利
城市发展规划及方法研究

1992年　穆尔哈德, 德国
2005用地规划与景观规划, 11个地区的区域发展规划

1992 -
1993年　施特拉尔松, 德国
以城市设计为重点的城市发展规划

1991年　波茨坦, 德国
城市东南地区与北部地区的发展研究

1990 -
1993年　波茨坦, 德国
以文物保护与城市设计为重点的城市发展规划

1980 -
1981年　内卡河畔的埃斯林根, 德国
城市发展规划与城市设计的初步研究

PROJECT LIST

REGIONAL AND URBAN DEVELOPMENT PLANNING

2011	**Quanzhou, China** Rersearch about urban integration around Quanzhou Bay **Chongqing, China** Tourism development planning of the Northern Hotspring **Guangzhou, China** Consulting of the general Masterplan for the new city Nansha **Xi`an, China** Concept strategy for the ruined ancient city of Changan **Taiyuan, China** Conceptual planning and urban planning competition - urban expansion for Taiyuan, competition 2. prize **Ningbo, China** Development planning of the Xiangshan Port **Longyou, China** Concept plan of the grotto tourismus area **Harbin, China** Site plan for China Town of Hainan Airlines
2010	**Quanzhou, China** Planning of the north new citypart Beiyi **Fuzhou, China** Conceptual masterplan of experimental planning area in Pingtan **Beijing, China** Monographic study of cultural area preservation in the surrounding area of Tanzhesi **Guangzhou, China** Urban development planning and urban design for the Huangpu district and the Changzhou island **Seoul, Korea** City image planning for the city of Seoul
2009	**Lijiang, China** Conceptual planning of Luguhu Lake **Taizhou, China** Concepteptual planning and city design of Lingang New Town's city center **Tangshan, China** Development planning and urban design for the south east area of Tangshan after a earthquake **Tangshan, China** Detailed planning of shopping area **Tangshan, China** Detailed regional planning for the Yangliu block **Luzhou, China** Conceptual masterplan of Luzhou **Chengdu, China** Conceptual masterplan and urban design of the core area in north of Chengdu **Beijing, China** Overall positioning and development strategies for Fengtai Science Park **Taiyuan, China** Urban redevelopment for the area along the river Fen, Competition 1st prize **Weil der Stadt, Germany** Development planning of the historical city center **Hangzhou, China** Conceptual (Strategical) planning for Qianjiang Economic Development Zone **Hangzhou, China** Masterplan for the area along the canal, competition 1. prize **Fuzhou, China** Preliminary study of Mawei and Kuai'an **Wenchang, China** Masterplan for Baohujiao area
2008	**Wenchang, China** Conceptual planning for Baohujiao area **Tianjin, China** Strategic planning for Moon Bay Tourism Resort in the district of Jixian **Dongshan, China** Development planning of Dongshan Island **Berlin, Germany** Process-related Urban development, Tempelhofer Feld - Columbiaquartier **Anapa, Russia** Urban planning preliminary studies for a tourism resort on the waterfront of Black Sea
2008	**Chongqing, China** Urban design of the riverfront city core Gailanxi, Chongqing **Zhuzhou, China** Waterfront and landscape planning for Xiang river of Zhuzhou City **Changsha, China** Regional Planning for the city of Changsha **Beijing, China** Tourism development planning of north part of Pinggu
2007	**Öresund, Denmark** "Öresundsvisioner 2040" **Yangzonghai, China** Site plan of Sailing Tourism Town
2006	**Yangzonghai, China** Conceptual planning of the nordwest part of Yangzonghai Resort **Yangzonghai, China** Regional planning and tourism planning for northwest part of Yiliang **Taiyuan, China** Urban development planning for the industrial area Taiyuan-North, int. competition 1. prize **Fuzhou, China** Urban development planning and architectural design of government area, int. competition 1. prize **Paris, France** Urban development study and competition preperation for the Paris-Region-Plateau de Saclay
2005	**Kunming, China** Urban development and landscape planning, int. competition 1.prize **Jiujiang, China** Urban development and new town planning, int. competition 1.prize
2004	**Binzhou, China** Tourist regional planning
2003	**Shanghai, China** Regional planning an new town planning for the new town "Chenjiazhen" on Chongming island **Shanghai, China** Urban development plan for Yachting
1996	**Illingen, Germany** Land use plan with integrated landscape plan
1995	**Heckengaeu, Germany** Land use plan with integrated landscape plan
1995-1996	**Illingen, Germany** Urban development planning with integrated development concept for ecology
1994	**Amtsberg, Germany** Land use plan, integrated development concept for ecology
1994-1995	**Jena, Germany** Urban development planning for the area of Lobeda
1993	**Heringsdorf, Germany** Land use plan for Bansin by Usedom **Talcahuano, Chile** Urban development planning for the whole city, developing of a methodology
1992	**Murrhardt, Germany** Landuse plan and landscape plan 2005, development concepts for 11 selected areas
1992-1993	**Stralsund, Germany** Urban development plan, special focus on urban design
1991	**Potsdam, Germany** Urban development study, southeastern and northern areas
1990-1993	**Potsdam, Germany** Urban development planning, special focus on urban conservation and urban design
1980-1981	**Esslingen am Neckar, Germany** Urban development and urban design, pilotstudy

从两种文化中学习　LEARNING FROM TWO CULTURES

新城规划和总体规划

2011 年　哈尔滨, 中国
　　　　阿城区东部山水新城概念性城市设计
　　　　广州, 中国
　　　　黄埔滨江新城城市设计
　　　　高雄, 台湾
　　　　海港码头城市设计竞赛
　　　　菲林跟, 德国
　　　　城市中心附近居住区的城市设计方案
　　　　西安, 中国
　　　　汉长安城遗址概念规划设计
　　　　太原, 中国
　　　　南部区域概念规划及重点地段城市设计, 竞赛二等奖
　　　　斯图加特, 德国
　　　　Olga 医院及其周边环境设计竞赛
　　　　奥芬堡, 德国
　　　　郊区城市设计邀请竞赛
　　　　泉州, 中国
　　　　百崎湖区域城市设计

2010 年　北京, 中国
　　　　门头沟新城南部地区规划设计
　　　　西柏坡, 中国
　　　　西北干部学院 (西柏坡行政学院) 规划设计

2010 -　迪琴根, 德国
2011 年　郊区城市设计邀请竞赛
　　　　福州, 中国
　　　　琅岐国际旅游度假区规划设计
　　　　大庆, 中国
　　　　乘风庄地块规划设计
　　　　临汾, 中国
　　　　汾河两岸城市设计

2009 -　维也纳, 奥地利
2010 年　"最喜爱的住宅 (Favoritenhoefe)" 整体设计, 规划区 D, 竞赛第 2 阶段
　　　　泉州, 中国
　　　　蟳埔中心区概念性城市设计与修建性详细规划

2009 年　唐山, 中国
　　　　涇阳新城控制性规划与城市设计
　　　　福州, 中国
　　　　海峡高科技园区规划
　　　　哈尔滨, 中国
　　　　群力新区 604 地块居住区规划设计
　　　　重庆, 中国
　　　　渝中区十八梯片区城市设计
　　　　唐山, 中国
　　　　矿山主体公园规划竞赛
　　　　卢森堡, 德国
　　　　施梅尔茨迪朗日空间结构规划, 竞赛三等奖
　　　　深圳, 中国
　　　　宝安中心区碧海片区城市设计

2008 年　深圳, 中国
　　　　宝安中心区碧海片区城市设计
　　　　于伯林根, 德国
　　　　"西部城市入口" 竞赛
　　　　唐山, 中国
　　　　南湖生态城概念性总体规划设计及起步区城市设计
　　　　哈尔滨, 中国
　　　　群力新区东区重点区域修建性详细规划意向设计及项目策划
　　　　北京, 中国
　　　　京东方科技集团恒通商务园区规划
　　　　大同, 中国
　　　　御东新区概念性总体规划及核心区概念性城市设计
　　　　广州, 中国
　　　　白云新城市设计

2007 年　太原, 中国
　　　　西山地区概念性总体规划
　　　　肇庆, 中国
　　　　城东新区概念设计

2007 年　雷克雅未克, 冰岛
　　　　Vatnsmyri 新城总体规划
　　　　北京, 中国
　　　　昌平科技园居住区及配套设计竞赛

2006 年　十堰, 中国
　　　　东部新城概念规划及城市设计
　　　　泉州, 中国
　　　　东海组团城市设计
　　　　拉巴特, 摩洛哥
　　　　新城规划及城市设计方案
　　　　大理, 中国
　　　　海东国际休闲度假小镇规划设计咨询
　　　　郑州, 中国
　　　　中州国际 Hotel 社区规划

2005 年　广州, 中国
　　　　南沙区行政中心概念规划和主体工程建筑设计, 竞赛一等奖
　　　　北京, 中国
　　　　昌平奥运自行车训练馆及周边地区城市设计, 竞赛一等奖
　　　　九江, 中国
　　　　八里湖新区概念规划, 竞赛一等奖
　　　　北京, 中国
　　　　冷泉居住区规划设计
　　　　福州, 中国
　　　　沃川, 韩国
　　　　新居住区规划

2004 年　东营, 中国
　　　　城市新区与城市副中心规划与城市设计
　　　　首尔, 韩国
　　　　恩平新城规划
　　　　温州, 中国
　　　　七都生态岛规划与城市设计
　　　　北京, 中国
　　　　中国国际航空公司后沙峪住区规划
　　　　情人岛, 中国
　　　　旅游度假区规划及建筑设计
　　　　烟台, 中国
　　　　信息产业园规划设计
　　　　丽江, 中国
　　　　玉龙新城规划, 竞赛二等奖

2004 -　北京, 中国
2005 年　上庄生态旅游新镇新城规划

2003 年　上海, 中国
　　　　崇明陈家镇东滩侯鸟保护区区域规划及城镇总体规划
　　　　上海, 中国
　　　　复兴岛及周边地区规划
　　　　上海, 中国
　　　　三林花园区住区规划与设计
　　　　上海, 中国
　　　　外高桥洲海国际生态社区总体规划
　　　　苏州, 中国
　　　　阳澄湖地区唯亭新城与旅游区概念规划

2002 年　上海, 中国
　　　　宝阳别墅保护性改造方案
　　　　凯尔, 德国
　　　　城市中心区更新规划

2002 -　上海, 中国
2003 年　崇明城桥新城概念总体规划及重点地区概念方案

2001 年　上海, 中国
　　　　新江湾城结构规划与详细设计

2000 年　莫斯巴赫, 德国
　　　　大自然中的居住区规划, 竞赛
　　　　普富林根, 德国
　　　　新居住区规划, 竞赛

1999 -　首尔, 韩国
2001 年　首尔新城 Sang Am 发展规划与总体规划

1997 年　首尔, 韩国
　　　　新居住区的开发——居住在高楼中, 竞赛

1996 年　魏玛, 德国
　　　　Ettersburger 街道体系规划

1996 年　巴黎, 法国
　　　　赛让新城市新区总体规划

1996 -　巴黎, 法国
1998 年　航天城赛让新城居住区设计

1993 -　内卡河畔罗腾堡, 德国
1995 年　Frommenhausen 地区发展概念规划

1994 -　古灵根, 德国
1995 年　Frauenzimmern 地区发展概念规划

1993 -　托尔格洛, 德国
1994 年　"Droegeheide" 木质房屋居住区改造与更新规划设计导则

1992 年　施特拉尔松, 德国
　　　　Ossenreyer 与 Heilgeist 大街间区域详细性规划

1992 年　施特拉尔松, 德国
　　　　Triebseer Vorstadt 工业区未来发展的详细性研究

1992 年　施特拉尔松, 德国
　　　　Knieper West 大规模居住区未来发展的详细性研究

1992 -　施特拉尔松, 德国
1993 年　Spuel 岛及火车站周边地区总体规划及城市设计

1991 年　魏斯巴赫, 德国
　　　　城市总体规划

1991 年　比格施塔特, 德国
　　　　城市中心区总体规划

1991 年　腓特烈斯, 德国
　　　　Westersielzug 内港区区域总体规划

1991 年　波茨坦, 德国
　　　　Drewitz Kirchsteigfeld 地区总体规划

1991 年　托尔格洛, 德国
　　　　城市及其中心区总体规划

1991 -　施特拉尔松, 德国
1992 年　港口区域总体规划
　　　　赫林斯多夫, 德国
　　　　班森总体规划
　　　　赫林斯多夫, 德国
　　　　阿卑克海滨总体规划
　　　　安克拉姆, 德国
　　　　建于上世纪 30 年代的 Gellendiner 居住区发展概念规划
　　　　费尔贝林, 德国
　　　　城市中心区发展概念规划

1991 -　施特拉尔松, 德国
1993 年　海港岛总体规划

1989 年　卡尔斯鲁厄, 德国
　　　　Vogelsand 科技园设计, 竞赛优胜奖

1989 -　斯图加特, 德国
1990 年　以城市设计为重点的 Feuerbach 东部工业区发展研究, 老工业区更新

1988 -　普富伦多夫, 德国
1989 年　历史城区总体规划

1986 年　奥夫特尔斯海姆, 德国
　　　　城市中心区总体规划

1986 -　阿尔伯斯多夫, 德国
1987 年　城市中心区总体规划

1986 -　舒特瓦尔德, 德国
1987 年　城市中心区总体规划

NEW TOWN PLANNING AND MASTERPLAN

2011 Harbin, China
Urban design of East New Town
Guangzhou, China
Urban design of the waterfront Huangpu New Town
Kaohsiung, Taiwan
Kaohsiung Port Station Urban Design Competition
Villingen, Germany
Urban design concept for a residential area near the city center
Xi`an, China
Concept strategy for the ruined ancient city of Changan
Taiyuan, China
Conceptual planning and urban planning competition - urban expansion for Taiyuan, competition 2. prize
Stuttgart, Germany
Planning competition for the area of the Olgahospital and its environment
Offenburg, Germany
Urban concept planning for the outskirts of the city of Offenburg, invited competition

2010 Quanzhou, China
Masterplan for the eastern area around baiqi lake
Beijing, China
Planning of the south part of the new town Mentougou
Xibaipo, China
Masterplan for a specialist training school in Xibaipo

2010 - 2011 Ditzingen, Germany
Urban concept planning for the outskirts of the city of Ditzingen, invited competition

2010 Fuzhou, China
Urban planning and tourism concept for Langqi island
Daqing, China
Urban concept for the planning area Chengfengzhuang
Linfen, China
Masterplaning for the area along the Fen River, competition 1st prize

2009- 2010 Vienna, Austria
Masterplan for the "Favoritenhoefe", Development area D, Competition 2. phase
Quanzhou, China
Masterplannning for a CBD district in the Xunpu area

2009 Tangshan, China
Regulatory plan and city design for the Gengyang New Town
Fuzhou, China
Planning for the High-Tech Park at Fuzhou Canal
Harbin, China
Planning and design for the Qunli new district on land Nr. 604
Chongqing, China
Urban design for the Shibati district of Yuzhong, competition 1st prize
Tangshan, China
Tangshan Main Mine Park Planning, Competition
Luxemburg, Germany
Spatial and structural concept for Schmelz Diddeleng, Competition 3. prize
Guangzhou, China
Concept planning and city design of high-tech park Guanggu in Nansha

2008 Shenzhen, China
Masterplanning for the Bi Hai area in the city center
Bao'an
Ueberlingen, Germany
"City Entrance West", Competition
Tangshan, China
Conceptual masterplan of Tangshan Nanhu Eco-City
Harbin, China
New Town Planning for the east part of Qunli, Competition 1. prize
Beijing, China
Planning of the Hengtong Business Park
Datong, China
New Town Master Planning
Guangzhou, China
Urban development planning of a new district "Baiyun New Town"

2007 Taiyuan, China
Conceptual Planning of the Xishan area of Taiyuan
Zhaoqing, China
New Town Planning for the Eastcity New District
Reykjavik, Iceland
New Town Master Planning for Vatnsmyri
Beijing, China
Dense Residential Area "Ermao District" in Changping, Competition

2006 Shiyan, China
Shiyan East New Town - conceptual planning and urban design
Quanzhou, China
Planning of new central district Donghai
Rabat, Morocco
Planning of a new town and urban design concept
Dali, China
Planning of an international tourist resort area
Zhengzhou, China
Planning of a new resindential area and hotel complex

2005 Guangzhou, China
New town planning and structural planning for the new goevernment area in Nansha district (int. competition 1. prize)
Beijing, China
Urban renewal and strutural planning for the new town of Changping, competition 1.prize
Jiujiang, China
urban development and new town planning, int. Competition 1.prize
Beijing, China
New garden residential area - Yicheng Lengquan
Okcheon, Korea
New residential area

2004 Dongying, China
New town planning
Seoul, Korea
New town planning of Eunpyeong district
Wenzhou, China
New town planning for Qidu Island, competition 1.prize
Beijing, China
Planning of a new residential area for Air China
Qingren Island, China
Planning of a new luxurious residential area
Yantai, China
Masterplan fuer the science and industry area
Lijiang, China
New town planning "Yulong New Town", competition 2.prize

2004- 2005 Beijing, China
Planning of a new town "Shangzhuangzhen"

2003 Shanghai, China
Regional planning an new town planning for the new town "Chenjiazhen" on Chongming island
Shanghai, China
Urban renewal planning of Fuxing Island
Shanghai, China
Renewal of the residential area Sanlin
Shanghai, China
Planning of the new city district Zhouhai
Suzhou, China
New town planning for Weiting at Yangcheng Lake

2002 Shanghai, China
Redevelopment and conversion of the historic villa district Baoyang and architectural design
Kehl, Germany
Urban renewal concept for the city center

2002- 2003 Shanghai, China
New town planning and urban design concept of Cheng Qiao New Town

2001 Shanghai, China
New town planning and design guidelines for "New Jiang Wan Town"

2000 Mosbach, Germany
New residential area in the nature, competition
Pfullingen, Germany
Planning of a new residential area, competition

1999- 2001 Seoul, Korea
Development plan and masterplan for a new city center "Sang Am"

1997 Seoul, Korea
Development of a new residential area - living in high-rise buildings, competition

1996 Weimar, Germany
Masterplan for the area "Ettersburger Strasse"
Paris, France
Masterplan for a new city district in Cergy-Pontoise

1996- 1998 Paris, France
Planning of a new residential district in Sattelite-City Cergy-Pontoise

1993- 1995 Rottenburg am Neckar, Germany
Conceptual masterplan for the district Frommenhausen

1994- 1995 Gueglingen, Germany
Conceptual Masterplan for the district Frauenzimmern

1993- 1994 Torgelow, Germany
Design guidelines for conversion and redevelopment of the wooden housing estate "Droegeheide"

1992 Stralsund, Germany
Masterplan for the Ossenreyer- and Heilgeiststreet
Stralsund, Germany
Detailed studies for the future development of "Triebseer Vorstadt", industrial area
Stralsund, Germany
Detailed studies for the future development of "Knieper West", mass housing estate

1992- 1993 Stralsund, Germany
Masterplan and urban design for the island of Spuel and the area around the train station

1991 Weissbach, Germany
Masterplan
Burgstaedt, Germany
Masterplan for the city center
Friedrichstadt, Germany
Masterplan for Westersielzug, Interior Harbor
Potsdam, Germany
Masterplan Drewitz - Kirchsteigfeld
Torgelow, Germany
Masterplan for the whole city and the city center

1991- 1992 Stralsund, Germany
Masterplan for the harbor area
Heringsdorf, Germany
Masterplan for the Seaside Bansin
Heringsdorf, Germany
Masterplan for the Seaside Ahlbeck
Anklam, Germany
Masterplan for a neighborhood of the 30ies "Gellendiner Siedlung"
Fehrbellin, Germany
Masterplan for the city center

1991- 1993 Stralsund, Germany
Masterplan for the harbour island

1989 Karlsruhe, Germany
Planning of "Technology Park Vogelsand" (urban design competition, merit award)

1989- 1990 Stuttgart, Germany
Development study for Feuerbach-Ost, urban renewal of a historic industrial area. special focus on urban design

1988- 1989 Pfullendorf, Germany
Masterplan for the historic area

1986 Oftersheim, Germany
Masterplan for the downtown

1986- 1987 Albersdorf, Germany
Masterplan for the downtown

1986- 1987 Schutterwald, Germany
Masterplan for the downtown

城市设计

2011 年	武汉，中国 二环线汉阳段（龙阳大道至滨江大道）城市设计 南京，中国 海峡两岸科技工业园城市设计	
2010 - 2011 年	普福尔茨海姆，德国 Jahn 大街城市规划	
2010 年	唐山，中国 大成山周边地区城市设计 福州，中国 晋安新城茶会核心区城市设计	
2009 - 2010 年	首尔，韩国 城市空间发展规划	
2009 年	南京，中国 浦口新城公共活动轴线及两侧区域城市设计 唐山，中国 丰润北部新区概念规划 昆明，中国 机场高速路两侧景观设计与城市设计 江阴，中国 中心城区总体城市设计研究 泉州，中国 晋江世纪大道片区控规与城市设计	
2008 - 2009 年	埃斯林根，德国 城市中心区太阳能电池设计导则	
2008 年	蒂宾根，德国 Steinlach 东部地区设计推荐	
2007 年	约勒松得，丹麦 "约勒松得发展愿景 2040" 泉州，中国 闽南（泉州）传统建筑文化在新区建设中的延续和发展研究	
2004 - 2006 年	研究项目 企业园区气候罩	
2003 年	青岛，中国 麦岛城居住区规划 上海，中国 游艇业发展规划	
2002 - 2004 年	泰克山上基尔夏伊姆，德国 城市中心区色彩设计导则	
2002 年	内卡河畔埃斯林根，德国 城市中心区城市入口广场设计导则	
2001 年	内卡河畔埃斯林根，德国 采尔地区城市公共空间——街道和广场设计导则	
2001 - 2007 年	内卡河畔埃斯林根，德国 城市中心区公共和私人街道小品设计导则	
2000 年	内卡河畔埃斯林根，德国 火车站大街公共和私人街道小品设计导则	
2000 - 2001 年	内卡河畔埃斯林根，德国 梅廷根地区城市公共空间——街道和广场设计导则	
1999 年	康斯坦茨，德国 Stromeyersdorf 工业园区开发和设计导则	
1999 - 2000 年	内卡河畔埃斯林根，德国 城市中心区街道和广场城市设计	
1997 - 1999 年	波茨坦，德国 世界遗产地区城市设计	
1998 - 2000 年	波茨坦，德国 Jaeger 郊区设计导则	
1998 - 2000 年	波茨坦，德国 Nauener 郊区设计导则	
1997 - 1998 年	波茨坦，德国 城市整体视廊规划	
1997 - 2000 年	新勃兰登堡，德国 城市形象规划	
1997 - 2000 年	波茨坦，德国 整个城区文物保护规划	
1996 年	魏玛，德国 北部地区大型居住区城市设计导则	
1996 - 1998 年	魏玛，德国 北部地区大型居住区城市导则	
1995 - 1996 年	格拉，德国 Lusan 大型居住区色彩设计导则	

1995 - 1996 年	耶拿，德国 洛卑达大型居住区色彩设计导则	
1995 - 1996 年	格里门，德国 历史城区设计规范	
1994 年	拉桑，德国 历史城区设计规范	
1993 - 1994 年	格里门，德国 历史城区设计导则手册	
1994 年	安克拉姆，德国 历史城区设计导则	
	博伊岑堡，德国 "窗，门，外门与屋顶"信息宣传手册	
	波茨坦，德国 仓库城的城市设计导则和视觉轴线规划	
1994 - 1995 年	福冈，日本 生态岛滨水地区生态规划以及城市形象规划	
1994 - 1999 年	康斯坦茨，德国 Stromeyersdorf 工业园区设计协调与实施工作	
1993 - 1994 年	奥拉宁堡，德国 Eden 地区设计导则手册	
1993 年	安克拉姆，德国 设计规范	
	格里门，德国 标识与宣传册的设计规范	
1992 - 1994 年	波茨坦，德国 整体城市景观规划	
1992 年	格里门，德国 历史城区——集市广场的城市设计导则	
1992 - 1993 年	埃格辛，德国 历史城区景观规划	
1992 - 1993 年	彭昆，德国 历史城区景观规划	
1992 - 1994 年	居茨科，德国 历史城区景观规划	
1992 - 1994 年	海尔布隆，德国 城市中心区城市设计导则	
1992 - 1994 年	亚尔门，德国 历史城区景观规划	
1992 - 1994 年	托尔格洛，德国 城市中心区景观规划	
1991 - 1992 年	赫林斯多夫，德国 班森总体规划	
	赫林斯多夫，德国 阿卑尔海滨总体规划	
	于克明德，德国 历史城区景观规划	
1991 - 1993 年	拉桑，德国 城市中心区景观规划	
	施特拉尔松，德国 历史城区设计规范	
1991 - 1994 年	安克拉姆，德国 城市中心区景观规划	
	博伊岑堡，德国 历史城区景观规划	
1990 年	阿尔布施塔特，德国 Ebingen 历史城区设计规范	
	康斯坦茨，德国 Stromeyersdorf 工业区概念性城市设计及公共空间设计导则	
	默尔恩，德国 历史城区设计规范实施	
1990 - 1992 年	施特拉尔松，德国 历史城区景观规划	
1990 - 1994 年	维尔茨堡，德国 城市景观分析与城市设计导则	
1989 - 1990 年	腓特烈斯港，德国 城市中心区景观规划	
	奥芬堡，德国 城市入口设计导则	
1988 - 1989 年	弗伦斯堡，德国 东部历史城区景观设计	

1988 - 1989 年	普富伦多夫，德国 历史区街道设计，一般概念设计及公共空间设计导则	
1987 年	斯图加特，德国 Berg 地区街道设计与城市设计导则	
1987 - 1988 年	斯图加特，德国 Feuerbach 东部工业区街道设计及城市设计导则	
1987 - 1989 年	腓特烈施塔特，德国 历史城区景观设计	
	腓特烈施塔特，德国 历史城区设计规约	
1986 - 1987 年	阿尔伯斯多夫，德国 中心城区设计规范	
	阿尔伯斯多夫，德国 中心城区景观设计	
	阿尔布施塔特，德国 Ebingen 历史城区城市设计	
1985 - 1987 年	海尔布隆，德国 城市中心区总体规划	
1984 年	迪琴根，德国 Schoeckingen 地区景观规划	
1984 - 1985 年	内卡河畔埃斯林根，德国 Oberesslingen 地区街坊改造，城市设计及城市设计导则	
1982 年	安格尔巴塔尔，德国 Michelfeld 中心区改造及城市设计导则	
1982 - 1983 年	比蒂希海姆，德国 历史城区城市设计及城市设计导则	
	默尔恩，德国 历史城区城市设计及城市设计导则	
	默尔恩，德国 历史城区城市设计	
1981 年	伦茨堡，德国 城市景观规划	
1980 - 1981 年	内卡河畔的埃斯林根，德国 城市发展规划与城市设计的初步研究	
	伦茨堡，德国 Neuwerk 历史城区设计规范	
1979 - 1980 年	乌尔姆，德国 Herdbrucker 大街和 Donaufront 历史城区城市设计导则	
1979 - 1981 年	路德维希堡，德国 城市中心区总体规划	
1977 - 1978 年	内卡河畔埃斯林根，德国 城市中心区设计规范	
1977 - 1978 年	内卡河畔埃斯林根，德国 城市景观分析与城市设计导则	
1974 - 1977 年	吕贝克，德国 城市中心区设计规范	
	吕贝克，德国 城市中心区景观规划	
1973 年	莱昂贝格，德国 总体城市规划	

URBAN DESIGN PLANNING

2011	**Wuhan, China** Urban design of the 2nd ring road **Nanjing, China** Urban design of the Strait High-Tech Industry Park
2010-2011	**Pforzheim, Germany** Urban planning workshop for Jahnstrasse
2010	**Tangshan, China** Urban design for the surrounding area of Dachengshan **Fuzhou, China** City design of the Chahui new town
2009-2010	**Seoul, Korea** City image planning for the city of Seoul
2009	**Nanjing, China** Urban design for the Pukou new town **Tangshan, China** Conceptual urban design of the north area of Fengrun District **Kunming, China** Landscape planning and urban design for both sides of Kunming airport highway **Jiangyin, China** Urban design study of Jiangyin city center
2008-2009	**Quanzhou, China** Structural planning and city design for districts at Jinjiang Century Avenue
2008	**Esslingen, Germany** Design Guidelines for solar cells in the city center
2007	**Tuebingen, Germany** Design commendations for the area "Östlich der Steinlach" **Öresund, Denmark** „Öresundsvisioner 2040"
2004-2006	**Quanzhou, China** Planning of the city´s silhouette
2003	**Research project** Climatic covers for industrial areas **Qingdao, China** Urban design concept for the Maidao Suburb Area
2002-2004	**Shanghai, China** Urban development plan for yachting
2002	**Kirchheim u. Teck, Germany** Colour guidelines for the city center
2001	**Esslingen am Neckar, Germany** Guideline planning of squares ans city entrances for the city center
2001-2007	**Esslingen am Neckar, Germany** Guideline planning for urban spaces - street and squares - for Zell District
2000	**Esslingen am Neckar, Germany** Design guidelines for the public and private street furnitures for the whole city center
2000-2001	**Esslingen am Neckar, Germany** Design guidelines for the public and private street furnitures in the Bahnhofstrasse
1999	**Esslingen am Neckar, Germany** Guide planning for urban spaces - street and squares - in Mettingen District
1999-2000	**Konstanz, Germany** Development of design guidelines for the industrial park of Konstanz-Stromeyersdorf
1997-1999	**Esslingen am Neckar, Germany** Urban design guidelines for the streets and squares in the city center
1998-2000	**Potsdam, Germany** Urban design plan for the area of the Unesco world heritage
1998-2000	**Potsdam, Germany** Design rules for Jaegervorstadt
1997-1998	**Potsdam, Germany** Design rules for Nauener Vorstadt
1997-2000	**Potsdam, Germany** Planning of the view axis for the whole city
1997-2000	**Neubrandenburg, Germany** City image plan
1996	**Potsdam, Germany** Monument preservation plan for the whole city
1996-1998	**Weimar, Germany** Urban design guideline for the mass housing estate in Weimar-North
1995-1996	**Weimar, Germany** Urban guidelines for mass housing estate in Weimar-North **Gera, Germany** Colour guidelines for the mass housing estate of Lusan **Jena, Germany** Colour guidelines for the mass housing estate of Jena-Lobeda
1994	**Grimmen, Germany** Design codes for the historic area
1993-1994	**Lassan, Germany** Design codes for the historic area
1994	**Grimmen, Germany** Manual of design guideline for the historic area **Anklam, Germany** Urban design guideline for the historic area **Boizenburg a. Elbe, Germany** Information flyer for "Windows, Doors, Gates" and "Roofs"
1994-1995	**Potsdam, Germany** Urban design guideline and view axis planning of Speicherstadt
1994-1999	**Fukuoka, Japan** Urban design planning with special focus on ecology for the waterfront area "Ecopark - Island City"
1993-1994	**Konstanz, Germany** Design coordination and realization of the industrial park Stromeyersdorf
1993	**Oranienburg, Germany** Manual of design guideline for Eden **Anklam, Germany** Design codes
1992-1994	**Grimmen, Germany** Design codes for the signage and brochure
1992	**Potsdam, Germany** City image plan for the whole city
1992-1993	**Grimmen, Germany** Urban design guideline for the historic area - Market Square **Eggesin, Germany** City image plan for the historic area
1992-1994	**Penkun, Germany** City image planning of the historic area **Guetzkow, Germany** City image plan for the historic area **Heilbronn, Germany** Urban design guideline for the city center **Jarmen, Germany** City image plan for the historic area
1991-1992	**Torgelow, Germany** City image plan for the city center **Heringsdorf, Germany** Masterplan for the Seaside Bansin **Heringsdorf, Germany** Masterplan for the Seaside Ahlbeck
1991-1993	**Ueckermuende, Germany** City image plan for the historic area **Lassan, Germany** City image plan for the city center
1991-1994	**Stralsund, Germany** Design codes for the historic area **Anklam, Germany** City image plan for the city center
1990	**Boizenburg, Germany** City image plan for the historic area **Albstadt, Germany** Design codes for the historic area of Ebingen **Konstanz, Germany** Conceptual urban design planning and design guideline for public spaces in industrial area of Stromeyersdorf
1990-1992	**Moelln, Germany** Actualisation of design codes for the historic area
1990-1994	**Stralsund, Germany** City image plan for the historic area
1989-1990	**Wuerzburg, Germany** City image analysis and urban design guideline **Freidrichshafen, Germany** City image plan for the city center
1988-1989	**Offenburg, Germany** Urban design guideline for city entrances **Flensburg, Germany** City image plan for the eastern historic area
1987	**Pfullendorf, Germany** Street design, general concept and design guideline for public spaces in the historic area
1987-1988	**Stuttgart, Germany** Street design and urban design guidelines for Stuttgart-Berg
1987-1989	**Stuttgart, Germany** Street design and urban design guidelines for the industrial area Feuerbach-East
1987-1989	**Friedrichstadt, Germany** City image plan for the historic area
1986-1987	**Friedrichstadt, Germany** Design codes of the historic area **Albersdorf, Germany** Design codes of the downtown **Albersdorf, Germany** City image plan for the downtown
1985-1987	**Albstadt, Germany** City image plan for the historic area of Ebingen
1984	**Heilbronn, Germany** Urban design masterplan for the city center
1984-1985	**Ditzingen, Germany** City image plan of Schoeckingen
1982	**Esslingen am Neckar, Germany** Neighborhood improvement, urban design planning and urban design guideline of Oberesslingen
1982-1983	**Angelbachtal, Germany** Reorganization and urban design guidelines for the center of Michelfeld
1982-1983	**Bietigheim, Germany** Urban design plan and urban design guideline for the historic area
1982-1983	**Moelln, Germany** Design codes of the historic area **Moelln, Germany** Urban design plan for the historic area
1981	**Rendsburg, Germany** City image plan
1980-1981	**Esslingen am Neckar, Germany** Urban development and urban design, pilotstudy
1980-1981	**Rendsburg, Germany** Design codes of the city center, "Historic Area"-
1979-1980	"Neuwerk" **Ulm, Germany** Urban design guideline for the historic area, Herdbrucker Street and Donaufront
1979-1981	**Ludwigsburg, Germany** Urban design masterplan for the city center
1977-1978	**Esslingen am Neckar, Germany** Design codes of the city center
1977-1978	**Esslingen am Neckar, Germany** City image analysis and urban design guidelines
1974-1977	**Luebeck, Germany** Design codes of the city center
1974-1977	**Luebeck, Germany** City image planning of the city center
1973	**Leonberg, Germany** Urban design masterplan

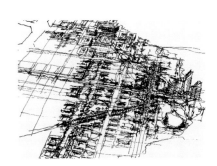

从两种文化中学习 LEARNING FROM TWO CULTURES

详细规划

2011 年	广州, 中国		1996 年	耶拿, 德国
	华南国际港航服务中心规划与建筑概念设计			洛伯达地区详细规划, 更新与协同
2009 年	西柏坡, 中国		1996 -	巴黎, 法国
	西北干部学院 (西柏坡行政学院) 规划设计		1997 年	赛让新城 Quatier de Liesse 地区详细规划
	南京, 中国		1995 -	阿姆茨贝尔格, 德国
	生命科技园控制性详细规划与详细规划		1996 年	城市新中心详细规划
2007 年	阳宗海, 中国			阿姆茨贝尔格, 德国
	柏联国际温泉度假小镇规划			Schloesschen 地区详细规划
	西安, 中国		1994 -	居格林根, 德国
	大明宫区域概念规划设计		1996 年	Frauenzimmern Gaessle 地区详细规划
	泉州, 中国			耶拿, 德国
	法石街区保护与整治规划及环境设计			洛伯达大型居住区详细规划
2006 年	阳宗海, 中国		1993 年	舒特瓦尔德, 德国
	西北片区控制性详细规划			基希贝格北部地区详细规划
	内卡河畔埃斯林根, 德国			穆尔哈德, 德国
	采尔南部地区公共空间详细规划			Kirchenkirnberg - Lettengasse 详细规划
	斯图加特, 德国			韦尔特海姆, 德国
	斯图加特西区 Olga 医院可行性研究			Dietenhan 地区初步设计与详细规划
	韦德尔, 德国		1992 -	格里门, 德国
	Schulauer 港口结构规划概念设计		1993 年	以城市设计为重点的详细规划
2004 -	丽江, 中国		1992 -	居茨科, 德国
2005 年	玉龙新城规划		1994 年	详细规划
2004 年	北京, 中国		1991 -	施特拉尔松, 德国
	十三陵明帝陵世界文化遗产保护与发展规划研究与咨询		1992 年	历史城区单栋详细规划
2004 -	厦门, 中国		1990 年	奥芬堡, 德国
2005 年	滨北金融商贸区城市设计, 竞赛一等奖			城市入口详细规划
2002 -	内卡河畔埃斯林根, 德国		1990 -	阿尔伯斯多夫, 德国
2003 年	Maille 地区详细规划及 Fabrik 和 Neckar 大街规划		1991 年	以生态规划为重点的中心城区详细规划
2001 年	内卡河畔埃斯林根, 德国		1987 年	斯图加特, 德国
	采尔地区详细规划			Berg 地区详细规划
	内卡河畔赖兴巴赫, 德国		1987 -	默尔恩, 德国
	公共空间详细规划		1989 年	中心城区详细规划
2001 -	康斯坦茨, 德国		1985 -	默尔恩, 德国
2006 年	Stromeyersdorf 工业园区详细规划		1986 年	火车站地区和 Berg 大街详细规划
1997 -	巴黎, 法国		1981 -	布龙斯比特尔, 德国
1998 年	赛让新城 Quatier de Liesse 地区第一阶段概念性详细规划		1982 年	城市设计与详细规划
1997 -	波茨坦, 德国		1981 -	布龙斯比特尔, 德国
1999 年	Marx 大街 Neu-Babelsberg 别墅区详细规划		1982 年	以 Suederbelmhusen B29 地区城市设计为重点的详细规划
1996 年	明斯特塔尔, 德国		1981 -	伦茨堡, 德国
	Parkhotel 地区概念性详细规划		1983 年	中心历史城区与 Neuwerk 地区详细规划
	安克拉姆, 德国		1979 -	乌尔姆, 德国
	历史城区详细规划, 更新与咨询		1980 年	城市中心 Herdbrucker 大街地区详细规划

街道, 广场与公共小品设计

2011 年	明登, 德国		2008 年	凯尔, 德国
	步行区重新设计竞赛			城市入口设计研究
2010 年	太原, 中国		2007 年	釜山, 韩国
	东三道巷历史文化街区规划设计			Haeundae 城市更新设计
	非斯, 摩洛哥		2007 -	霍夫, 德国
	Lalla Yeddouna 地区设计		2009 年	城市中心区提升城市品质的街道设计
	埃斯林根, 德国		2007 年	斯图加特, 德国
	新中央车站覆顶设计竞赛			Helfferich 大街重新设计
	卡尔斯鲁厄, 德国		2006 -	内卡河畔埃斯林根, 德国
	主要消防站和控制站设计竞赛		2007 年	采尔地区 Wilhelmstr./Bachstr. - Kennerplatz 公共空间重新设计
2009 -	锡根, 德国		2006 年	弗赖堡, 德国
2010 年	Sieg 河的开发与周边地区重新规划竞赛			Platz Der Alten Synagoge 设计竞赛
2009 年	北京, 中国			威斯巴登, 德国
	轨道交通房山线长阳镇站及周边用地一体化设计			Synagoge, Michelsberg 历史地区概念性结构规划与城市设计
	埃尔旺根, 德国			伯布林根, 德国
	集市广场重新设计竞赛			Elbenplatz 重新设计, 多方委托
	慕尼黑, 德国			郑州, 中国
	Prinz-Eugen-Kaserne 地区城市和景观重新设计竞赛			航海东路景观规划
	考夫博伊伦, 德国		2005 年	广州, 中国
	Kemptener Tor 街道与广场空间重新设计竞赛			珠江新城核心区地下空间及中央广场建筑方案设计, 竞赛四等奖
2008 年	碛口, 中国			内卡河畔埃斯林根, 德国
	古镇中市街恢复性规划方案设计			梅廷根城市更新区公共空间设计
	埃斯林根, 德国			比尔, 德国
	Waeldenbronner 大街重新设计初步设计方案			结构规划与城市设计竞赛
2008 -	埃斯林根, 德国		2004 年	科恩韦斯特海姆, 德国
2009 年	Am Backhaeusle 广场公共空间设计			集市广场灯光规划与照明要素设计
2008 年	泉州, 中国		2004 -	内卡河畔埃斯林根, 德国
	城市新区公共空间控制规划方法研究		2005 年	视觉保护墙设计

STRUCTURE PLAN AND BUILDING PLAN

2011	**Guangzhou, China** — Planning and conceptional architecture design of Huanan international port and ship service center
2009	**Xibaipo, China** — Concept plan for the Xibaipo Institut
	Nanjing, China — Development planning and architectural design for the Bio-High-Tech Park
2007	**Yangzonghai, China** — Implementation planning of Bolian Yachting holiday resort at Yangzonghai Lake
	Xi`an, China — Urban Concept Planning and Design Project of Da Ming Palace Region
	Quanzhou, China — Rehabilitation of an area of conservation of ancient monuments
2006	**Yangzonghai, China** — Regulatory detailed planning for a part of Yangzhonghai District
	Esslingen am Neckar, Germany — Structure planning for the public spaces for the southern part of the Zell District
	Stuttgart, Germany — Feasibility study for Olgahospital, Stuttgart-West
	Wedel, Germany — Conceptual structure planning of Schulauer Hafen
2004-2005	**Lijiang, China** — Yulong New Town - structure planning and focus planning
2004	**Beijing, China** — Structure planning concept for the Ming-Graves area
2004-2005	**Xiamen, China** — Urban structure planning for the central district of Xiamen, competition 1. prize
2002-2003	**Esslingen am Neckar, Germany** — Structure planning of Maille and district planning for the Fabrik-/ Neckarstrasse area
2001	**Esslingen am Neckar, Germany** — Structure planning Zell District
	Reichenbach am Neckar, Germany — Structure planning concept for the public spaces
2001-2006	**Konstanz, Germany** — Building plan for industrial park Konstanz-Stromeyersdorf
1997-1998	**Paris, France** — Conceptual building plan for the first phase of realization Quartier de Liesse in Cergy-Pontoise
1997-1999	**Potsdam, Germany** — Building plan for the villa district Neu-Babelsberg Karl-Marx-Strasse
1996	**Muenstertal, Germany** — Conceptual building plan for the area of Parkhotel
1996	**Anklam, Germany** — Structure plan for the historic area, update and consulting
	Jena, Germany — Structure plan Jena-Lobeda, update and coordination
1996-1997	**Paris, France** — Structure plan for the district "Quartier de Liesse" in Cergy-Pontoise
1995-1996	**Amtsberg, Germany** — Building plan for the new center
	Amtsberg, Germany — Building plan for the Schloesschen district
1994-1996	**Gueglingen, Germany** — Building plan for the district Frauenzimmern Gaessle
	Jena, Germany — Structure plan for the housing estate of Jena-Lobeda
1993	**Schutterwald, Germany** — Building plan Kirchberg - Nord
	Murrhardt, Germany — Building plan Kirchenkirnberg - Lettengasse
	Wertheim, Germany — Preliminary design and building plan Dietenhan
1992-1993	**Grimmen, Germany** — Structure plan with special focus on urban design
1992-1994	**Guetzkow, Germany** — Structure plan
1991-1992	**Stralsund, Germany** — Planning for single blocks in the historic area
1990	**Offenburg, Germany** — Structure plan for city entrances
1990-1991	**Albersdorf, Germany** — Structure plan for the downtown with special focus on ecological planning
1987	**Stuttgart, Germany** — Structure plan for Stuttgart-Berg
1987-1989	**Moelln, Germany** — Structure plan for the downtown
1985-1986	**Moelln, Germany** — Structure plan for "Train Station Area" and "Bergstrasse"
1981-1982	**Brunsbuettel, Germany** — Urban design and building plan
1981-1982	**Brunsbuettel, Germany** — Structure plan with special focus on urban design for Suederbelmhusen Area B29
1981-1983	**Rendsburg, Germany** — Structure plan for the downtown "Historic Area" and "Neuwerk"
1979-1980	**Ulm, Germany** — Building plan of the city center, Herdbrucker Street area

STREET, PLACE AND PUBLIC FURNITURE DESIGN

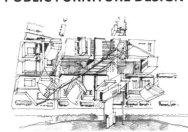

2011	**Minden, Germany** — Competition for redesign of the pedestrian zone
2010	**Taiyuan, China** — Planning of the Dongsanxiang historic conservation area
	Fes, Morocco — Design of Place Lalla Yeddouna
	Esslingen, Germany — Design of a shelter for the new central bus station, Competition
	Karlsruhe, Germany — New planning of the main fire station and control station, Competition
2009-2010	**Siegen, Germany** — Unearthing of the river Sieg, reorganisation of the urban surroundings, Competition
2009	**Beijing, China** — City design of the Metro-Stadtion Changyangzhen in Fangshan
	Ellwangen, Germany — Redesign of the market square, competition
	Munich, Germany — Redesign of the area of Prinz-Eugen-Kaserne, urban and landscape competition
	Kaufbeuren, Germany — Redesign of street- and square spaces at Kemptener Tor in Kaufbeuren, competition
2008	**Qikou, China** — Restoration design of Zhongshi street
	Esslingen, Germany — Preliminary design for redesign of the Waeldenbronnerstrasse
2008-2009	**Esslingen, Germany** — Public square design of " Platz am Backhaeusle"
2008	**Quanzhou, China** — Urban development of Quanzhou, research project
	Kehl, Germany — Workshop "Stadteingang Kehl"
2007	**Busan, Korea** — Stadterneuerungskonzept Haeundae
2007-2009	**Hof, Germany** — Street design for revaluation of the city center
2007	**Stuttgart, Germany** — Redesign of the Helfferich Street
2006-2007	**Esslingen am Neckar, Germany** — Public space redesign: Wilhelmstr./Bachstr. - Kennerplatz, Zell District
2006	**Freiburg, Germany** — Place design "Platz Der Alten Synagoge", competition
	Wiesbaden, Germany — Conceptual structure planning and place design "Ehemalige Synagoge, Michelsberg", competition
	Boeblingen, Germany — Redesign of the Elbenplatzes, Mehrfachbeauftragung
	Zhengzhou, China — Street design of the Ost Hanghai Street, int. competition
2005	**Guangzhou, China** — Central Plaza in Zhujiang (int. competition 4.prize)
	Esslingen am Neckar, Germany — Public spaces design for the urban renewal area, Mettingen district
	Buehl, Germany — Structural planning concept and place design, competition

从两种文化中学习　LEARNING FROM TWO CULTURES

街道，广场与公共小品设计

2004 - 2005 年	内卡河畔埃斯林根，德国 Ritter 大街重新设计
2003 年	伯布林根，德国 Wolfgang Brumme Allee 购物街与购物中心重新设计
	常州，中国 城西地区概念性详细规划及城市设计方案
	内卡苏尔姆，德国 步行区与集市广场和集市大街重新设计竞赛
	纳戈尔德，德国 Vorstadtplatz 设计竞赛
2002 年	内卡尔滕茨林根，德国 城市中心区与工作坊更新
2001 年	天津，中国 新城市中心市政广场设计
2001 - 2004 年	斯图加特，德国 Hoelderlinplatz 街道及广场设计与实施
2001 - 2005 年	内卡河畔埃斯林根，德国 采尔中心区广场设计与实施
2000 年	内卡河畔埃斯林根，德国 统一广场设计
2000 - 2002 年	内卡河畔埃斯林根，德国 Am Kronenhof 街道及广场设计与实施
2000 - 2003 年	内卡河畔埃斯林根，德国 梅廷根地区 Kirchplatz 设计与实施
2000 - 2005 年	内卡河畔埃斯林根，德国 梅廷根地区 Oberturkheimer 大街设计与实施
1999 - 2000 年	内卡河畔埃斯林根，德国 城市中心区公交车站设计
1998 年	巴黎，法国 赛让新城 Cergy - Le Haut 中心轴线街道与广场设计方案
1998 - 2000 年	埃尔旺根，德国 埃尔旺根步行区设计与实施，竞赛一等奖
	内卡河畔埃斯林根，德国 火车站大街多功能街道设施设计
1997 - 2000 年	内卡河畔埃斯林根，德国 火车站大街设计与实施
1996 年	内卡河畔罗滕堡，德国 两个广场的设计与建设规划
1996 - 1998 年	卢肯瓦尔德，德国 火车站广场概念性设计
1995 - 2000 年	耶拿，德国 大型居住区公共空间设计
	耶拿，德国 11 层大型居住区地区入口重新设计
1994 - 1996 年	费尔滕，德国 新 Post 大街设计，4 车道街巷最终建设
1994 - 1998 年	耶拿，德国 Stauffenberg 大街 Karl - Marx - Allee 概念设计
1993 - 1996 年	安克拉姆，德国 Peene 河岸大桥设计与实施
1992 年	海尔布隆，德国 Neckarufer 售票亭设计与实施
1992 - 1993 年	彭昆，德国 集市广场设计与实施
	施特拉尔松，德国 Frankendamm 街道设计
1992 - 1995 年	海尔布隆，德国 凯撒大街和集市广场街道设计与设计导则
1990 年	奥芬堡，德国 Engler 大街 B3 段设计
	瓦尔登布赫，德国 集市广场和公共行政大楼重新设计竞赛
1990 - 1991 年	穆尔哈德，德国 历史城区公共卫生间设计
1990 - 1994 年	阿伯斯多夫，德国 Berg 大街 -Kapellen 大街设计与建设规划
1989 年	内卡河畔埃斯林根，德国 内卡岛设计与照明要素建设规划
1989 - 1990 年	舒特瓦尔德，德国 街道和公共空间设计与建设规划
	内卡河畔埃斯林根，德国 与公共交通系统协同的照明要素设计与建设规划
1989 - 1990 年	穆尔哈德，德国 Am Oberen Tor 广场设计与实施，喷泉设计与街道小品
1988 年	阿伯斯多夫，德国 照明要素设计
1988 - 1989 年	普富伦多夫，德国 历史城区街道设计，一般概念设计与公共空间设计导则
1988 - 1990 年	阿伯斯多夫，德国 街道，地区与照明小品设计与实施
	穆尔哈德，德国 历史城区公共空间设计与部分实施工作，竞赛一等奖
1987 年	斯图加特，德国 Berg 地区街道设计与城市设计导则
1987 - 1988 年	斯图加特，德国 Feuerbach 东部工业区街道设计与城市设计导则
1986 - 1990 年	海尔布隆，德国 城市西南中心区 Deutschhofplatz，Kirchenplatz, Neckarufer, Kirchbrunnenstr. 的公共空间设计
1982 - 1983 年	伦茨堡，德国 Neuwerk 巴洛克地区重新设计

城市灯光规划

2005 - 2006 年	科恩韦尔斯特海姆，德国 Dr. Siegfried Pflugfelder 广场灯光规划
2004 年	科恩韦尔斯特海姆，德国 集市广场灯光规划及照明要素设计
2004 - 2005 年	辛德尔芬根，德国 城市中心区灯光规划
2001 - 2003 年	内卡河畔埃斯林根，德国 城市灯光规划
2000 年	内卡河畔埃斯林根，德国 火车站大街照明设计
1999 - 2000 年	埃尔旺根，德国 步行区照明设计
1996 年	埃门丁根，德国 集市广场灯光规划
	坦佩雷，芬兰 集市广场设计及灯光规划，竞赛一等奖
1995 年	波茨坦，德国 城市灯光规划
1992 年	布伦瑞克，德国 Altstadtmarkt 灯光规划
1989 年	内卡河畔埃斯林根，德国 内卡岛照明要素设计与建设规划
	图尔库，芬兰 集市广场设计及灯光规划
1989 - 1990 年	内卡河畔埃斯林根，德国 与公共交通系统协同的照明要素设计与建设规划
1988 年	阿伯斯多夫，德国 照明要素设计
1987 年	奥卢，芬兰 城市中心区灯光规划
1987 年	赫尔辛基，芬兰 城市中心区灯光规划
1983 年	坦佩雷，芬兰 滨河灯光规划，竞赛一等奖
1978 年	内卡河畔埃斯林根，德国 Im Heppaecher 地区灯光规划
1976 年	吕贝克，德国 城市中心区灯光规划

生态规划研究

2009 年	唐山，中国 矿山公园规划，竞赛
2008 年	株洲，中国 湘江风光带滨水公共空间与环境景观设计
	石家庄，中国 滹沱河市区段生态开发整治区城市设计
2005 年	昆明，中国 绿地系统概念性规划，竞赛一等奖
1996 年	伊林根，德国 土地利用规划与景观规划整合
1995 年	黑肯高，德国 土地利用规划与景观规划整合
1995 - 1996 年	伊林根，德国 城市发展规划与生态发展概念规划整合
1994 年	阿姆茨贝格，德国 土地利用规划，生态发展概念规划整合
	波茨坦，德国 Babelsberg 区高层建筑环境影响研究
	拉芬斯堡，德国 Niederbiegen-Egelsee 北部 330 米高层环境影响研究
1994 - 1995 年	福冈，日本 滨水地区"岛城生态园"城市设计与重点地区生态规划
1993 年	Blaustein，德国 B28 沿线火车道，过路桥搬迁
1992 年	穆尔哈德，德国 土地利用规划与景观规划 2005,11 个待选区域发展概念规划
1991 年	康斯坦茨，德国 "商务园区"建筑设计
1991 - 1992 年	韦尔特海姆，德国 环境影响报告书及重点地区城市设计，村庄发展规划与 506 沿线交通规划

项目概览 PROJECT LIST

STREET, PLACE AND PUBLIC FURNITURE DESIGN

2004	**Kornwestheim, Germany**	
	Light planning for the Marktplatz and design of the lighting elements	
2004-2005	**Esslingen am Neckar, Germany**	
	Visual protection wall	
	Esslingen am Neckar, Germany	
	Redesign of the Ritterstreet	
2003	**Boeblingen, Germany**	
	Redesign of the shopping street and shoppping mall area Wolfgang Brumme Allee. Workshop	
	Changzhou, China	
	City entrance and highway sequance planning	
	Neckarsulm, Germany	
	Conversion of the pedestrian precinct and market square/Market street. Competition	
	Nagold, Germany	
	Place design of the Vorstadtplatz. Competition	
2002	**Neckartenzlingen, Germany**	
	Renewal city center and workshop	
2001	**Tianjin, China**	
	Place design of a shopping center and museum	
2001-2004	**Stuttgart, Germany**	
	Street and square design and realisation of Hoelderlinplatz	
2001-2005	**Esslingen am Neckar, Germany**	
	Square design and realisation of Center Zell	
2000	**Esslingen am Neckar, Germany**	
	Place design of Platz der deutschen Einheit	
2000-2002	**Esslingen am Neckar, Germany**	
	Street and square design and realisation of Am Kronenhof	
2000-2003	**Esslingen am Neckar, Germany**	
	Place design and realisation of Kirchplatz in Mettingen District	
2000-2005	**Esslingen am Neckar, Germany**	
	Street design and realisation of Obertuerkheimerstrasse in Mettingen District	
1999-2000	**Esslingen am Neckar, Germany**	
	Design of a busterminal in the city center	
1998	**Paris, France**	
	Street and square concept for the central axis of Cergy - Le Haut, Cergy-Pontoise	

1998-2000 **Ellwangen, Germany**
Design and realization of the pedestrian precinct Ellwangen, competition 1. prize
Esslingen am Neckar, Germany
Design of multifunktional construction as street equipment for the Bahnhofstreet
1997-2000 **Esslingen am Neckar, Germany**
Design and realization of the Bahnhofstreet
1996 **Rottenburg am Neckar, Germany**
Design and construction planning of two squares
1996-1998 **Luckanwalde, Germany**
Conceptual design of the square at the train station
1995-2000 **Jena, Germany**
Public space design in a mass housing estate, District 1
Jena, Germany
Conversion of the entrance areas of 11 storey mass residential buildings
1994-1996 **Velten, Germany**
Design of the new Poststrasse, final construction of a 4-lane alley
1994-1998 **Jena, Germany**
Conceptual street design of Karl - Marx - Allee, Stauffenbergstrasse
1993-1996 **Anklam, Germany**
Design and realization of Peene riverside - Peene bridge
1992 **Heilbronn, Germany**
Design and realization of ticket kiosk Neckarufer
1992-1993 **Penkun, Germany**
Design and realization of Market Square
Stralsund, Germany
Street design of Frankendamm
1992-1995 **Heilbronn, Germany**
Street design and design guidelines for Kaiserstrasse and Market Square
1990 **Offenburg, Germany**
Street design of the intersection Englerstreet, B3
Waldenbuch, Germany
Redesign of the Market Square and the Public Administration Building. urban design competition

1990-1991 **Murrhardt, Germany**
Design of public toilets for the historic area
1990-1994 **Albersdorf, Germany**
Design and construction planning of Bergstrasse - Kapellenstrasse
1989 **Esslingen am Neckar, Germany**
Design and contruction planning of lighting elements for Neckar Island
1989-1990 **Schutterwald, Germany**
Design and construction planning of streets and places
Esslingen am Neckar, Germany
Design and contruction planning of lighting elements in coordination with the public O-Bus system
Murrhardt, Germany
Design an realisation of the square Am Oberen Tor, fountain system and street furniture
1988 **Albersdorf, Germany**
Design of lighting elements
1988-1989 **Pfullendorf, Germany**
Street design, general concept and design guideline for public spaces in the historic area
1988-1990 **Albersdorf, Germany**
Design and realisation of streets, places and lighting furniture
Murrhardt, Germany
Public space design with partial realization in the historic area. competition 1st prize.
1987 **Stuttgart, Germany**
Street design and urban design guidelines for Stuttgart-Berg
1987-1988 **Stuttgart, Germany**
Street design and urban design guidelines for the industrial area Feuerbach-East
1986-1990 **Heilbronn, Germany**
Public space design for the south-west city center: Deutschhofplatz, Kirchenplatz, Neckarufer, Kirchbrunnenstr.
1982-1983 **Rendsburg, Germany**
Place redesign in the baroque area "Neuwerk"

CITY LIGHT PLANNING

2005-2006 **Kornwestheim, Germany**
Light planning for the Dr. Siegfried Pflugfelder Platz
2004 **Kornwestheim, Germany**
Light planning for the Marktplatz and design of the lighting elements
2004-2005 **Sindelfingen, Germany**
City light planning for the city center
2001-2003 **Esslingen am Neckar, Germany**
City light planning
2000 **Esslingen am Neckar, Germany**
Lighting design for the Bahnhofstrasse
1999-2000 **Ellwangen, Germany**
Lighting design for the pedestrian area
1996 **Emmendingen, Germany**
City light planning for the Marktplatz
Tampere, Finnland
Square design and lighting design for the Markt Square, competition 1.prize
1995 **Potsdam, Germany**
City light planning
1992 **Braunschweig, Germany**
Light planning of the Altstadtmarkt
1989 **Esslingen am Neckar, Germany**
Design and contruction planning of lighting elements for Neckar Island
Turku, Finnland
Square design and light planning of the Market Square
1989-1990 **Esslingen am Neckar, Germany**
Design and contruction planning of lighting elements in coordination with the public O-Bus system
1988 **Albersdorf, Germany**
Design of lighting elements
1987 **Oulu, Finnland**
Light planning for the city center

1987 **Rovaniemi, Finnland**
Light planning for the city center
1983 **Tampere, Finnland**
Light planning for the riverside, competition 1. prize
1978 **Esslingen am Neckar, Germany**
Light planning for Im Heppaecher
1976 **Luebeck, Germany**
Light planning for the city center

ENVIRONMENTAL IMPACT STUDIES

2009 **Tangshan, China**
Tangshan Main Mine Park Planning, Competition
2008 **Zhuzhou, China**
Waterfront and landscape planning for Xiang river of Zhuzhou City
Shijiazhuang, China
Ecological concept and planning for the riverside Hutuo
2005 **Kunming, China**
Urban development and landscape plannig, int. competition 1.prize
1996 **Illingen, Germany**
Land use plan with integrated landscape plan
1995 **Heckengaeu, Germany**
Land use plan with integrated landscape plan
1995-1996 **Illingen, Germany**
Urban development planning with integrated development concept for ecology
1994 **Amtsberg, Germany**
Land use plan, integrated development concept for ecology
Potsdam, Germany
Environmental impact study of scyscrapers - DEFA Babelsberg
Ravensburg, Germany
Environmental Study of the north highway 330, Niederbiegen-Egelsee
1994-1995 **Fukuoka, Japan**
Urban design planning with special focus on ecology for the waterfront area "Ecopark - Island City"
1993 **Blaustein, Germany**
Relocation of the train tracks, train overpasses and bypass B28
1992 **Murrhardt, Germany**
Landuse plan and landscape plan 2005, development concepts for 11 selected areas
1991 **Konstanz, Germany**
Building project "Business Park"
1991-1992 **Wertheim, Germany**
Environmental impact report for Reichholzheim with special focus on city planning, village development and traffic bypass L 506

从两种文化中学习　LEARNING FROM TWO CULTURES

建设咨询与公共关系管理

2008 - **2009 年** 埃斯林根, 德国 城市中心区太阳能改造设计导则	**1997 年** 新勃兰登堡, 德国 建筑项目设计咨询	**1992 年** 博伊岑堡, 德国 建筑设计许可申请与多方案比较建议书
2007 年 首尔, 韩国 方背洞粉红岛街区广场	**1996 年** 波茨坦, 德国 Neu-Babelsberg 别墅区建筑项目设计咨询	**格里门, 德国** 结构规划信息宣传手册
2007 - **2008 年** 埃斯林根, 德国 城市小品概念规划与公共关系咨询	**恩根, 德国** 居住区建筑设计咨询	**格里门, 德国** 建筑设计许可申请与多方案比较建议书
2006 年 宁波, 中国 新海港城市梅山岛规划	**1995 年** 魏玛, 德国 魏玛北部住宅群落建筑设计咨询	**亚尔门, 德国** 市民参与城市形象设计信息宣传手册
2006 - **2008 年** 埃斯林根, 德国 采尔地区 Wilhelmstr./Bachstr. - Kennerplatz 公共空间改造	**耶拿, 德国** 城市总览与信息宣传手册	**亚尔门, 德国** 建筑设计许可申请与多方案比较建议书
2006 年 阿伦, 德国 工业区气候防护罩可行性研究	**1994 年** 耶拿, 德国 洛伯达地区建筑设计咨询与色彩咨询	**拉桑, 德国** 建筑许可申请咨询
2002 年 巴黎, 法国 巴黎区 Plateau de Saclay 城市发展研究与竞赛	**博伊岑堡, 德国** "窗, 门, 外门与屋顶"信息宣传手册	**波茨坦, 德国** 城市发展规划信息宣传手册
2001 - **2003 年** 埃斯林根, 德国 车站大街展示改造	**海尔布隆, 德国** 城市设计竞赛	**施特拉松, 德国** 建筑设计许可申请与多方案比较建议书
2001 - **2007 年** 埃斯林根, 德国 城市夜景灯光规划	**康斯坦茨, 德国** Stromeyersdorf 工业区建筑设计咨询	**托尔格洛, 德国** 建筑设计许可申请与多方案比较建议书
2000 年 埃斯林根, 德国 城市中心区公共与私人街道小品设计导则	**波茨坦, 德国** Babelsberg 区"欧洲传媒中心大厦"展示与公众参与工作组	**1992 -** **1993 年** 海尔布隆, 德国 凯撒大街 50 号设计建议书
埃斯林根, 德国 城市中心区独栋建筑设计咨询	**1993 年** 波茨坦, 德国 城市发展规划与城市景观规划展览 –Podium discussion Aedes 画廊	**1988 -** **1990 年** 普富伦多夫, 德国 建筑设计咨询
1999 年 埃斯林根, 德国 车站大街公共与私人街道小品设计导则	**安克拉姆, 德国** 马市城市设计竞赛	**1987 -** **1990 年** 腓特烈施塔特, 德国 建筑设计咨询
1998 - **1999 年** 康斯坦茨, 德国 Stromeyersdorf 工业园区发展与设计导则	**格里门, 德国** 设计编码标识与手册	**1986 -** **1987 年** 阿尔伯施塔特, 德国 建筑设计咨询
1998 - **2003 年** 腓特烈港, 德国 城市中心区独栋建筑设计咨询	**波茨坦, 德国** 城市景观规划展示	**1986 -** **1990 年** 阿尔伯斯多夫, 德国 建筑设计咨询
1997 年 埃斯林根, 德国 车站大街建筑项目设计咨询	**1993 -** **1996 年** 腓特烈港, 德国 建筑咨询与多方案设计	**1985 -** **1987 年** 海尔布隆, 德国 建筑设计咨询
腓特烈港, 德国 城市信息宣传手册	**1992 年** 安克拉姆, 德国 建筑设计许可申请与多方案比较建议书	**1979 -** **1980 年** 乌尔姆, 德国 建筑设计咨询

建筑设计

2011 年 拉萨, 中国 柳梧城市综合体及沿河景观带概念方案	**2006 年** 郑州, 中国 中州国际酒店社区规划设计	**1996 年** 腓特烈港, 德国 人民银行建筑设计与工程咨询指导
斯图加特, 德国 阿卡萨公寓设计	**2005 年** 广州, 中国 南沙行政中心概念规划和主体工程建筑设计, 竞赛一等奖	**1995 年** 施特拉松, 德国 新市建筑设计与工程咨询指导
2010 年 汉堡, 德国 IBA 地区威廉斯堡桥, 竞赛一等奖	**北京, 中国** 亿城冷泉花园住区规划设计	**1995 -** **1997 年** 耶拿, 德国 11 层高层住宅立面改造设计
2009 年 大加纳利岛, 西班牙 度假别墅发展规划	**沃川, 中国** 新居住区规划	**1993 -** **1994 年** 托尔格洛, 德国 Droegeheide 木质房屋改造翻新指导准则
林根, 德国 Ross 内卡运河人行桥设计, 竞赛一等奖	**2004 年** 首尔, 韩国 大学食堂改造室内设计	**1991 年** 奥芬海姆, 德国 市政厅扩建, 学校扩建, 多功能中心及高档居住区建筑设计
2008 年 北京, 中国 京东方恒通商务园规划	**滨州, 中国** 邹平大学园区设计	**腓特烈施塔特, 德国** Westersielzug 步行桥设计
太原, 中国 东隅国际金融街高层建筑初步设计	**釜山, 韩国** 五金建材中心设计	**波茨坦, 德国** Waldstadt 购物中心选址咨询
2007 年 圣莱昂罗特, 德国 高尔夫俱乐部展示中心室内设计	**韩国** 私家宅邸"申宅"设计	**斯图加特, 德国** 韦兴根地区"惠普商务园区"总体规划, 竞赛
约勒松得, 丹麦 约勒松得发展愿景 2040	**2004 -** **2005 年** 丽江, 中国 玉龙新城规划与中心区设计	**1990 年** 海尔布隆, 德国 腓特烈艾伯特桥修缮设计
2007 - **2009 年** 埃斯林根, 德国 Hermann-Sohn 广场设计与施工	**2004 年** 舟山, 中国 情人岛高级度假区规划	**穆尔哈德, 德国** 历史居住区改造与扩建咨询
2007 年 太原, 中国 运城黄河文化博物馆建筑规划设计方案及各单体建筑方案设计	**2004 -** **2006 年** 研究项目 企业园区气候罩	**瓦尔登布赫, 德国** 集市广场与公共行政建筑改造设计, 城市设计竞赛
2007 - **2008 年** 首尔, 韩国 Gangseo Gu 地区城市发展规划	**2004 -** **2005 年** 舟山, 中国 情人岛高级度假区建筑设计	**1988 年** 伯布林根, 德国 购物中心重建
2007 年 北京, 中国 通盈中心酒店公寓概念方案设计, 竞赛	**2003 年** 上海, 中国 世博会游艇天穹设计	**海尔布隆, 德国** Fa.Maizena 行政建筑设计, 竞赛优胜奖
2006 年 斯图加特, 德国 斯图加特西区 Olga 医院可行性研究	**上海, 中国** 崇明城桥工业区设计	**伯布林根, 德国** 克劳斯时尚中心设计
弗莱堡, 德国 "犹太人广场"场地设计, 竞赛	**2002 年** 上海, 中国 宝阳别墅区保护性改造与建筑设计	**1986 -** **1987 年** 海尔布隆, 德国 "海尔布隆之声"高层建筑改造设计及咨询
巴特黑斯费尔德, 德国 工业区气候防护罩可行性研究	**2001 年** 上海, 中国 浦东高档住宅区建筑设计	**1981 -** **1983 年** 路德维希堡, 德国 联邦德国邮政服务中心建筑设计与施工
阿伦, 德国 工业区气候防护罩可行性研究	**天津, 中国** 购物中心与博物馆广场设计	**1976 -** **1977 年** 罗伊特林根, 德国 购物中心概念规划与建筑设计
福州, 中国 东部新城中心城市设计与概念建筑方案, 竞赛一等奖	**1997 年** 首尔, 韩国 高层住宅设计, 竞赛	

项目概览　PROJECT LIST

DESIGN CONSULTING AND PUBLIC RELATIONS

2008-2009	Esslingen, Germany — Design Guidelines for solar cells in the city center	1996	Neubrandenburg, Germany — Design consulting for building projects	1992	Boizenburg, Germany — Consulting on building permit applications and alternative design proposals

2008-2009　Esslingen, Germany
　　Design Guidelines for solar cells in the city center
2007　Seoul, Korea
　　Pink Islands place Bangbae-Dong
2007-2008　Esslingen, Germany
　　City Furniture Concept and public relations
2006　Ningbo, China
　　Masterplan for the Eco-Harbour-City Meishan Island
2006-2008　Esslingen am Neckar, Germany
　　Public space redesign: Wilhelmstr./Bachstr. - Kennerplatz, Zell District
2006　Aalen, Germany
　　Feasibility study for climatic covers for industrial area
2002　Paris, France
　　Urban development study and competition preperation for the Paris-Region-Plateau des Saclay
2001-2003　Esslingen am Neckar, Germany
　　Exhibition of the Bahnhofstreet
2001-2007　Esslingen am Neckar, Germany
　　City light planning
2000　Esslingen am Neckar, Germany
　　Design guidelines for the public and private street furnitures for the whole city center
　　Esslingen am Neckar, Germany
　　Architectural consulting for single building projects in the city center
1999　Esslingen am Neckar, Germany
　　Design guidelines for the public and private street furnitures in the Bahnhofstrasse
1998-1999　Konstanz, Germany
　　Development of design guidelines for the industrial park of Konstanz-Stromeyersdorf
1998-2003　Friedrichshafen, Germany
　　Building consulting for single building projects in the city center
1997　Esslingen am Neckar, Germany
　　Design consulting for building projects in the Bahnhofstreet
　　Friedrichshafen, Germany
　　Information flyer for the citizens

1996　Neubrandenburg, Germany
　　Design consulting for building projects
　　Potsdam, Germany
　　Design consulting for building projects in Neu-Babelsberg
1995　Engen, Germany
　　Building design consulting for the residential areas
　　Weimar, Germany
　　Building design consulting for the mass housing estate in Weimar-Nord
1994　Jena, Germany
　　Citizen Interview and informtaion flyer for the citizens
　　Jena, Germany
　　Building design consulting and colour consulting for the Lobeda district
　　Boizenburg a. Elbe, Germany
　　Information flyer for "Windows, Doors, Gates" and "Roofs"
　　Heilbronn, Germany
　　Preparing of an urban design competition
　　Konstanz, Germany
　　Building design consulting for the industrial park Stromeyersdorf
　　Potsdam, Germany
　　Exhibition "Euromedia Building" and public participation workshop Babelsberg
1993　Potsdam, Germany
　　Exhibition about the urban development planning and city image planning - podium discussion Aedes Gallery
　　Anklam, Germany
　　Preparing of an urban design competition Anklam-Pferdemarkt
　　Grimmen, Germany
　　Design codes for the signage and brochure
　　Potsdam, Germany
　　Exhibition about the city image plan
1993-1996　Friedrichshafen, Germany
　　Building consulting and alternative designs
1992　Anklam, Germany
　　Consulting on building permit applications and alternative design proposals

1992　Boizenburg, Germany
　　Consulting on building permit applications and alternative design proposals
　　Grimmen, Germany
　　Information flyer about structure plan
　　Grimmen, Germany
　　Consulting on building permit applications and alternative design proposals
　　Jarmen /Germany
　　Information flyer about city image planning for the citizen participation
　　Jarmen, Germany
　　Consulting on building permit applications and alternative design proposals
　　Lassan, Germany
　　Consulting on building permit applications
　　Potsdam, Germany
　　Information flyer about urban development planning
　　Stralsund, Germany
　　Consulting on building permit applications and alternative design proposals
　　Torgelow, Germany
　　Consulting on building permit applications and alternative design proposals
1992-1993　Heilbronn, Germany
　　Design proposals for the building Kaiserstr. 50
1988-1990　Pfullendorf , Germany
　　Design consulting for building projects
1987-1990　Friedrichstadt, Germany
　　Design consulting for building projects
1986-1987　Albstadt, Germany
　　Design consulting for building projects
1986-1990　Albersdorf, Germany
　　Design consulting for building projects
1985-1987　Heilbronn, Germany
　　Design consulting for building projects
1979-1980　Ulm, Germany
　　Design consulting for building projects

ARCHITECTURE

2011　Lhasa, China
　　Design of a new modern district of the city Lhasa
　　Stuttgart, Germany
　　Acasa Apartments
2010　Hamburg, Germany
　　Bridges Wilhelmsburg IBA Area, Competition 1st prize
2009　Gran Canaria, Spain
　　Development of a new holiday mansion
　　Esslingen, Germany
　　Design of the pedestrian bridge Rossneckarkanal, Competition 1. prize
2008　Beijing, China
　　Planning of the Hengtong Business Park
　　Taiyuan, China
　　Preliminary design for a high-rise building complex in the financial district
2007　Sankt Leon Roth, Germany
　　Interior design of the demonstration center, Golfclub
　　Öresund, Denmark
　　„Öresundsvisioner 2040"
2007-2009　Esslingen, Germany
　　Design and realisation of the Hermann-Sohn-Square
2007　Taiyuan, China
　　Regional and Urban Development Planning and Architecture for the industrial zone Yuncheng museum
2007-2008　Seoul, Korea
　　Urban Development planning for the district Gangseo Gu
2007　Beijing, China
　　Residential High Rise Buildings - Tongying Plaza, Competition
2006　Stuttgart, Germany
　　Feasibility study for Olgahospital, Stuttgart-West
　　Freiburg, Germany
　　Place design "Platz Der Alten Synagoge", Competition
　　Bad Hersfeld, Germany
　　Feasibility study for climatic covers for industrial area
　　Aalen, Germany
　　Feasibility study for climatic covers for industrial area
　　Fuzhou, China
　　Urban development planning and architectural design of government area (int. competition 1. prize)
　　Zhengzhou, China
　　Planning of a new resindential area and hotel complex

2005　Guangzhou, China
　　New town planning and structural planning for the new goevernment area in Nansha district (int. competition 1. prize)
　　Beijing, China
　　New garden residential area - Yicheng Lengquan
　　Okcheon, Korea
　　New residential area
2004　Seoul, Korea
　　Interior design - conversion of a university canteen into a bar
　　Binzhou, China
　　University campus design in Zouping
　　Busan, Korea
　　Design of hardware store complex
　　Korea
　　Planning of a private villa "House Shin"
2004-2005　Lijiang, China
　　Yulong New Town - structure planning and focus planning
2004　Zhoushan, China
　　Planning of a new luxurious residential area Qingren Island
2004-2006　Research project
　　Climatic covers for industrial areas
2004-2005　Zhoushan, China
　　Design for luxurious villas Qingren Island
2003　Shanghai , China
　　Sailing Dome
　　Shanghai , China
　　Design of a coverd industrial area for Chengqiao, Chongming Island
2002　Shanghai , China
　　Redevelopment and conversion of the historic villa district Baoyang and architectural design
2001　Shanghai , China
　　Design of luxury apartments in Pudong
　　Tianjin, China
　　Place design of a shopping center and museum
1997　Seoul, Korea
　　Development of a new residential area - living in high-rise buildings, competition
1996　Friedrichshafen, Germany
　　Design and design consulting for the building project Volksbank

1995　Stralsund , Germany
　　Design and design consulting for new buildings "am Neuen Markt"
1995-1997　Jena, Germany
　　Fassade-renovations of 11-storey mass residential buildings
1993-1994　Torgelow, Germany
　　Design guidelines for conversion and redevelopment of the wooden housing estate "Droegeheide"
1991　Oftersheim, Germany
　　Design of town hall extension, municipal center, school extension and seniors residence
　　Friedrichstadt , Germany
　　Design of the pedestrian bridge Westersielzug
　　Potsdam , Germany
　　Consulting on the position of the shopping center Waldstadt
　　Stuttgart, Germany
　　Masterplan for the "Hewlett Packard Business Park" in Vaihingen district, competition
1990　Heilbronn, Germany
　　Design for renovation of the Friedrich-Ebert-Bridge
　　Murrhardt, Germany
　　Consulting on the conversion and expansion of a residential building in the historic area
　　Waldenbuch, Germany
　　Redesign of the Market Square and the Public Administration Building. urban design competition
1988　Boeblingen, Germany
　　Remodelling of a shoppingcenter
　　Heilbronn, Germany
　　Design of the administrative building Fa. Maizena, Competition (merit award)
　　Boeblingen, Germany
　　Design of Modezentrum Krauss
1986-1987　Heilbronn, Germany
　　Design and design consulting for the remodelling of a small high-rise building "Heilbronner Stimme"
1981-1983　Ludwigsburg, Germany
　　Building design and realization of the Federal German Post Service
1976-1977　Reutlingen, Germany
　　Urban design concept and building design for a department store

编著

Seog-Jeong Lee   Michael Trieb

张亚津   鲁西米   董元明   Anja Goehringer   米婉

英文校对

Anthony Cannon Walker

版式设计及排版

Kew-Yearn Chung   Young-Gook Park   Jin-Kyu Choi

德国 ISA 意厦国际设计集团驻北京独资公司

意厦国际规划设计（北京）有限公司

地址：北京市海淀区学院路甲 5 号 768 创意园 A 座 3-005

邮编：100083

电话：86-10-62698680

传真：86-10-62698689

邮箱：contact@isa-design.cn

www.isa-design.cn

Editors

Seog-Jeong Lee   Michael Trieb

Zhang Yajin   Lu Ximi   Dong Yuanming   Anja Goehringer   Mi Wan

English Proofreader

Anthony Cannon Walker

Format Design and Layout

Kew-Yearn Chung   Young-Gook Park   Jin-Kyu Choi

ISA-Internationales Stadtbauatelier

Add: Furtbachstrasse 10

D-70178 Stuttgart, Germany

Tel: 49-711-6403031

Fax: 49-711-6403032

E-Mail: kontakt@stadtbauatelier.de

www.stadtbauatelier.de

Title

LEARNING FROM TWO CULTURES

Urban Development, Renewal, Preservation and Management in Europe and Asia

Author

ISA Internationales Stadtbauatelier

Publisher

China Architecture & Building Press

ISBN 978-7-112-15052-6

图书在版编目（CIP）数据

从两种文化中学习——欧亚城市发展、更新、保护及管理理论与实践 / 德国ISA意厦国际设计集团编著. —北京：中国建筑工业出版社，2013.1
ISBN 978-7-112-15052-6

Ⅰ. ①从… Ⅱ. ①德… Ⅲ. ①城市规划-建筑设计-作品集-世界 Ⅳ. ①TU984

中国版本图书馆CIP数据核字（2013）第009464号

责任编辑：费海玲　张幼平
责任校对：姜小莲　陈晶晶

从两种文化中学习
——欧亚城市发展、更新、保护及管理理论与实践
德国 ISA 意厦国际设计集团　编著

*

中国建筑工业出版社出版、发行（北京西郊百万庄）
各地新华书店、建筑书店经销
北京中新华文广告有限公司制版
北京宏信印刷厂印刷

*

开本：889×1194毫米　1/12　印张：29⅔　字数：685千字
2014年1月第一版　　2014年1月第一次印刷
定价：**480.00元**
ISBN 978-7-112-15052-6
（23122）

版权所有　翻印必究
如有印装质量问题，可寄本社退换
（邮政编码　100037）